AF541614

COLLOIDAL CHEMISTRY

By

Dr. A. Goel

DISCOVERY PUBLISHING HOUSE
NEW DELHI-110 002

First Published – 2006

Reprinted – 2026

ISBN: 978-81-8356-171-6

© Author

Colloidal Chemistry

Published by:

DISCOVERY PUBLISHING HOUSE

4383/4B, Ansari Road, Darya Ganj

New Delhi-110 002 (India)

Phone: +91-11-23279245; 23253475; 43596065

Mobile: +91 9811179893 / +91 9871656464

E-mail: discoverybooksindia@gmail.com

orderdphbooks@gmail.com

namitwasan9@gmail.com

web: www.discoverypublishinggroup.com

Printed at:

Infinity Imaging Systems

Delhi

Preface

This book has been written for the students of under-graduate and post-graduate level of the various universities in India. A special feature of the book is that the text has been illustrated with a large number of line diagrams and the data presented in the form of numerous tables for reference and comparison. In the preparation of text standard works and review by renowned author have been freely consulted and the reference given chapter wise. At the end of the book will be found useful by those who wish to make a more detailed study of the topics discussed.

We are extremely grateful to our respected Managing Director Shri Tilak Wasan, Discovery Publishing House, for valuable and active cooperation. We humble request our students and chemistry teachers to send us their constructive criticism and suggestions which we shall be using in the publication of the next edition.

Author

CONTENTS

1

COLLOID SCIENCE

COLLOIDAL STATE

Thomas Graham (1861), from his investigation on the diffusion of various substances is a liquid medium, classified substances as :

(i) *Crystalloids* : They diffuse rapidly in solution and can rapidly pass through animal or vegetable membranes, *e.g.*, urea, sugar, salts and other crystalline substances.

(ii) *Colloids* : They diffuse very slowly in solution and cannot pass through animal or vegetable membranes, *e.g.*, starch, gelatin, silicic acid, proteins etc. This class of substances generally exist in amorphous or gelatinous condition, and hence the name *colloid* meaning "glue form".

The original meaning attached to the words colloids and crystalloids do not hold good. In recent years, Graham's view about this classification of substances has undergone a great change, *because it has been shown that every substance irrespective of its nature can be a crystalloid or colloid under suitable conditions*. For example:

(i) NaCI though a crystalloid in water behaves like a colloid in benzene.

(ii) Soap is a colloid in water, while it behaves like a crystalloid in benzene.

(iii) Insoluble metals like Cu, Ag, Au and Pb can also be transformed by suitable methods in colloidal form.

From the above examples, it is clear that the concepts of crystalloids and colloids do not hold good. Thus, we now talk of colloidal state which may be defined as follows :

"*A substance is said to be in the colloidal state, when it is dispersed in another medium in the form of very small particles having diameter between* 10^{-4} to 10^{-7} cm (100 mμ to 1 mμ).

Thus, the size of particles in a colloidal state may be regarded as an intermediate between molecular size (*i.e.*, 10^{-7} to 10^{-8} cm.) and particles of coarse suspension (10^{-3} to 10^{-4} cm), *i.e.*,

Molecular size in true solution	***Colloidal particle size***	***Coarse suspension particle size***
10^{-7} to 10^{-8} cm	10^{-5} to 10^{-7} cm	10^{-3} to 10^{-5} cm

COLLOIDAL SOLUTIONS

Colloidal solution may be considered as a heterogeneous system consisting of the following three essential components :

1. *A dispersed phase :* It is also known as discontinuous or inner phase. It consists of discrete particles significantly larger than ordinary molecules. The dispersed phase is sometimes referred to as internal phase and usually constitutes the smiler fraction of the colloidal solution.
2. *A dispersion medium or continuous phase or the outer phase :* It is the medium in which dispersed phase is present. This consists of continuously interlinked molecules. This phase is sometimes referred to external phase and this phase usually constitutes the larger fraction of the colloid.
3. *A stabilising agent :* This is a substance which tends to keep the colloidal particles apart. Some colloids are self-stabilizers.

Dispersed phase + Dispersion medium = Dispersion system (Colloidal solution)

The colloidal particles may be single molecules particularly of those substances which are big enough to be of colloidal dimensions, *e.g.*, starch, gelatin, albumins, cellulose etc. or may be aggregate of large number of molecules, *e.g.*, a solution of NaCl in benzene. The size of the particles of the dispersed phase and dispersion medium is not same. Thus, while the former are usually between 10^{-4} to 10^{-7} cm., the latter are less than 10^{-7} cm. in diameter. Thus *a colloidal solution is one in which a definite substance is distributed in the form of very small* *particles of diameter between 10^{-4} to 10^{-7} cm as a dispersed phase in another substance called the dispersion medium.*

Each of the two phases constituting a colloidal system may be a gas, a liquid or a solid. In milk, for example, the fat globules are dispersed in water. Hence, fat globules form a dispersed phase and water is the dispersion medium. The term sol is applied to the dispersion of a solid in a liquid, solid or gaseous medium. The dispersion of a solid (dispersed phase) in a liquid (dispersion medium) is termed as *colloidal solution.* The dispersion of a solid (dispersed phase) in a gas (dispersion medium) is termed as a solid aerosol. If the dispersed phase is a liquid and the dispersion medium a gas, the resulting sol is called a liquid aerosol. When a liquid is dispersed in another liquid the resulting system is called *an emulsion.* If colloidal system becomes fairly rigid, it is termed as a gel.

CLASSIFICATION OF COLLOIDS

There are a number of basis for the classification of colloids.

1. Depending upon the nature of the dispersed phase and that of dispersion medium the colloidal solutions are divided into the following eight categories.

S.N.	*Dispersed phase*	*Dispersion medium*	*Name*	*Examples*
1.	Solid	Solid	Solid sol	Coloured glass, gems, alloys.
2.	Solid	Liquid	Sol	Paints, inks, white of eggs, mud.
3.	Solid	Gas	Aerosol	Smoke, dust.
4.	Liquid	Solid	Gel	Curds, pudding, cheese, jellies.
5.	Liquid	Liquid	Emulsion	Milk, cream, butter, oil in water.
6.	Liquid	Gas	Liquid aerosol	Clouds, mist, fog (water in air).
7.	Gas	Solid	Solid foam	Cake, broad, lava, pumice stone.
8.	Gas	Liquid	Foam	Soap lather, froth on beer, whipped cream.

Since the two gases are completely miscible with each other, they always form a true solution.

2. *Depending upon the appearance of colloids :* On this basis colloids are divided into the following two main categories.

(1) *Sol :* When a colloidal solution appears as fluid, it is termed as sol. Sols are named after dispersion medium. For example when dispersion medium is water, they are called *hydrosols* when the dispersion medium is alcohol they are called *alcosols* and so on.

(2) *Gels :* When a colloid has a solid-like appearance, it is termed as gel. The rigidity of gel varies from substance to substance.

Some substances may occur both as sols as well as gels. This depends upon the relative concentrations of the dispersed phase and the dispersion medium. For example at high temperatures and low concentrations of gelatin it occurs as a hydro sol in water. But if the temperature is lowered and the concentration of the gelatin is high, the colloid takes the form of a gel. It is because under these conditions, water acts as dispersed phase and solid gelatin particles act as dispersion medium.

3. *Depending upon the interaction of the two phases :* According to Perrio and Freaadlich, colloids may be classified into lyophobic and lyophilic.

(a) *Lyophobic or solvent-hating :* When the dispersed phase has less affinity for the dispersion medium, the colloids are termed as *lyophobic.* But when the dispersion medium is water, they are given the name *hydrophobic.* Substances like metals, etc. which have particles of size bigger than the colloidal particles or NaCI which has particles of size smaller than the colloidal size, fall in this category. Such substances are brought into colloidal state with difficulty.

(b) *Lyophilic or solvent loving :* When dispersed phase has a greater affinity for the dispersion medium, the colloids are termed as lyophilic and when the dispersion medium is water, they are given the name *hydrophilic.* They are also called natural colloids. Substances like proteins, starch and rubber etc. are grouped under this category. Such substances easily pass into the colloidal state when brought in contact with the dispersion medium. The lyophobic colloids are called *suspensoids* when they present a fluid appearance and are not very stable. Most inorganic colloids are hydrophobic, most organic colloids are lyophilic.

It was believed at one time that viscosity is the satisfactory measure of the lyophilic character, *i.e.*, the degree of solvation. But this is not always true because viscosity depends chiefly on particle shape. There are various colloids of low viscosity which are hydrophilic such as haemoglobin, albumin, glycogen etc. This type of classification will be used throughout the book.

4. *Depending upon the electrical charge on the dispersed phase* : On this basis the colloids may be divided into :

   (a) *Positive colloids* : The dispersed phase carries the positive charge. The particles of $Fe(OH)_3$ sol in water arc positively charged. Examples of this type are also methylene blue and TiO_2 sols.

   (b *Negative colloids* : The dispersed phase carries the negative charge. For example the particles of As_2S_3 sol in water are negatively charged. The other examples are copper or gold sol and certain dye-stuffs like eosin, congo red etc.

5. *Depending on the structure of colloid particles* : According to *Lumiere and others*, colloids can also be classified into molecular and micellar colloids. The particles of *molecular colloids* are single macromolecules and their structure is similar to that of small molecules. The particles of micellar colloids are aggregates of many molecules or groups of atoms which are held together by cohesive or van der Waal's forces. The examples of molecular colloids are albumin, silicones, rubber, etc. while that of micellar colloids are sulphur, gold, soap, detergents, etc.

6. *Based on particle shape* : Colloids may also be classified into *sphere* and *linear colloids*. The former is composed of more or less compact globular particles while the latter have long fibrous units. Astbury actually employed the terms globular and fibrous particles. Native albumin is an example of sphere colloid but after denaturation it becomes a fibrous colloid.

7. *Based on chemical composition* : The colloidal solutions are also divided into inorganic and organic colloids depending upon the chemical composition. The inorganic colloids may be metals (*e.g.*, Ag, A.U sols), non-metals (graphite sulphur), oxides and hydroxides [$Fe(OH)_3$. TiO_2] and salts (AgBr, AS_2S_3). The organic colloids may be

(i) Homopolar sols like rubber in benzene.

(ii) Hydroxy sols like starch,

(in) Heteropolar sols like proteins, soaps in water.

LYOPHILIC AND LYOPHOBIC COLLOIDS

Lyophilic Colloids are those colloids which have great affinity for the dispersion medium. These substances directly pass into colloidal solution when put in contact with the dispersion medium. When water is the dispersion medium, the name given is *hydrophilic*. For example, gelatin, albumin, etc.

Lyophobic colloids are those colloidal solutions which cannot be prepared by bringing them in contact with a solvent. When water is the dispersion medium, they are given the name hydrophobia.

Sometimes the names Emulsoids and Suspensoids are also used for hydrophilic and hydrophobic colloids respectively.

In general lyophilic sols are more stable than lyophobic sols. The additional stability is due to the presence of an envelop of the solvent (say water) around the colloidal particle. The process is known as hydration. To coagulate a hydrophilic sol we have to add a dehydrating agent in addition to an electrolyte. Main points of differences between the two types are given in the Table 1.1.

Table 1.1

Property	*Lyophilic (Intrinsic sol.)*	*Lyophobhic (Extrinsic sol.)*
1. Preparation	They are easy to prepare. Only contact with the dispersion medium is needed to stabilise them.	They are difficult to prepare. Special methods are used. Addition of stabilisers is essential for their stability.
2. Size of the particles.	The particles arc just bigger molecules.	The particles are aggregates of thousands of molecules.
3. Effect of electrolytes.	Small quantities of the electrolytes have no effect. Larger quantities,	Even small quantities of the electrolytes cause coagulation.

	however, cause coagulation.	
4. Nature	Reversible; once precipitated easily pass back into the colloidal state by contact with dispersion medium.	Irreversible, once precipitated does not easily pass into colloidal state.
5. Density, Refractive index, etc.,	Physical properties do riot follow the law of mixtures.	Density, refractive index follow the law of mixtures.
6. Conductivity.	With lyophilic salts high conductivities can generally be measured.	Owing to their sensitivity in electrolytes the conductivity of lyophobic sol can rarely be measured over a considerable range of concentration.
7. Colligative Properties.	They have relatively high osmotic pressure, depression in freezing point and high lowering of vapour pressure.	They have high osmotic pressure, small depression in freezing point, less elevation of boiling point and less lowering of vapour pressure.
8. Gelatinisation.	They may be gelatinised very easily and show the isoelectric point.	Do not gelatinise rapidly but they are unstable on the reversal of charge.
9. Tyndall effect.	Less distinct.	More distinct.
10. Viscosity.	Higher than that of water.	Almost same as that of water.
11. Surface Tension.	Lower than that of water.	Almost same as that of water.
12. Hydration.	Particles are heavily hydrated.	Particles are poorly hydrated.
13. Stability,	Very stable, coagulated with difficulty.	Less stable, coagulated easily.
14. Charge.	Depends on the pH of the medium. It can be even zero.	Have characteristic charge (positive or negative).
15. Concentration of the	Higher concentrations	Only low concentrations

dispersed phase	of dispersedphase are possible.	of the dispersed phase are possible.
16. Cataphoresis.	The measurement of cataphoresis is more difficult than lyophobic colloids.	The property of cataphoresis can be easily observed and easily measurable.
17. Examples.	Albumin, Glycogen. Rubber, Silicic acid etc.	Au, Ag, some emulsions etc.

PREPARATION OF SOLS

Preparation of lyophilic sols. Many organic substances like gelatin, starch, agar, egg albumin, glycogen etc., dissolve readily in water either in cold or on warming to give colloidal solutions directly. These are the lyophilic colloids. For example, sols of egg albumin or glycogen can be prepared by dissolving 1–2 grams of the finely ground substance in 100 ml of distilled water and then allowing it to stand for two hours after constant stirring. After two hours, the solutions are filtered.

Gelatin may be regarded as a typical lyophilic linear colloid. If two grams of gelatin are placed in distilled water and kept there for several hours, it has been observed that unlike egg albumin and glycogen, gelatin does not dissolve in cold water although it does swell. The swellen gelatin may be dissolved by heating with water at 80–90°C. If two grams of gelatin are dissolved in 400 ml of distilled water, a clear sol is obtained on cooling.

Preparation of lyophobic sols. Such sols can be prepared by the two general ways.

1. *By dispersion of coarse particles (Dispersion method). Here we start with bigger particles and break them down to the colloidal size.*

2. *By inducing molecular particles to form large aggregates (Condensation method. Here we start with particles of molecular dimensions and condense them to the colloidal dimensions.*

Dispersion Methods

(a) Mechanical Dispersion

Here the substance is first finely powdered and a coarse suspension is made by shaking the powdered substance with the dispersion medium.

This suspension is then passed through a colloid mill consisting of two discs, moving in opposite directions at a very high speed (Fig. 1.1). The particles of the suspension are subjected to a great shearing force and break down to the colloidal dimension. The space between the two discs controls the size of the colloidal particles to be obtained. Rubber, ink, paints and varnishes are prepared by this method.

A ball mill shown in Fig. 1.2 has been employed to get a colloidal solution from a coarse suspension. Due to a high speed rotation of the mill the coarse ball-like particles roll over one another and drop down at accretion position of the mill, thereby grinding the preliminary wetted mixture. This gives rise to particles of size 6Å. If some quantity of the dispersion medium is added, it becomes possible to get a colloidal solution.

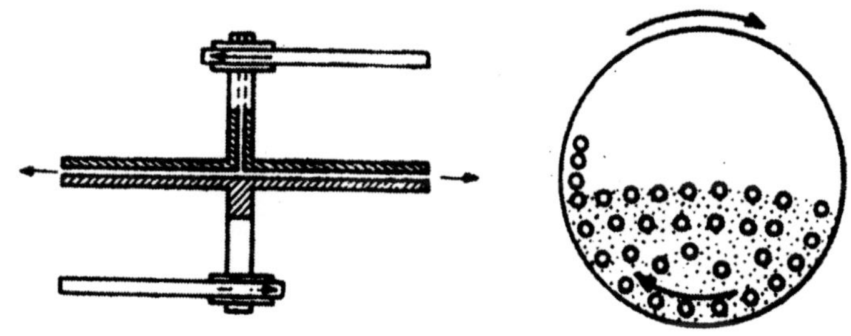

Fig. 1.1 **Fig. 1.2 : A colloid ball mill.**

(b) Electrical Dispersion

Bredig's arc methods : This is commonly used method for preparing the colloidal solutions of metals. An electric arc is struck between two metallic rods kept under the liquid (dispersion medium). A current of 10 amperes and a voltage of 100 to 300 volts is generally employed. The liquid is kept cooled by surrounding it with a cooling mixture. Tiny particles of the metal break away from the roads and disperse in the liquid. Gold, platinum, silver, copper and such other metals can thus be obtained in the colloidal form (Fig. 1.3).

Svedberg modified this method and prepared sols in non-aqueous media like pentane, diethyl ether, by striking an arc with high frequency alternating current which greatly diminishes the decomposition of the liquid (Fig. 1.4).

Svedberg has shown that the electrical methods are suitable not only for preparing the hydro sols of metals like gold, silver, platinum etc., but also for sols of strongly electropositive metals, such as sodium in benzene.

(c) Peptization

The process of bringing a precipitated substance back into the colloids state is known as population. It is carried .out by the addition of an electrolyte. The electrolyte added is termed as *peptising* or *dispersing agent.* It involves the adsorption of a suitable ion supplied by the electrolyte added by the particles of the precipitate. Peptisation may be carried out by the following ways :

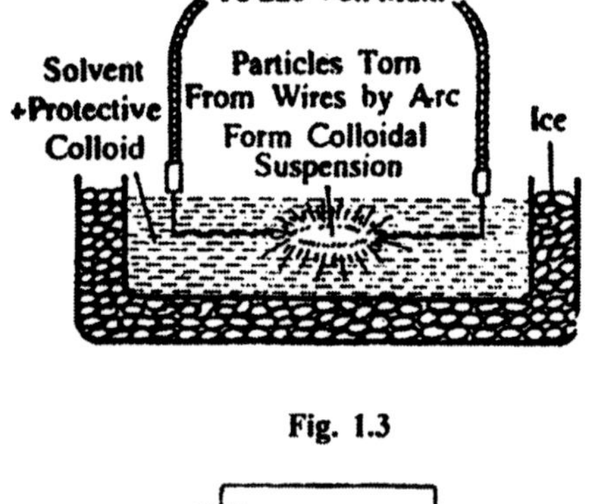

Fig. 1.3

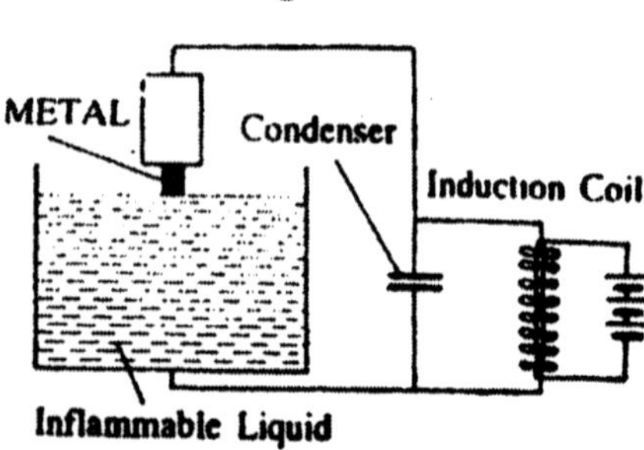

Fig. 1.4 : Svedberg's method of preparing metal sols in inflammable liquids.

(i) *By electrolyte :* Freshly prepared precipitate of Fe $(OH)_3$ can be changed into colloidal state when precipitate is treated with a small amount of $FeCl_3$ solution. The sol thus obtained is positively charged due to the preferential adsorption of Fe^{+++} ions (from $FeCl_3$) on sol particles of Fe $(OH)_3$ as [Fe $(OH)_3$] Fe^{+++}.

It should be noted that only freshly prepared precipitates can be peptized.

(ii) *By washing a precipitate :* Peptisation sometimes can be brought about by repeated washings of a precipitate. For example if the precipitate of $BaSO_4$ is washed continuously, a state is reached when the washings carry some of the particles of the substances in the form of colloidal solution.

The process of peptisation can be best explained by adsorption of the electrolyte. In most of the cases, the chelate compound is formed between the precipitate and the peptising agent.

Ostwald and Buzagh (1927) investigated the effect of the relative amount of the peptising agent and the substance to be peptised. They concluded that the amount peptised depends upon the amount of the reactant. The amount peptised has been found to increase either with the increasing amount of the precipitate or there occurs a maximum peptisation at a medium amount of the reactant. The plot of amount peptised versus the amount of the precipitates to be dispersed has been shown in Fig. 1.5.

From Fig. 1.5, it is evident that the amount peptised depends upon the amount of the substance to be peptised. In certain cases such as high polymers, the amount peptised has been found to increase with the increasing amount of the precipitate (curve 1). The increase of the value has been explained by the fact that the substance to be peptised is composed of various components which have the various resistances towards the dispersing action of the peptiser.

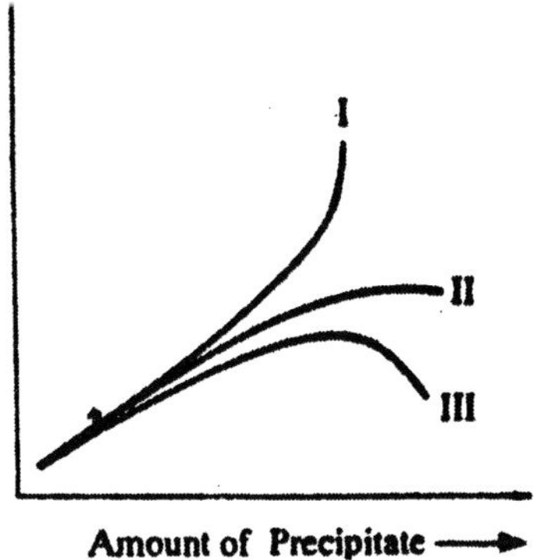

Fig. 1.5 : Dependence of the amount peptised upon the quantity of precipitate.

From curve II : it is evident that the amount peptised in certain cases increases with the increasing amount of the solid only when the quantities of the latter are small. Here the amount peptised becomes constant at a certain quantity of the solid.

From curve III : it is evident that the peptisation is maximum with a medium quantity of the precipitate and then begins to decrease. This may be explained as the maximum peptisation occurs at a certain medium concentration of the peptising agent Curve III is shown by the precipitated hydroxide, sulphides or other amorphous solids.

(d) *Electrolytic disintegration :* This type of disintegration is carried out in an electrolytic cell, the cathode of which is made of the metal to be dispersed. The electrolyte consists of a solution of sodium hydroxide and a current of high density is applied. Sodium is discharged at the cathode and forms an alloy with the metal. The sodium in the alloy is attacked by water in the solution and we get the metal in the form of sol.

(e) *Dispersion by ultrasonic waves :* The application of ultrasonic waves for the preparation of colloidal solution was first introduced by Wood and Loomis. Substances like oils, sulphur, graphite, sulphides and oxides of metals can be dispersed very easily with the help of ultrasonic waves (sound waves of very high frequency of the order of 20000 cycles/sec). The most important advantage of this method is that the colloids can be prepared without adding foreign substances to the system.

Claus (1936) prepared mercury sol by subjecting mercury to sufficiently high frequency ultrasonic vibrations as shown in Fig. 6. Here the generator produces the ultrasonic vibrations which spread through the oil and hit the vessel having mercury under water. Then, the ultrasonic vibrations travel through the walls of the mercury container and produce clouds of mercury which form the mercury sol.

(f) *Removal of agglomerating agent :* When a mass is obtained by the precipitating action of certain substances, there is a possibility of regaining the colloidal state by the removal of these substances. In various analytical procedures, a precipitate is formed instead of a colloidal suspension by the adsorption of ions or molecules. On washing with water sometimes it happens that the adsorbed material is not removed. If, however, certain acids are added to the wash water, the agglomerating agent may be displaced or removed and colloidal solution will again be obtained.

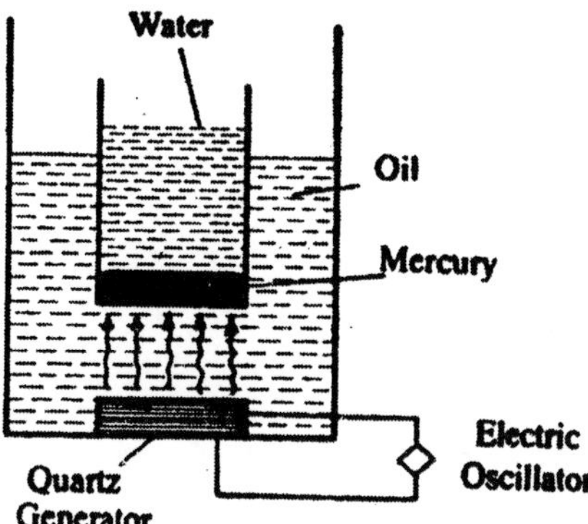

Fig. 1.6 : Ultrasonic dispersion of mercury in water.

Condensation Methods

The materials from which the sol is prepared are originally present in true solution in the form of either ions or molecules. If a chemical reaction occurs between these ions of molecules, particles of colloidal size can be obtained. The type of chemical reaction and general principles which have been employed will be described shortly. P.P. Von Weimarn and his coworkers in 1908 have attempted to ascertain the various factors determining the condensation of molecules and ions into particles of appreciable size and according to them,

$$\text{Initial rate of condensation (V)} = k \frac{\text{condensation pressure}}{\text{condensation resistance}}$$

Here k is a constant. According to *Von Weimarn*, the normal solubility (S) of the substance separating out from the solution may be regarded as a measure of condensation resistance, whereas, the difference between the total concentration (C) momentarily produced as a result of chemical change and the normal solubility (S) may be regarded as equal to the condensation pressure. Thus the above equation may be put as,

$$V = k \frac{C - S}{S} \qquad ...(1)$$

From this equation it follows that if V, the rate of condensation is large, the substance will separate from solution in the form of very small particles and if V is of the order of 100,000 to 200,000, with k equal to one, the particles separating are of colloidal dimensions. Thus there arc two general methods of increasing the rate of condensation.

(1) *By making C, the total concentration large* : In order to make C large the solution or solutions (as the case may be), from which the particles separates should be very concentrated.

(2) *By making S, the normal solubility small* : In order to retain the particles in sol form it is preferable to keep C small and to reduce S as far as possible.

Sols of substances which are sparingly soluble in water, *e.g.*, metals, sulphur and certain metallic sulphides, (*i.e.*, those substances having S very low), can be obtained readily by starting with a relatively concentrated solution of a suitable compound.

Following are the methods of preparing colloidal solutions.

Physical Methods

(a) *Condensation of hot vapours under a liquid phase* : Sols of certain elements can be obtained by passing their vapours into a cooled dispersion medium. A mercury or sulphur sol is obtained by boiling mercury or sulphur and passing the vapours into cold water containing a stabilising agent.

(b) *Exchange of solvent* : When a substance in a solution is shaken with another solvent in which it is less soluble, it changes into colloidal form *e.g.*, phenolphthalein in alcohol on shaking with water gives a colloidal solution.

Substances such as sulphur, phosphorus, etc., which are more soluble in alcohol than water are first dissolved in alcohol and then diluting them with water in which they are less soluble. The substance in alcohol is present in the molecular state which on transference to water precipitates out to form particles of colloidal size.

(c) *Excessive cooling* : Colloidal ice is obtained by chilling a mixture of water and the organic solvent as chloroform, pentane or ether. The preparation of ice cream is based on this principle.

(d) *Cathodic reduction* : Certain sols have been prepared by the rapid electrolysis of their solutions. For example a sol of lead is obtained at the cathode by electrolysing lead salts. Similarly, platinum black for electrodes is prepared by electrolysing platinum (IV) chloride on a gold or platinum electrode.

(e) *Controlled condensation :* Colloidal sols of certain insoluble substances are prepared by their precipitation in the presence of peptiser or protective colloid, *e.g.*, gum, starch, gelatin or agar-agar etc. A sol of Prussian blue may be prepared by this method.

Chemical Methods : All chemical changes giving rise to insoluble reaction products can be used for the formation of sols, particularly when suitable, stabilizers are also present–

(a) *Oxidation :* Sols of some non-metalş are obtained by oxidation. For example, sulphur is obtained in colloidal form by passing H_2S gas through bromine water or dil. HNO_3 solution.

$$H_2S + Bn_2 \rightarrow 2HBr + \underset{\text{colloidal}}{S}$$

Similarly, sol. of iodine is obtained by oxidising hydriodic acid with iodic acid as HIO_3 + 5HI ® 3H3O + $3I_2$, it can be made stable by adding a small amount of gelatin.

(b) *Reduction :* Sols of some metals are obtained by the reduction of their salts. Gold sols are made by reduction of $HAuCl_4$ solution with reducing agents like tannic acid. HCHO. H_2O_2, phosphorus, hydrazine etc. *Carey Lee's silver sole* is obtained by reducing solution of $AgNO_3$ containing alkaline dextrin as stabilizer with a variety of reducing agents. Stabilised colloidal suspension of *graphite* in water is known as *aquadag* and one in oil is known as *oildag.*

(c) *Hydrolysis :* This method is generally employed for the preparation of sols of number of a hydroxides and hydrous oxides. $Fe(OH)_3$ and $Al(OH)_3$ sols are obtained by boiling solution of the corresponding chlorides.

$$FeCl_3 + 3HOH \rightarrow Fe(OH)_3 + 3HCl$$

A beautiful red sol of ferric hydroxide is prepared by boiling ferric acetate in a beaker having distilled water (500 ml).

$$(CH_3COO)_3\,Fe + 3H_3O \rightarrow Fe(OH)_3 + 3CH_3COOH$$

The excess of ferric acetate may be removed by electrodialysis because its presence renders the sol unstable–

Organic esters of silicon, such as ethyl silicate, will hydrolyse in water to form colloidal silicic acid, $Si(OH)_4$ and alcohol.

(d) *Double decomposition* : This is the usual way of forming sols of insoluble salts. If the solutions containing the component ions of an insoluble substance are mixed, a precipitate will result. If the substance has low solubility, the precipitate will be colloidal.

Colloidal sol of prussian blue may be prepared by mixing very dilute solutions of $FeCl_3$ and $K_4Fe(CN)_6$.

$$3K_4Fe(CN)_6 + 4FeCl_4 \rightarrow Fe[Fe(CN)_6]_3 + 12KCl$$

Sols of arsenius sulphide and mercurie sulphide are obtained by passing H_2S into the saturated solutions of corresponding soluble salts of their oxides,

$$As_2O_3 + 3H_2S \rightarrow As_2S_3 + 3H_2O$$

$$Hg(CN)_2 + H_2S \rightarrow H_2S + 2HCN$$

(e) *Oxidation-reduction methods* : Many colloidal sols are prepared by oxidation-reduction reactions. An example of this method is the preparation of the colloidal molybdenic oxide which may be prepared by the reduction of ammonium molybdate with hydrogen sulphide, The sol of hydrated MnO_2 can be prepared by the reduction of 0.01 M $KMnO_4$ with ammonia. This sol can be made stablized by adding a definite quantity of gelatin.

PURIFICATION OF SOLS

Excessive quantities of electrolytes and some other soluble impurities remain in a sol as a result of the method selected for preparation, particularly in chemical condensation methods. The presence of small quantities of the electrolytes is sometimes essential for the stability of the colloidal systems, but their presence in excess results in the precipitation of the colloidal solutions on standing. Following methods are commonly used for the purification of colloidal solutions:

1. *Dialysis* : This method is based on the fact that colloidal particles are retained by animal membrane or a parchment paper while electrolytes pass through them. The sol is taken in a parchment or cellophane bag, which itself is placed in running water in a

trough. Gradually the soluble impurities diffuse .out leaving a pure sol behind (Fig. 1.7). Dialysis is a slow process and it takes several hours and sometimes even days for complete purification.

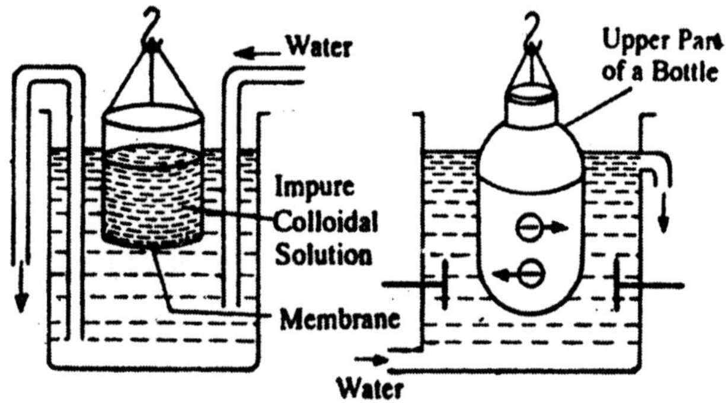

Fig. 1.7 : Dialyser.

2. *Electrodialysis :* Dialysis can be fastened by applying an electric field if the substance in true solution is an electrolyte. This process is then called electrodialysis Fig. (1.8). By means of electrodialysis it is possible to get a colloid in pure state in short time, although the electric current does not affect non-conducting impurities such as alcohol, sugar, etc. Staaffer (1956) noted that the current may also cause electrolytic decomposition and changes the activity of sol. Normal dialysis is generally quicker, when the sol contains much electrolyte.

3. *Hot dialysis :* The various factors upon which dialysis depends, are :

   (a) Nature of particles present in the solution.

   (b) Nature of the membrane used.

   (c) Difference in the densities of the two liquids.

   (d) Stirring both the sol and water.

   (e) Difference in temperature of the two liquids.

   (f) Area of the membrane in contact with the liquids.

Colloid Science

The rate of dialysis can be increased:

(a) by taking distinctly semipermeable membrane

(b) increasing the effective area of the membrane

(c) by stirring both the sol and water

(d) by increasing the temperature of the solution which is being dialysed as much as possible without the decomposition or coagulation of the colloid.

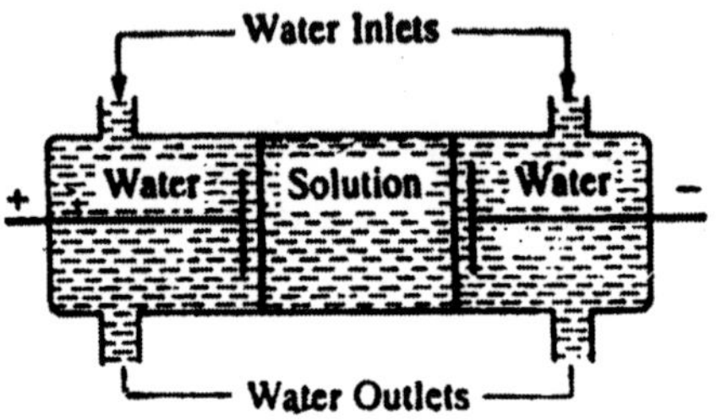

Fig. 2 : Electrodialysis.

Hence the process of diffusion can be hastened tc some extent by using hot water in place of cold water as was used in case of ordinary dialysis. This process is called hot dialysis. It has been observed that a temperature of 65–90°C increases the rate of dialysis by about three times. Heating may be carried out by placing an electric heater in the dialyser.

4. *Ultra filtration* : This is a method not only for purification of the sol but also for concentrating the sol. The pores of the ordinary filter paper are large enough (1030 mμ) for the colloidal particles (203 mμ) to pass through. But if the pores are made smaller the colloidal particles may be retained on the filter paper. This process is known as ultra filtration,

Preparation of ultra filters : Colloidal particles are more or less completely retained by ultra filters. There are various ultra filters which are available with different average pore size varying form 100Å to 200Å *i.e.,* from 10 mμ to 200 mμ.

(a) *Cellophane* is the material which is generally used for the ultrafiltration. This ultra filter may be prepared in the following way : From a sheet of thin cellophance a disc is cut and folded into the shape of a dish. By using a rubber-Ting, it is fastened to a funnel as shown in the Fig. 1.0.

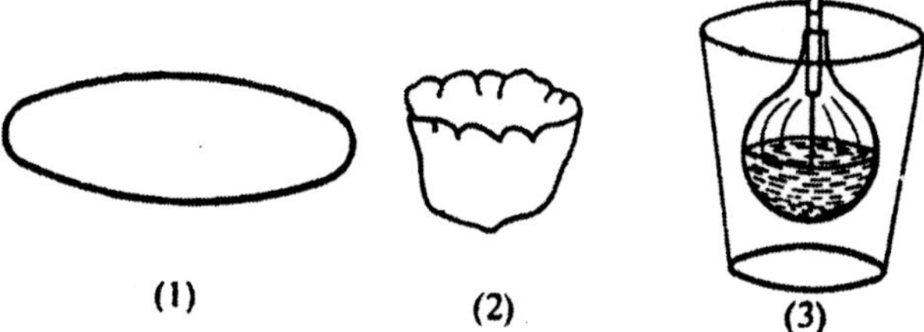

Fig. 1.9 : Formation of cellophane filter.

If the sol is poured, in the beginning the filtration would be quite slow but after a certain time the liquid would penetrate into the membrane.

(b) Another material for the preparation of ultrafilters is *collodion* which is 4% solution of nitrocellulose in the mixture of alcohol and ether. The collodion is generally poured into a sintered glass filter or filter crucible to a thickness of several m.m. It is then kept until it solidifies In the other method the filter paper is impregnated with collodion.

The pore size of these collodion ultrafilters depends upon the degree of the collodion layer. If it is kept for some time the nitrocellulose layer becomes so hard that it does not allow water through this membrane. Therefore after certain time of drying, these ultrafilters are kept on water in order to prevent further drying and the pore size; Bachhold (1907) separated certain virus by using this method.

Uses of Ultrafilters

(i) They are used for separating a coarser colloid from a finer one *e.g.*, prussian blue from haemoglobin.

(ii) They are used in bacteriology for removing the bacteria from solutions.

(iii) Ultrafilters are used for the determination of water of hydration from colloidal particles.

Preparation of f ultrafilters : Ultra filtrers are prepared by impregnating the filter paper with collodion (nitro-cellulose) solution and subsequently hardening by dipping it in a solution of HCHO and finally drying The size of the pores is varied by changing the concentration of the collodion solution.

Use of pressure on the sol side or the filtrate side facilitates such filtration.

A biological example of ultrafiltration is the removal of water and other substances from the blood through the glomeruli in the kidneys. *Elford* investigated the particle size of virus proteins by the help of ulrafiltration.

5. *Electro decantation :* This is a method not only for purification of the sol but also for concentrating the sol. If electrodialysis is carried out without stirring the sol, the lower layer becomes more concentrated whereas the top layer becomes dilute. This process is known as electro decantation and was introduced by Pauli (Fig. 1.9)

Pauli prepared a number of sols. The sol is introduced through the tube A. After applying 110 volts for sometimes, the negative colloid particles move towards the positive electrode and then collected in the container S. The dispersion medium flows upwards along m_1. This is finally withdrawn through the upper tube A. The new portion of the sol is supplied through the tube B.

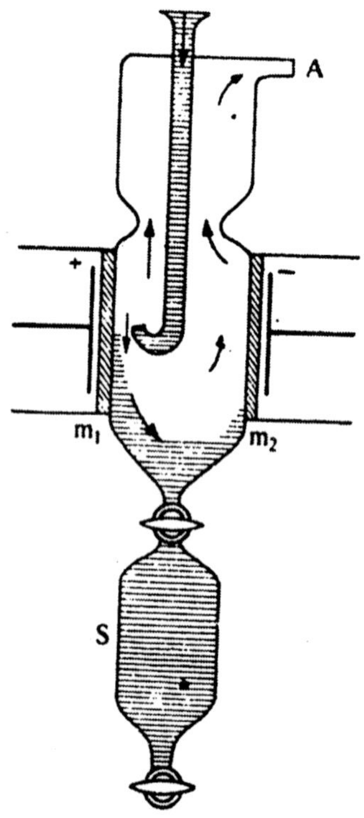

Fig. 1.10 : Paulis, process of electrodecantation.

The process of decantation is slow in the initial stages of electrodialysis. However in the later stages the electrodecantation is not helpful for removing the electrolyte because the electrolytes are also decanted. This process is applicable in the purification and the concentration of unstable cold sols.

With the process, a sol of 10% concentration can be obtained. *Kirkwood* (1990) developed a more refined form of electrodecantation.

It is to be noted that continuous electrodecantation is a very mild treatment, because the particles remain under the influence of current only for a short while. Even the very unstable gold sols can be purified and concentrated easily. For example, Pauli prepared a gold sol having 70g of gold per 100ml.

6. *Ultracentrifuging* : The colloidal particles share the motion of the molecules of the dispersion medium and are in a state of continuous zigzag motion, called Brownian movement. Sol particles are prevented from settling by this continuous haphazard zigzag motion. The sol is kept in a high speed centrifuging machine revolving at a very high speed (about 15,000 revolutions per minute) so that the colloidal particles settle quickly. The slime can be suspended in water so as to get a sol.

PROPERTIES OF COLLOIDS

General properties of colloids may be studied under several heads:

1. *Heterogeneous nature* : Colloidal solutions are heterogeneous in nature consisting of two distinct phases viz., the *dispersed phase* and the *dispersion medium*. Experiments like dialysis, ultrafiltration and ultracentrifuging, clearly indicate the heterogeneous character of colloidal system. Recent investigations have, however, shown that colloidal solutions are neither obviously homogeneous nor obviously heterogeneous.
2. *Non-settling nature* : Colloidal solutions are quite stable. The suspended colloidal particles remain suspended in the dispersion medium indefinitely. In other words there is no effect of gravity on the colloidal particles. A gold sol was prepared by Michael Faraday in 1857. This sol is still present in the museum of London without any detectable settling.
3. *Filtrability* : Colloidal particles readily pass through ordinary filter papers. It is because the size of the pores of the filter paper is larger than that of the colloidal particles. Colloidal particles can often be removed by especially prepared filters of extremely small and fine porosity, called ultrafilters. The latter are usually prepared by soaking the ordinary filter paper in a solution of collodion in acetic acid and subsequently hardened by soaking in formaldehyde. This process of separation is termed as ultra filtration.
4. *Concentration and density of sols* : It is a rather tricky process. Most sols are fairly stable when dilute but when the volume of

the dispersion medium is decreased, the colloidal particles get closer together and often combine together first to form larger particles and then finally to a coarse suspension. Collodion has shown that the density of sols increases linearly with the concentration. According to him, the density of a sol is given by

$$\rho_s = \rho + C\,\frac{\rho_0 - \rho}{\rho}$$

where ρ_s = density of sol

ρ = density of dispersed substance

ρ_0 = density of the solvent

C = number of grams of the dispersed substance in one ml.

The density of the particles can be calculated by the above equation. The concentration of most of the colloids is small. Metal sols usually contain 0.1–0.5% metal, the sols of hydroxides and sulphides 1–5 percent of the solid. The concentration of fibrous molecular colloids is also small and usually lies between 0.2 to 1%.

It is possible, however, to prepare concentrated sols of hydrophilic sphere colloids, for instance 10–15 percent sols of albumin or casein. Quite concentrated ferric hydroxide or arsenious sulphide sols have also been prepared. For example, Boutaric and Viuallume (1924) prepared an arsenious sulphide sol having 300 gms of As_2S_3 per litre. However, since the sulphide particles are larger than that of micromolecular units, the molar concentration, even of this concentrated sol., is very low.

5. *Size of the particles of the dispersed phase :* Colloidal dispersions generally range in particle size from 1 mμ to 1μ in diameter. The properties which distinguish them from true solution are mainly due to their large size.

6. *Diffusibility :* Colloidal suspensions unlike true solutions do not readily diffuse through fine membranes and have a little power of diffusion. This is due to the large size of the colloidal particles as compared to ordinary solute particles. In fact, it was this property which led Thomas Graham to lay the foundation of colloid chemistry, as already discussed.

7. *Colour :* The colour of the sol is not always the same as the colour of the substance in the bulk. The colour of the colloidal

solution changes as the size and shape of the particles change. It may exhibit different colours when seen by reflected and transmitted light. For example diluted milk gives a bluish tinge in reflected light and reddish tings in transmitted light.

In certain cases the colour of a sol also depends upon the shape of the particles, but the theory of the phenomenon is very complicated. Frequently, the specific absorption of light by colloidal particles is of greater importance than the scattering in determining the colour.

The various factors on which the colour of a sol depends are—

(a) Size and shape of colloidal particles.

(b) Wavelength present in the source of light used.

(c) The specific selective absorptive power of the dispersed phase and dispersion medium.

(d) Method of preparation of the sol.

(e) The way an observer receives the light, *i.e.*, whether by reflection or transmission.

8. *Shape of the colloidal particles* : Different sol particles have different shapes, for example, red gold sol, silver sol, platinum sol. As_2S_3 sol has spherical particles whereas $Fe(OH)_3$ sol and blue gold sol have disc or platelet-like particles. Rod-like particles appear in V_2O_5 sol, tungstic acid sol, etc.

   Colloidal particles may possess very different shapes and so it is very difficult to give a complete classification of colloids based on particle shape.

9. *Visibility* : Most of the sols appear to be true solutions with a naked eye, but the colloidal particles can be seen through an ultramicroscope.

10. *Surface area* : The total surface area of the particles in colloidal system is very large, compared to an equal mass of matter in the useful coarse grained size. This can be demonstrated very readily by taking a centimetre cube of material and dividing it into an increasing number of smaller cubes. The increase in surface area with decreasing particle size may be made clear by the following table.

Table 1.2 : Effect of sob-division on surface area

No. of cubes	*Length of Edge*	*Total surface area*
1	1 cm	6 cm^2 0.93 sq inch
10^3	1.0 mm	60 cm^2 93 sq inch
10^6	0.1 mm	600 cm^2 93.0 sq inch
10^9	0.01 mm	6000 cm^2 6.5 sq ft.
10^{12}	1.0 μ	6 m^2 64.6 sq ft.
10^{15}	0.1 μ	60 m^2 645 8 sq ft.
10^{18}	0.01 μ	600 m^2 717.6 sq. yd.
10^{21}	1.0 mμ	6000 m^2 7176.0 sq. yd. (about 3/2 acre)

From the above table it can be seen that last three lines include particles of colloidal size. It should be noted that the above data are calculated by assuming each particle as a solid cube. Where the particle is spongy, a much larger surface area will be obtained.

Due to this large surface area, colloidal systems show extensive adsorption of foreign materials. If the particles which are in the form of ions, then the colloidal particle is likewise charged.

11. *Colligative properties* : Sol particles because of their free suspensions amongst the molecules of the medium, share kinetic energy with them in the same manner as molecules of regular solutes do. This gives rise to colligative properties *like osmotic pressure lowering, depression of freezing point and elevation of boiling point.*

    Of these, the *osmotic pressure* alone has measurable values and its measurement has been used for finding average particle weight in colloidal suspensions. The reason for this may be put as follows:

    As colloidal particles are not simple molecules and are bigger particles (*i.e.*, physical aggregations of about 1000 molecules) their number in the colloidal solutions are comparatively small. As the magnitude of colligative properties depends upon the number of solute particles present in the solvent, their values are smaller.

12. *Viscosity of sols* : It is an important property of sols because it is directly related to the shape of particles. Therefore, some useful conclusions may be drawn from the effect of concentration, temperature, shear stress, and of different additives on the viscosity of colloids.

Dependence of viscosity on concentration : The dependence of viscosity on concentration has been thoroughly investigated for a number of sphere colloids. Einstein (1906) correlated these experimental facts and gave the following empirical relation on the basis of laws of hydromechanics.

$$\eta_{rel} = 1 + 2.5\,\phi \qquad ...(1)$$

where η_{rel} = The relativ^ viscosity and

ϕ = The relative volume-concentration of the colloid particles.

From Eq. (1) it is evident that the viscosity of a sphero-colloid increases linearly with its concentration.

Eirich, Bunzl Margaretha (1936) verified Eq. (1) experimentally by measuring the viscosity of suspensions of glass particles and of yeast cells by different methods. Staudinger modified Eq. (1) as follows :

$$h_{sp}\,\frac{\rho}{c} = k = 0.0025 \qquad ...(2)$$

where c = The grams of the dispersed substance per litre.

Later on, various experiments were carried out which revealed that two kinds of deviations occurred in Eq. (2).

(i) The constant k in equation (2) may also be greater than 0.0025. If k is greater than 0.0025, it indicates that the particles arc solvated or hydrated. If k is 0.0025, the hydration is negligible.

(ii) The linearity between the viscosity and concentration holds good at low and moderate concentration of the sphero-colloids.

Viscosity and particle size : As both Eqs. (1) and (2) do not involve any term for particle size, it follows that the viscosity of sphero-colloids is independent of the particle size. Buzagh and Erenyl (1940) argued that this statement is only correct for the systems having comparatively coarse particles like finely ground quartz but not for smaller particles tor which viscosity increases with the decrease of particle size. This increase may be explained by the fact that with increasing subdivision of the same amount of the material, new surfaces are formed. At these new surfaces more of the solvent is adhered so that the value of k is increased.

Viscosity of linear colloids. Fikantscher gave an empirical relation between relative viscosity and concentration of the linear colloids:

$$\log \eta_{rel/c} \frac{0.75\, k^2}{1 + 1.5kc} . + k \qquad ...(3)$$

where k is a constant which is characteristic of linear colloids. This relation holds good for different concentrations of many linear colloids.

Viscosity of linear colloids having different particle size : According to *Staudinger* (1928), there exists a quantitative relation between the molecular weight of truely linear macromole-Colloid Science cules and intrinsic viscosity of their solutions. .Staudinger gave the following relation;

$$\underset{c \to 0}{Lt} \frac{\eta_{sp}}{c} = K_m P \qquad ...(4)$$

where P denotes the number of radicals in the macromolecule and Km is a constant characteristic of each molecular colloid in a definite solvent. The very important condition is that the measurements must be made with solutions as dilute as possible. Another condition is that the calculated values of reduced viscosity must be extrapolated to zero concentration. The plot of η_{sp} against the concentration is depicted in Fig. 1.11. in which reduced viscosity is extrapolated to zero concentration.

From the measurement of viscosity of molecular colloids. The value of constant Km .can be calculated. The value of M can be measured by the light scattering experiment. If P is known, then M can be calculated by multiplying P with the weight of the radical. For instance, the radical weight of cellulose is 162. Then. M will be P × 162.

13. *Surface tension :* It is an important characteristic of inorganic and organic colloids. In the case of inorganic hydrophobic colloidal solutions, their water/air surface tension is not much different from that of pure water. Thus, hydrophobic sols Al_2O_3, V_2O_2, etc., and even hydrophilic sols of silica and stannic oxide do not exhibit significant difference in surface tension as compared with water.

As most of the organic colloids are lyophilic in nature, they have a low surface tension. If protein is dissolved in water, it will lower its surface tension moderately. But if surface active substances like soap, detergent, fluorochemicals, etc. are dissolved in water, they affect the surface tension appreciably.

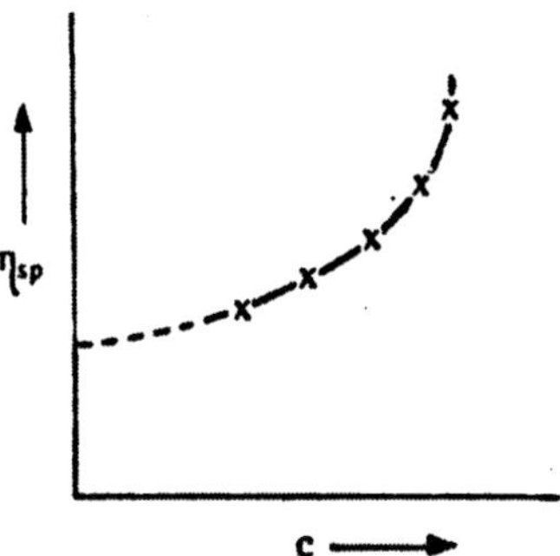

Fig. 1.11 : Extrapolation of the reduced viscosity to zero concentration.

Du Nouy observed that the surface tension of many colloidal sols changes with time after the addition of surface active agents.

This phenomenon is known as Du Nouy phenomenon (Fig. 1.12) of the surface tension. For example, the surface tension of a soap sol at first decreases and then rises rapidly to its original value.

14. *Optical properties* : Colloidal sols exhibit the following noted optical properties :

    (a) Optical rotation by solution of high polymers. The molecules of some organic colloids are optically active. Hence they will rotate the plane of polarised light either towards left or right.

For instance if a starch solution is kept in polarised light, the effect of light can be measured by rotating the analyser prism of a polarimeter to a degrees. The specific rotation [a] may be expressed in the form of following expression :

$$[\alpha] = \frac{\alpha.100}{c.l}$$

where a is the observed angle of rotation, c is the concentration in grams per ml and l is the length of the polarimeter tube.

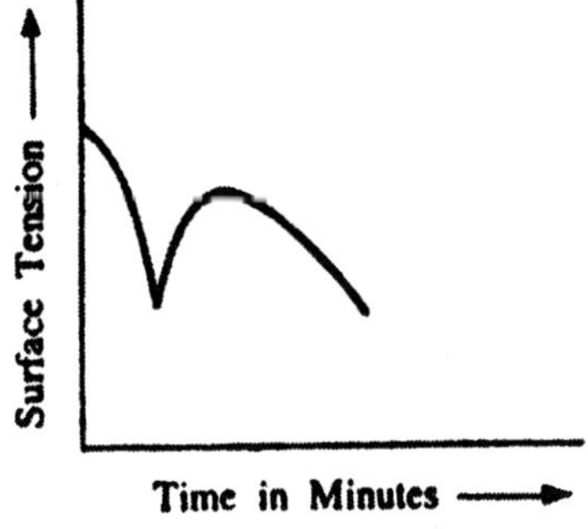

Fig. 1.12 : The surface tension dependence on time in colloidal soap solutions.

Drudge gave a general relationship between the specific rotation and wavelength. Yang *et.al.* found that the proteins obey linear equation

$$\lambda^2 [a] = \lambda_c^2 [a] + K$$

where λ is the wave length, $[\alpha]$ is the specific rotation and λ_c and K are constants. When $\lambda^2 [\alpha]$ is plotted against $[\beta]$, a straight line shown in Fig. 1.13 is obtained.

(b) *Optical anisotropy in colloids : Optical isotropic* substances are those substances whose optical properties like refractive index, absorption are same in all directions. Examples are liquids, glass and cubic crystals. The *optically anisotropic* substances, however, exhibit different optical properties. Colloidal sols are optically anisotropic. It is because the particles are comparatively large and asymmetric and hence the optical properties are different from the dispersion medium.

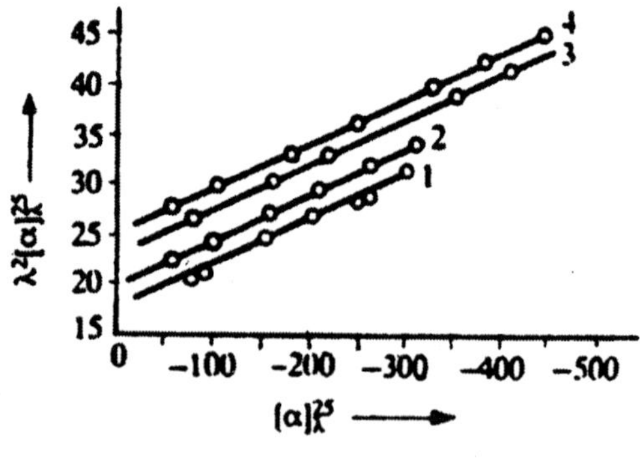

Fig. 1.13

The particles of ferric hydroxide and vanadium pentoxide sols are smaller and rod-like structures. Therefore, they form the anisotropic sols.

The spontaneous anisotropy can be seen in coarse systems having distinctly asymmetric molecules. For example, the unstirred solutions of ferric hydroxide and tungsten tridxide show spontaneous anisotropy. The asymmetric flat particles thus get oriented when they lie in regular layers. This orientation is not disturbed in Brownian movement. Such spontaneously oriented systems are known as tactoids. The formation of tactoids has been shown in Fig. 1.14.

(c) *Tyndall effect* : If a homogeneous solution is observed in the direction of light, it appears clear and when it is observed from a direction at right angles to the direction of the light beam, it appears perfectly dark. But when a beam of light passes through colloidal solutions, it is scattered, the maximum scattered intensity being in the plane at right angles to the path of light. The path beam becomes visible. The effect was first observed by *Faraday* but was studied in detail by *Tyndall* and the effect is now commonly known as *Tyndall effect.*

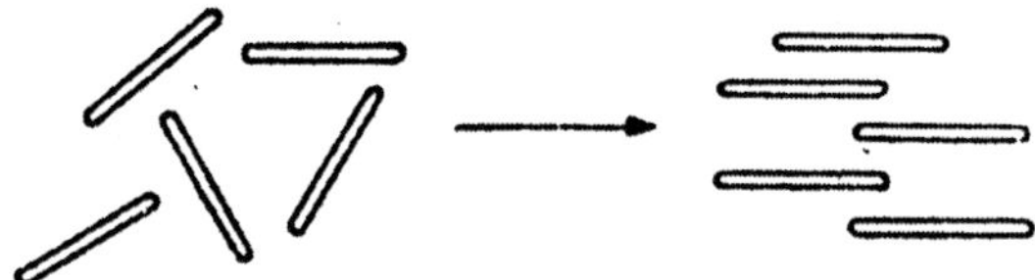

Fig. 1.14

Scattering also occurs in true solutions but the amount of scattering is extremely weak and there is a fundamental difference between the two types of scattering. "Tyndall effect is characteristic of large particles (Rayleigh scattering) and there being no difference between the wavelengths of incident and scattered light, solutions exhibit Raman scattering which is characteristic of molecules.

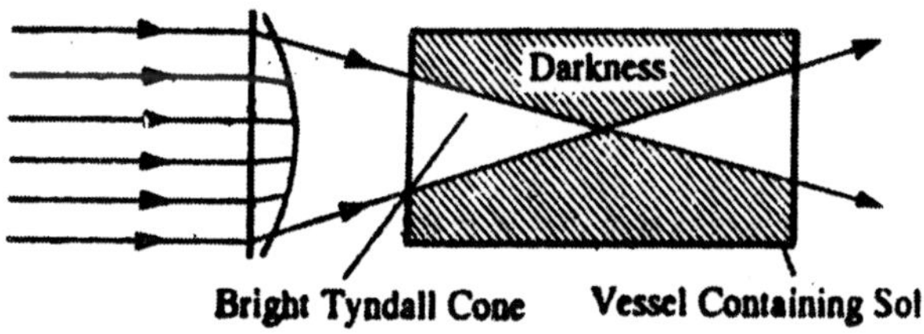

Fig. 1.15 (A) : The Tyndall effect.

The randomness of scattering of light by colloidal particles (Fig. 1.15A) was shown by Tyndall as a bright cone of light. It is known as *Tyndall cone.*

Explanation : The Tyndall effect is due to scattering of light by the colloidal particles. The colloidal particles first absorb light and then a part of the absorbed light is scattered from the surface of the colloidal particles. Maximum scattered intensity being in a plane at right angles to the plane of incident light, the path becomes visible when seen from that direction.

As the intensity of scattered light depends on the difference between the refractive indices of the dispersed phase and dispersion medium, the Tyndall phenomenon can be observed to an appreciable extent when refractive indices of the two phases differ considerably. For example, in lyophobic colloids, the difference is appreciable and so, the effect is well defined. But in lyophilic sols, the particles are largely solvated and so the difference is small and the Tyndall effect is very weak. Because of this reason, sols of silicic acid, albumin, blood serum etc. show a little or no Tyndall effect. According to Rayleigh, the intensity of scattered light can be expressed by the following expression:

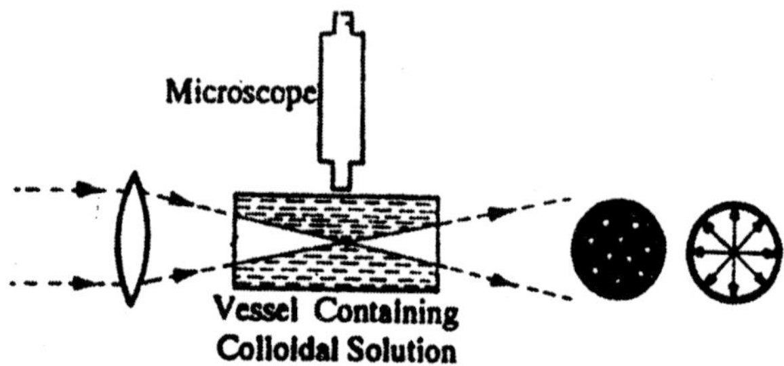

Fig. 1.15(B) : Ultra microscope.

Fig. 1.15(C) : Halo of scattered light round sol particle.

$$I = k\left(\frac{v^2}{d^2\lambda^4}\right)$$

where I = intensity of light scattered, v = volume of the particle.

d = distance between particle and observer.

λ = wavelength of scattered light.

This formula is true for insulating materials and is not applicable to metal particles. It is evident from the above equation that,

(1) Why sols like sulphur, arsenious sulphide etc. show a bluish Tyndall effect, why the term λ^4 appears in the denominator?

(2) Small wavelengths are generally favoured?

In conclusion it can be noted that in order for a system to show the Tyndall effect, two conditions must be satisfied.

(1) The diameter of the particles of the dispersed phase must not be much smaller than the wavelength of light used.

(2) The refractive indices of the dispersed phase and dispersion medium must differ considerably. This may be demonstrated by inserting a glass rod into some Canada balsam. The two

substances have practically the same refractive index, hence the rod becomes invisible where it enters the balsam.

Applications of Tyndall Effect : Tyndall effect has been used by *Zsigmody* and *Scidentonpf* for making an *ultramicroscope.*

Ultramicroscope is a microscope arranged so that light illuminates the object from the side instead of from below. In ultramicroscope, incident light does not strike the eye of the observer and thus observes the scattering produced by the sol particle against a dark background. *An actual image formation is not obtained but the presence of particle can be seen.*

There are three major types of ultramicroscopes.

(a) *Slit ultramicroscope :* It consists of an are lamp as source of light. The light from the source passes through the slit, and is condensed by a system of lenses so as to focus in a cell con-laming colloidal solution. The scattered light (Tyndall beam) is then viewed through a microscope which is placed at right angles to the beam. In this way, the colloidal particles can be seen as bright spots of light moving irregularly.

(b) *Immersion ultramicroscope :* This mircroscope was built by *Zsigmondy*. In this microscope, the illuminating lens and microscope lens are brought as close as possible, *i.e.,* and no special cells are used. Here the sol to be examined remains as a tiny drop between the two lenses.

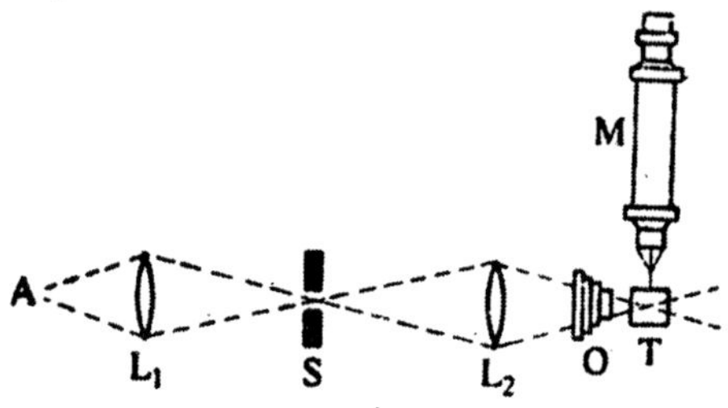

Fig. 1.16 : Slit microscope.

A drop of colloidal sol is placed between the lenses of the objectives. In this instrument the rays of light have to travel only a short distance. Here the resolving power and intensity of light are greatly increased. It is possible to study the coloured as well as colourless colloidal solutions.

(c) *Cardiod ultramicroscope :* Cardiod ultramicroscops is based on the principle of coaxial illumination. The light from an arc lamp is made parallel by means of a lens which is then reflected by the mirror under the microscope. The reflected light enters the cardiod condenser through the narrow slit. The light is finally passed through a vessel containing the sol in a quartz vessel. A very thin layer of the sol is taken in two glasses. Both of the glasses should be thoroughly cleaned, otherwise reflected light from dust particles enters the micro- scope. By cardiod condenser a very good illumination against a dark back ground is obtained. Now-a-days, ultra microscope has been replaced by means of electron microscope. It provides a true image of the colloidal particles.

ELECTRICAL PROPERTIES OF COLLOIDS

Electro-kinetic effects : It is now well established that colloid particles carry an electric charge. When the sols are placed in the electric field, certain special effects are noticed, which are termed as electro-kinetic effects. Such effects involve the relationship between the movement of one phase with respect to another in the exhibition of some electrical properties. Such effects are of four types :

1. Electrophoresis or cataphoresis
2. Electro osmosis or Electroendosmosis
3. Streaming potential
4. Sedimentation potential or Dorn Effect. Let us discuss these one by one.

1. *Electrophoresis or Cataphoresis :* The colloidal particles carry an electric charge which was shown first of all by Linder and Picton by placing the colloidal particles in electric field. The colloidal solution is taken in a U-tube and two platinum electrodes are dipped in the sol as shown in the figure. The current is then switched on. On closing the circuit it is found that colloid particles move to the oppositely charged electrode and on reaching that electrode they get discharged. As soon as the charge of the particles is neutralized, they aggregate and settle down.

This movement of colloid particles in electric field is known as electrophoresis, and corresponds to electrolysis in case of the true solutions.

Since the current in the colloidal solution must be carried by both positive and negative particles, ions of the diffused layer must be moving in a direction opposite to the direction of the movement of colloid particles. In a $Fe(OH)_3$ sol which is positively charged, the sol particles move to the negative electrode where their charge is neutralised and they aggregate and finally precipitate out. Thus, the entire colloidal matter settles down at the bottom.

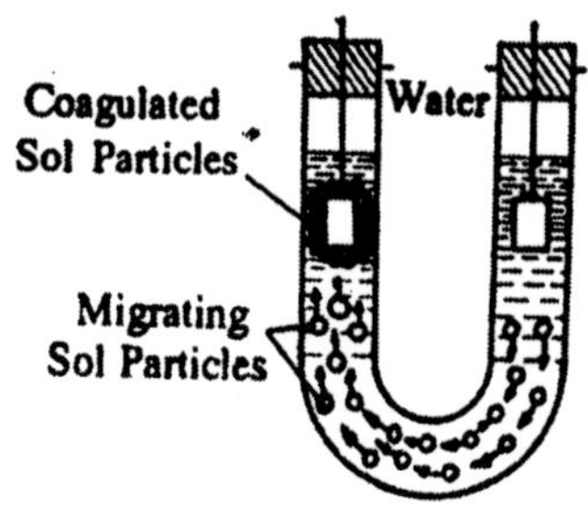

Fig. 1.17 : Cataphoresis

The electric charge carried by the colloidal particles and the direction of their displacement under the influence of an electric field depends, however, on the method of preparation of the sol. Moreover, addition of some electrolytes can also change the sign of the colloidal particles. Hardy (1899) has shown that particles of proteins dissolved in water are almost electrically neutral, but if a small amount of alkali is added to it the particles become --vely charged and if acid is added, the protein particles become positively charged, *i.e.,* they move towards the cathode. The common hydroxide sols are usually +vely charged, but it has been found that metal hydroxides having negatively charged particles can also be prepared.

The mobility of colloidal particles under a potential gradient of one volt./cm. is 2 to 4 μ per second. This is of the same order as that of potassium ion which is 6.6/μ/sec considering the size of the colloidal particles and the mobility to be very great. This is because the charges on colloidal particles is very high.

According to Svedberg and Tiselius fluorescence photography can be used for determining the velocity of migration of protein particles. In this technique, the ultraviolet light is used for causing the fluorescence of the protein. The intensity of the fluorescence was determined at different points and the position of the moving boundary could be estimated quantitatively.

Importance : This phenomenon of electrophoresis is made use of in the following ways :

1. *Determining the charge on the colloidal particles* : Direction of movement of the colloidal particles in the electric field shows the charge on them.
2. It can also be used to determine the rate at which colloidal particles migrate under the influence of an electric field.
3. It is also used in the identification and determination of homogeneity.
4. It is of great importance for the preparative separations of colloidal substances.

Both homogeneity determinations and separations have been based on the fact that the mobilities of particles depend on how highly charged they are, as well as how large they are, and how they are shaped. Electrophoretic separation may be effected in a monodisperse system if particles with different charges are present, since then they will have different mobilities as well as different zeta potentials. Separation can also be effected in the case of polydispersed linear colloids, but it is not possible in the case of polydispersed sphere colloids, such as gold, since the smaller particles also carry small charge.

Inspite of these limitations the phenomenon of electrophoresis is of great importance in chemistry, medicine, and industry. *Theorell* (1934) isolated the flavine enzyme from yeast using a very cleverly built apparatus of his own construction.

2. *Electro osmosis* : It is also known as electro-endosmosis and was discovered by *Reoss* (1809) and *Porret* (1816). If in the above experiment, a partition is made by animal membrane or parchment paper in between two electrodes (Fig. 1.18), so that only the dispersion medium can move through it and not the colloidal particles. When potential difference is set up between the electrodes, *then the dispersion medium is seen to move in a direction opposite to the direction of movement of the colloidal particles*. This movement of the dispersion medium relative to the dispersed phase under the influence of the electric field is known as *electro-osomosis*. This is indicated by the rise of water level in one limb of the U-tube.

The theoretical interpretation of electrosmosis has been given by Helmholtz, Berrin and Von-Smoluchowski.

Weidermann (1852) made a quantitative measurement of electro-osmosis. According to him, the difference in pressure maintained between the two sides of a porous diaphragm is found to be proportional to applied E.M.F. but is independent of the dimension of diaphragm, *i.e.,*

$$P = 2\xi\ DEl/\pi r^2$$

where D is the dielectric constant of the medium, E*l* is the voltage and ξ is the electro-kinetic voltage.

Measurement of electro osmosis : The simple apparatus used for the measurement of electro-osmosis is shown in Fig. 2a. This consists of a middle compartment C which is separated from the two compartments A and B by the two dialysing membranes L and L'. The compartment C is filled with colloidal suspension while the compartments A and B are filled with water. The water compartments A and B are fitted with the two side tubes T and T' to observe the level of the water. The function of membranes L and L' is to prevent the movement of colloidal particles.

When a potential difference is set up between the two electrodes, the water will begin to rise in tube T or T' depending upon the following conditions :

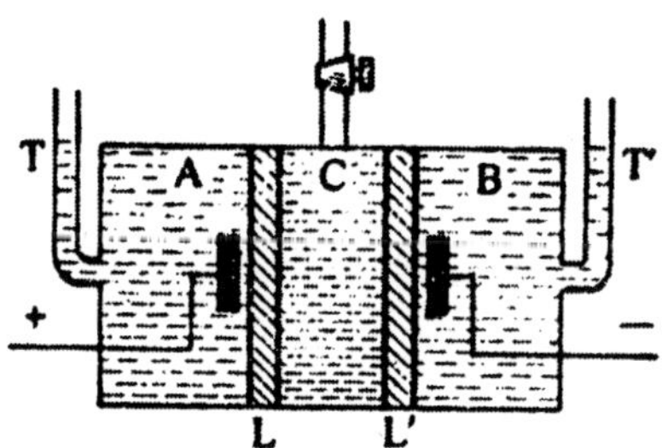

Fig. 1.18(a) : Measurement of electro-osmosis.

(i) If the colloidal particles are carrying positive charge, water will carry negative charge. In such a situation, the water will move towards the anode and the level of the water will rise in tube T.

(ii) If the colloidal particles are carrying negative charge, water will carry positive charge. In such a situation, the water will move towards the cathode and the level of the water will rise in tube T'.

The difference in water level h is measured and the pressure due to electro-osmosis may be calculated by the following relation :

$$P = hdg$$

where d = density of the particles

g = acceleration due to gravity.

Numerous measurements were made with similar devices in the middle of the last century and it was found that in most cases water moved towards the cathode, thus evidently carrying positive charge. It moves, however, towards the positive electrode when the porous material is weakly basic.

The phenomenon of electro-osmosis is the counterpart of electrophoresis and is modified in a similar way by adding acids, bases and neutral salts. It has been observed that negative diaphragms become more negative in alkaline solution while in acidic solutions they become less negative, electrically neutral and finally positive. Similarly, positive diaphragms become more positive in acidic solution and in basic solutions they become less positive; electrically neutral and finally negative.

Two most important conclusions have been arrived at by the endo-osmosis experiments in three years. These are :

(1) *For a given diaphragm, the greater the conductivity of the solution due to the presence of electrolyte ions, the less is the endo-osmosis flow.*

(2) *Under comparable conditions, fine particles with greater absorptive power, e.g., colloidal grains, making up the diaphragm cause greater endosmotic flow than if the diaphragm is made up of crystalline particles of small absorptive power.*

Applications : The phenomenon of electro-osmois is of great importance:

(i) In the preparation of pure colloids,

(ii) In the drying of peat and clay.

(iii) In the tanning of hides.

(iv) In the manufacture of gelatin for photographic emulsions, and high grade glue.

3. *Streaming potential* : The existence of electro-omosis has suggested that when liquid is forced through a porous material

or a capillary tube, a potential difference is set up between the two sides. This is called *streaming potential*. In other words, in electro-osmosis, the liquid flow across a solid surface of diaphragm is set up by an applied electric field. The reverse of electro-osmosis is the streaming potential.

Streaming potential can easily be measured by forcing a liquid under examination through a porous material or fine capillary tube and the potential difference across the electrodes is measured.

4. *Sedimentation potential or Dorn effect* : It is the reverse of electrophoresis. In electrophoresis, the colloidal particles are forced to move under the applied potential. The sedimentation potential is set up when a particle is forced to move in a resting liquid. The phenomenon was discovered by Dorn and is known as Dorn effect. The relation of the four effects described above are illustrated below:

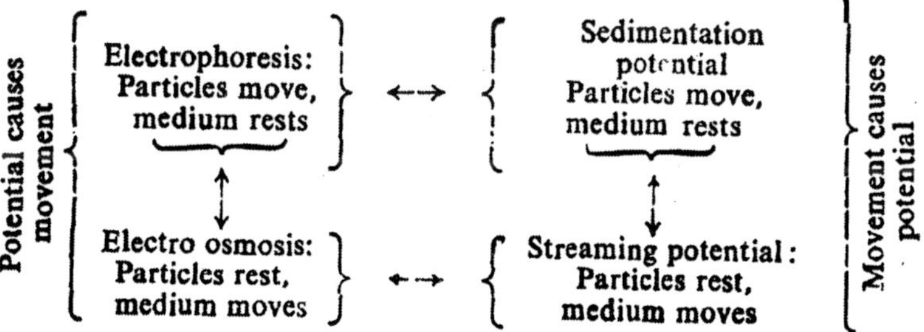

Fig. 1.18 (b)

ELECTRICAL DOUBLE LAYER

The electrical properties of colloids can be qualitatively explained on the concept of electrical double layer at solid-liquid interface proposed by Helmholtz (1879). Helmholtz considered that electrical double layer is formed at the surface of separation between two phases, *i.e.*, solid and liquid. He treated the double layer like a parallel plate condenser, with a molecular distance apart. Furthermore, Helmholtz proposed that the potential difference at the interface should be sharp.

However, recent views are in favour of diffused double layer, given by Gouy (1909) *i.e.*, there is only one layer instead of two. Stern developed views combining the essential characters of both.

Gouy (1909) calculated the thickness of the diffused double layer by applying Debye-Huckel theory to it.

According to Stern, double layer consists of two parts:

(i) One of the part remains almost fixed to the solid surface. It is known as the fixed part of the double layer having positive or negative ions. There is sharp fall of potential.

(ii) The other part extends some distance into the liquid phase. It is known as diffuse part. The layer contains ions of both sign. Its net charge is equal and opposite to that on the fixed part of the double layer. There is a gradual fall of potential into the bulk of the liquid where the charge distribution is not uniform.

In case of the diffuse part of the double layer, thermal agitation results the free movement of the particles, but the –ve and +ve ions are not uniformly distributed. The electrostatic field at the surface will result preferential attraction of those of opposite sign.

The existence of charges of opposite signs on the fixed diffuse parts of the double layer develops a potential between the two layers. This potential is known as elect rokinetic potential or zeta potential and is represented by ζ. *It is thus the electromotive force which is created between the fixed layer and the dispersion medium.*

We will now consider various theories about the structure of double layer.

1. *Helmholtz-Perrin theory :* According to this theory, the electric double layer is a flat condenser, as it were, one plate of which is connected directly to the surface of a solid (a wall) whereas the other plate which carries the opposite charge is in a liquid at a very small distance away from the first plate. The potential in such a double layer, like that in the plane condenser, should apparently fall very sharply (along the straight line), and the value of the surface charge density a will be determined by the formula

$$\sigma = \frac{\in}{4\pi\delta}\varphi_0 \qquad ...(1)$$

where $\in$ = absolute dielectric constant of the medium that fills up the space between the condenser plates;

φ_0 = potential difference between the dispersed phase and the solution;

δ = distance between the plates.

Fig. 1.19 schematically illustrates the structure of such an electric double layer: the crosshatched part is the solid phase, and the remaining part is the solution; positive and negative ions forming the double layer are denoted by "+" and "–", respectively. Free ions of an electrolyte which are always present in a liquid are not shown so as not to complicate the illustration. The potential falls as the distance x from the surface of the solid body grows; the total potential jump in such a double layer is at the same time a jump in potential between the solid phase and the solution.

The given illustration of the structure of the electric double layer does not explain several features of electrokinetic phenomena. At present, it is only of historical importance to colloid chemistry. The main shortcoming of the illustration is the fact that the thickness of the Helmholtz-Perrin double layer is very small and approximates molecular dimensions. At the same time hydrodynamic investigations have shown that the place of discontinuity (slipping plane) is always in the liquid phase at a comparatively great distance away from the interface when the solid and liquid phases move relative to each other. The thickness of the liquid layer that has "adhered" to the solid surface under these conditions is in any case greater than the thickness of the Helmholtz-Perrin electric double layer.

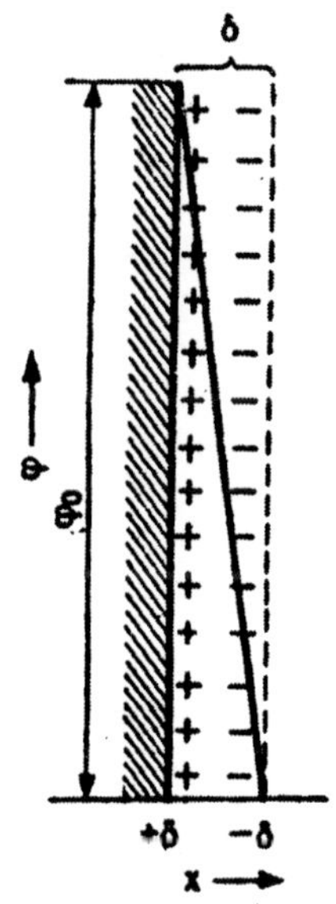

Fig. 1.19 : Electric double layer according to Helmholtz and Perrin and the potential jump in it.

If the Helmholtz-Perrin theory were correct, the Dorn effect or the streaming potential should not be observed when colloidal particles sediment in a liquid or when the liquid is pressed through a capillary, also eleetrophoresis and electro-osmosis would be impossible. However, even if we assume that as it has been done earlier, the slipping plane is between the two plates of the electric double layer, the Helmholtz-Perrin concepts create a contradiction. Indeed, if such an allowance is made, the electrokinetic potential of *eleetrophoresis should* correspond o the potential difference between all the potential-determining ions and all the counterions, *i.e.,* it should be equal to the total potential jump. However, the electrokinetic potential is, as a rule, smaller than the total potential jump, and changes quite differently under the action of various factors. For example, the total potential jump does not depend, to any considerable extent, on indifferent electrolytes that do not contain ions which are capable of completing the construction of the crystal lattice; at the same time, such electrolytes greatly affect the electrokinetic potential.

The foregoing exposition shows that the total potential jump and the electrokinetic potential are different quantities, and therefore Helmholtz's and Perrin's concepts of the flat electric layer are not enough for explaining electrokinetic phenomena.

2. *Gouy-Chapman theory* : A significant step forward was the theory of the electric double layer with a diffuse counterion layer, proposed independently by Gouy (1910) and Chapman (1913). This theory eliminated many of the shortcomings of the Helmholtz-Perrin theory. According to the Gouy-Chapman theory, counterions cannot concentrate only at the interface and form a monoionic layer, but are scattered in the liquid phase at a certain distance away from the interface. Such a structure of the double layer is determined, on one hand, by the electric field of a solid phase which attracts an equivalent amount of oppositely charged ions as close as possible to the wall and on the other, by the thermal motion of ions as a result of which counterions scatter in the liquid volume.

The action of the electric field prevails in the immediate proximity of the interface. As the distance from the interface increases, the intensity of the field gradually decreases and counterions of the double layer scatter more and more owing to thermal motion, as a result the concentration of counterions diminishes and becomes equal to that of ions which are connected with the solid, originates. Of course, the equilibrium of this layer is dynamic.

On the other hand, the ions in the liquid that bear the same sign as the potential-determining ions absorbed by the wall are repulsed from the solid phase by electric forces and go deep into the solution. This causes the distribution of potential-determining ions and counterions in the diffuse part of the electric double layer, as is illustrated in Fig. 1.20.

The structure of the electric double layer according to Gouy and Chapman and he potential drop in it are schematically illustrated in Fig. 1.21. The potential drop follows no a straight line, but a curve, because the counterions which compensate the wall charge are distributed non-uniformly. The curve drops steeper in places where there are more compensating counterions, and, conversely, it is smoother in places where there is a small amount of compensating counterions.

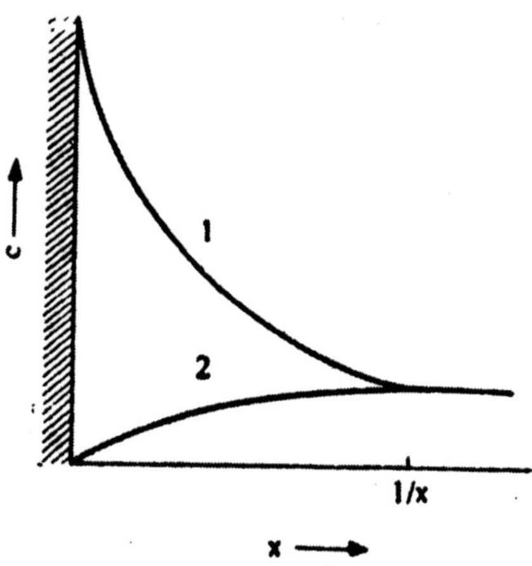

Fig. 1.20 : Dependence of the concentration c of counterions (curve 1) and potential-determining ions (curve 2) on the distance x from the plane wall (l/x is the thickness of the electric double layer)

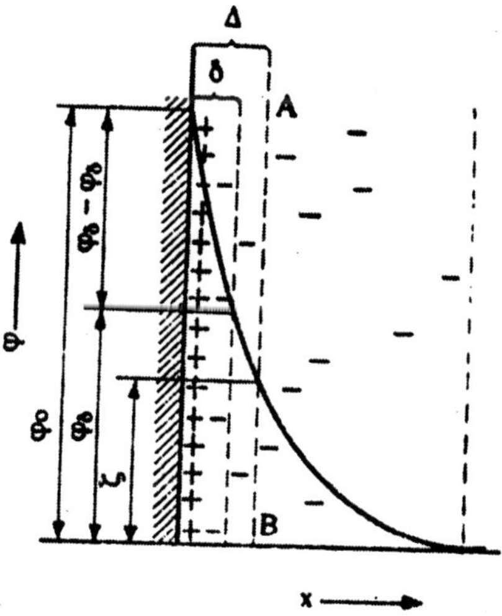

Fig. 1.21 : Electric double layer according to Gouy and Chapman and the potential drop in it.

Since the extent of diffusion of the counterion layer is determined by thermal scattering, all the counterions should be at the solid surface at a temperature of absolute zero. Consequently, the theoretical concepts of Helmholtz and Perrin turn out to be a particular case of the Gouy-Chapman theory.

The equilibrium distribution of ions established at the solid wall is similar to that of gas molecules established in the atmosphere under the action of gravity, the only difference is that the gravitational field does not depend on the distribution of molecules whereas in the case of the electric double layer, the electric field is a function of the distribution of charged ions. The number of counterions at the charged surface of a solid phase decreases according to the Boltzmann distribution law as the distance from the interface into the solution grows whereas the number of potential-determining ions increases according to the same law. It thus follows that if the concentration of positive and negative ions in a point, whose potential is φ_x, is denoted by c_+ and c_- (in moles per unit volume), then for distance $x = \infty$

$$C_+ = C_- = F\infty \quad \text{...(2)}$$

and for a distance x which is not too great

$$c_+ = c_\infty \exp(-Fz\varphi_x/RT) \quad \text{...(3)}$$

$$c_- = c_\infty \exp(Fz\varphi_x/RT) \quad \text{...(4)}$$

where c_∞ = electrolytic concentration at an infinitely great distance from the solid phase at $\varphi_\infty = 0$;

F = Faraday number (96.540 coulombs);

2 = valency of the ion;

R = gas constant;

T = absolute temperature.

According to the Boltzmann distribution equation, the product $Fz\varphi_x$, in the exponential factor of Eqs. (3) and (4) is the electric work of transfer of one mole of the appropriate type of ions from the volume of the solution (where $\varphi = 0$) to the point having the potential of φ_x.

It is clear that, at $x = \infty$, $\varphi_x + \varphi_\infty = 0$ and, at $x = 0$, $\varphi_x = \varphi_0$, *i.e.*, the potential is equal to zero at an infinitely great distance away from the interface, and is equal to the total surface potential jump at the interface.

The concepts evolved by Gouy and Chapman explain some electrokinetic phenomena. For example, since the slipping plane AB, when the solid and liquid phases move relative tome another, lies in the liquid at a small distance Δ away from the interface where the potential still does not drop to that of the liquid (Fig. 1.21), the difference between it and the potential in the liquid in this place corresponds to the charge of this part of the diffuse layer. It is this potential that determines the movement of phases when the electric field is applied, *i.e.,* the potential causes electrophoresis or electro-osmosis. The electrokinetic potential, or, as it is often called, the *ζ-potential,* is a part of the total jump of potential φ_0. Hence, it is clear why the electrokinetic potential differs from zero, but is not equal to the total potential jump. Moreover, the structure of the electric double layer proposed by Gouy and Chapman explains why different factors affect the two potentials differently.

Let us see, for example, how the introduction of an indifferent electrolyte into a system affects both of these potentials. In this case, the total potential jump barely changes. Things are quite different with the electrokinetic potential. As the concentration of the electrolyte being introduced increases, the thickness of the diffuse layer decreases because the same (equivalent) number of charges of the opposite sign is always required for compensating the potential-determining ions. The electric double layer is said to contract. As a result, the distribution of the potential in it also changes. Moreover, there will be a change in the ζ-potential which corresponds to the slipping plane of a liquid in electrophoresis or electro-osmosis.

The change in the ζ-potential whenever increasing quantities of an indifferent electrolyte are introduced is shown in Fig. 1.22. When the concentrations of the electrolyte arc large enough, the diffuse layer may contract to a monoionic layer, and the electric double layer thus becomes the Helmholtz-Perrin layer. Of course, ζ-potential will be equal to zero because this layer will be closer than the slipping plane to the wall.

The Gouy-Chapman theory is based on the same physical concepts as the Debye-Hückel theory of strong electrolytes, incidentally, the former originated earlier than the latter.

According to the strong electrolyte theory the thickness of the ionic atmosphere l/x is determined by the equation

$$\frac{1}{x} = \sqrt{\frac{\epsilon kT}{8\pi e^2 N_A \Sigma c_i z^2}} = \sqrt{\frac{\epsilon RT}{8\pi e^2 N_A^2 \Sigma c_i z^2}}$$

$$= \sqrt{\frac{\in kT}{8\pi F^2 \ \Sigma c_i z_i^2}} \qquad ...(5)$$

where $\in$ = absolute dielectric constant of the liquid;

k = Boltzmann constant;

T = absolute temperature;

e = electron charge;

N_A = Avogadro number;

c_i = bulk concentration of different ions;

z_i = valency of different ions;

R = gas constant;

F = Faraday number.

Eq. (5), which connects l/x, c_i and z_i, can be simplified to

$$\frac{1}{x} = k \frac{1}{\sqrt{\Sigma c_i z_i^2}}$$

where k = equation constant.

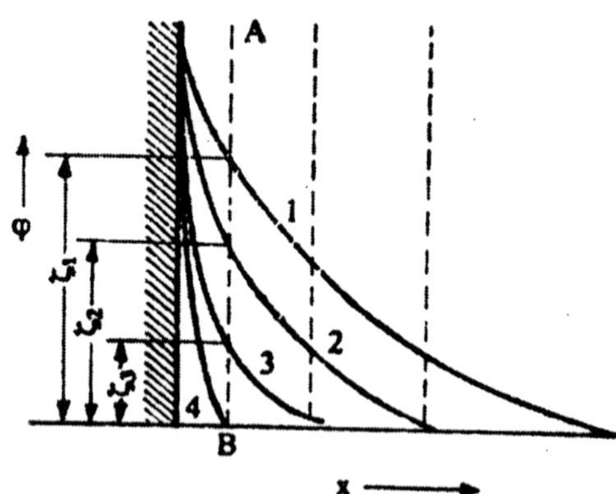

Fig. 1.22 : Effect of an indifferent electrolyte on both the thickness of the electric double layer and the electrokinetic potential (the amount of the electrolyte increases from curve 1 to curve 4)

It follows from the given equation that the ionic atmosphere should be the smaller, the higher is the concentration of ions in the solution and the greater is their valency. Naturally, a similar correlation is also maintained when the ionic atmosphere is formed not around the ion, but near the interface. However, it is only at small surface potentials that the value of the ζ-potential is determined merely by the thickness of the ionic atmosphere, and it is only under these conditions that it will be equally affected by the charges of counterions and those of secondary ions Which are present in a system. As theory shows, when surface potentials are high (over 50 mV), the decrease in the ζ-potential is affected much more strongly by the charge of a counterion, especially when it is large. The physical meaning of this phenomenon is that a strongly charged counterion of an electrolyte is attracted to the surface and greatly screens it. The growth in the ability of the counterion to reduce the ζ-potential with an increase in its valency can be clearly seen from the following experimental data:

Counterion valency	IV	III	IT	I
c_t	1	10-20	60-100	500-1000

In this case, the value of c_ζ is the relative counterion concentration that is required for reducing the electrokinetic potential to the same value, such as 50 mV. As we see, this series does not give any exact relationships. In fact, this cannot be expected because the ability of the ion to compress the electric double layer depends not only on its valency, but also on its size, polarizability, the ability to be hydrated, and so forth. Proceeding from Gouy's and Chapman's concepts, Müller showed purely mathematically that the ability of counterions to reduce the ζ-potential should grow rapidly with their valency; in addition, for a flat intçrfacc, he calculated the series

Counterion valency	IV	III	II	I
c_ζ	1	7.5	57	540

This is in good agreement with experimental data.

The Gouy-Chapman theory explains well a decrease in the ζ-potential when the concentration of the counterion and its valency increase. The untenability of the Gouy-Chapman theory can be shown in a way. If an electrolytic solution is not too dilute (say 0.1 N) and the potential near the wall is high (for example, 200 mV), for the theoretical and empirical values of capacity to coincide, the counterion concentration near the wall should be 300 N. and this is impossible. The experimentally and theoretically found values of capacity of the electric double layer differ because the Gouy-Chapman theory does not take the dimensions of ions into account and regards them as point charges which can closely approach the wall; this determines the higher values of theoretical quantities as compared to empirical ones.

Another shortcoming of the Gouy-Chapman theory is that it does not explain the reversal of charge, *i.e.*, a change in the sign of the electrokinetic potential when an electrolyte having a polyvalent ion whose charge is opposite in sign to that of the dispersed phase is introduced into a system.

Furthermore, the Gouy-Chapman theory does not explain why different counterions of the same valency produce different action on the electric double layer. According to the theory, the introduction of an equivalent amount of different counterions of the same valency should compress the electric double layer and reduce the ζ-potential to the same extent. However, experiments show that this is not so. The effectiveness of the action of ions of the same valency on the electric double layer grows with the ionic radius.

The Gouy-Chapman theory is relatively well applicable to sufficiently diluted colloidal solutions, but not to concentrated ones.

All these difficulties are overcome to a considerable extent in the theory of the structure of the electric double layer proposed by Stern.

3. *Stern's Theory :* In 1924, Stern proposed a scheme for the structure of the electric double layer in which he combined the Helmholtz-Perrin and Gouy-Chapman schemes. In elaborating the theory of the electric double layer. Stern used two prerequisites. First, he presumed that ions have certain finite dimensions and consequently ionic centres cannot be closer to the solid surface by a distance smaller than the ionic radius. Secondly, Stern took account of the specific, non-electric interaction of ions with the solid surface. This interaction is caused by a field of molecular (adsorption) forces at a small distance away from the surface. As we will see when we discuss the cause of the stability and coagulation of colloidal systems, the molecular forces which act between bodies consisting of many molecules are long-range ones owing to their additivity. However, when ion interacts with the solid surface, their action rapidly decreases with distance and therefore should be taken

into account only at the very surface of a solid (at a distance of several angstroms).

According to Stern, the first layer or even the first several layers of counterions are attracted to the wall under the action of both electrostatic and adsorption forces. As a result, a part of counterions is retained by the surface at a very close distance which is of the order of one-two molecules, forming a plane condenser having a thickness of δ, as envisaged by the Helmholtz-Perrrin theory. This layer in which the electric potential falls sharply is termed the *Helmholtz layer*, or the *Stern layer*, or the *adsorption layer*. The remaining counterions which are needed for compensating the potential-determining ions form, as result of thermal scattering, a diffuse part of the double layer: they are distributed in it according to the same laws as in the diffuse Gouy-Chapman layer. This part of the double layer in which the potential falls gradually is sometimes called the *Gouy layer*. The scheme of the electric double layer according to Stern and the electric potential drop in it are illustrated in Fig. 1.23.

It follows from the illustration that the total potential drop φ_0, consists of the potential drop y, in the diffuse part of the double layer and the difference in potentials $(\varphi_0 - \varphi_\delta)$ between the condenser plates. It still remains unclear where the slipping plane is in such a layer. Some authors believe that it coincides with the boundary between the Helmholtz layer and the Gouy layer. However, in general, we may suppose it to be in the Gouy layer, as is represented in Fig. 1.23 (the slipping plane is denoted by the dotted line AB). Hence, the potential at the boundary between the Helmholtz layer and the Gouy layer must not necessarily be equal to the ζ-potential.

Of course, when electrolytes are introduced into a system, the diffuse layer will contract and an increasing number of counterions will appear into the adsorption layer. According to Stern, the electric double layer in this case draws closer and closer to the layer envisaged in the Helmholtz-Perrin theory, while the ζ-potential decreases, gradually approaching zero. Conversely, when a system is diluted the diffuse layer expands and the ζ-potential grows.

According to Stern's theory, the distribution of ions in the electric double layer is strongly affected by the nature of counterions. If counterions have different valencies, the thickness of the diffuse layer and the number of counterions in the adsorption layer are determined mainly by the ionic valency and, consequently, depend on electrostatic forces. Of course, the diffuse layer is the thinner and the ζ-potential is the smaller, the greater is the valency of counterions. In this case, account should be taken of the same considerations as those in explaining the effect of the counterionic valency on the ζ-potential according to the Gouy-Chapman theory.

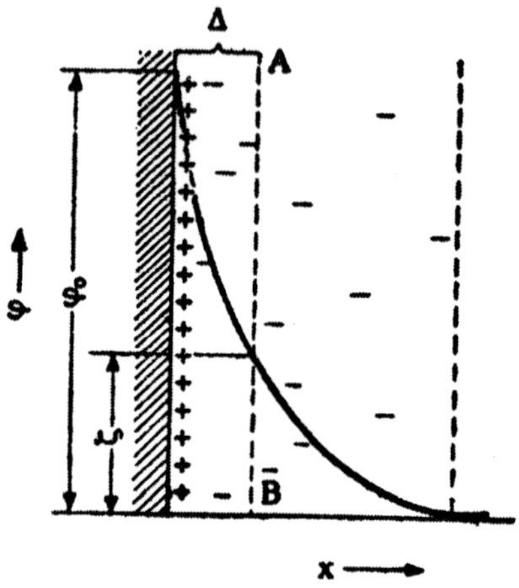

Fig. 1.23 : Electric double layer according to Stem and the potential drop in it.

If counterions have the same valency, the thickness of the electric double layer and the number of counterions in the diffuse layer are determined by the specific adsorbability of ion, which, depends on their polarizability α and hydration. These properties of ions are determined by their true radius or, what is the same, by the position of the elements in Mendeleyev's Table.

Cation	*$\alpha \times 10^{24}$, cm²*	*Anion*	*$\alpha \times 10^{24}$ cm³*
Li^+	0.03	F^-	0.96
Na^+	0.19	Cl^-	3.60
K^+	0.89	Br^-	5.00
Rb^+	1.50	I^-	7.60
Cs^+	2.60		

The great polarizability of the ion should, of course, favour a decrease in the thickness of the electric double layer because then additional adsorption forces originate between a solid and the induced dipole; in addition, the ion can come closer to the surface. Anions are usually polarized to a greater extent than cations because the ion deforms more as its dimensions grow and since the anionic radii are, in general, considerably greater than the cationic radii :

That incidentally explains why the negative surface charge is found in nature far more often than the positive one. Ionic hydration decreases as the true ionic radius increases. A decrease in ionic hydration should promote the compression of the electric double layer because the hydrate membrane reduces electrostatic interaction between the counterions and the solid surface.

It becomes clear from the foregoing why the ability to compress the electric double layer and reduce the ζ-potential grows in the series of cations from Li^+ to Cs^+ and m the series of anions, from F^- to I^-. Fig. 1.24 shows the dependence of the ζ-potential, of the negatively charged day particles in an aqueous suspension, which is saturated with various monovalent and bivalent counterions, on the size of ions. As we may see, the dependence of the ζ-potential on the radius of bivalent cations is the same as that of monovalent cations.

The values of the capacity of the electric double layer, calculated according to the Stem theory with regard to ionic radii, are similar to the empirical values, and therefore the theory overcame one of the shortcomings of the Gouy-Chapman theory. In addition, unlike the Gouy-Chapman theory, the Stern theory can explain why the sign of the electrokinetic potential changes when polyvalent ions whose charge is opposite in sign to that of the dispersed phase are introduced into a system. Such polyvalent ions enter the adsorption layer owing to both strong electrostatic interactions and great adsorbability connected with the polarizability of such ions. Ions can be adsorbed in so great an amount that they not only neutralize the charge of the solid surface, but also reverse its charge. Then, as may be seen from Fig. 1.25, the pattern of the potential drop in the electric double layer changes whereas the φ_{δ} and ζ-potentials which earlier had the same sign as the φ-potential ($\varphi_{\delta 1}$ and ζ_1 in the scheme) change to the opposite sign ($\varphi_{\delta 2}$, and ζ_2 in the scheme). Of course in this case, the potential φ_0 remains constant because extraneous ions are incapable of completing the construction of the crystal lattice of a solid. Hence, φ_{0-} sod ζ- potentials have different signs. Of course, a further increase in the concentration of an electrolyte in a system should cause the diffuse layer to contract and the ζ-potential to drop.

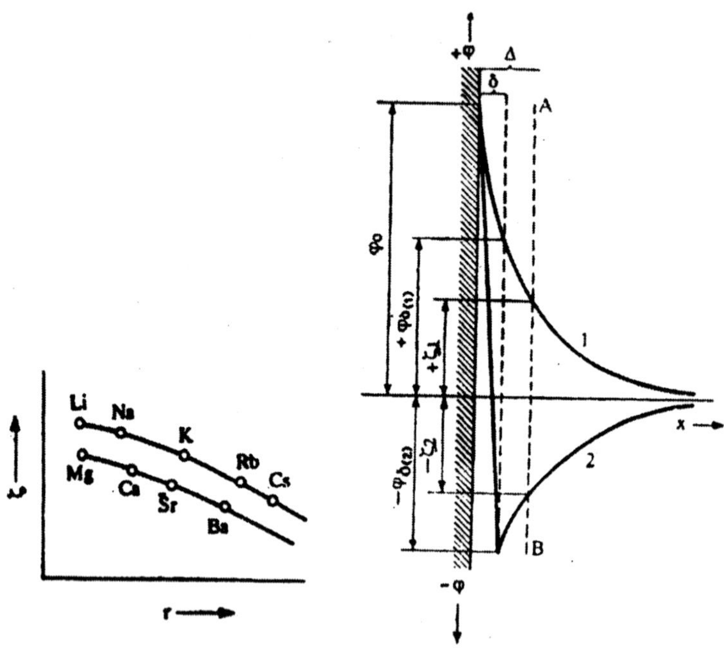

Fig. 1.24 : Dependence of the ζ-potential of negatively charged clay particles on the counterion radius r.

Fig. 1.25 : Potential drop in the electric double layer changed after a recharge by strongly adsorbed ions : 1–potential drop before a recharge; 2–potential drop after a recharge.

Polyvalent ions of aluminium and thorium are often used for recharging the particles which have a negative ζ-potential. In fact, monovalent ions can also cause a recharge if they have a great adsorption potential. There are ions of many alkaloids, such as strychnine and quinine, and basic dyes such as crystal violet and methylene blue. This is because such large ions not only can be polarized, but also have constant dipole moments.

It follows from the foregoing that the Stem theory corresponds to experimental results better than the Gouy-Chapman theory. Since it specifies the role of the ion dimensions and introduces the concept of the adsorption potential, it can explain several features of the action of electrolytes on the electric double layer and the electrokinetic potential. However, this theory is also not perfect because it is based on several assumptions and has many ambiguities, *e.g.*, it allows the adsorption potential to be independent of concentration, which is hardly probable. The concepts of the slipping plane in the electric double layer are not accurate. The slipping plane as conceived by Stern apparently does not exist; there is the slippage of liquid from the interface deep into the solution, and it begins at a certain distance from the solid surface. Thus, the value of the ζ-potential, an important characteristic of any colloidal system, is determined in electro-kinetic phenomena not only by the nature of the potential drop in the electric double layer, but also by the nature of liquid motion near the solid surface that depends on its rheological properties.

Several studies were later carried out in order to specify Stern's assumed structure of the electric double layer. Among them, of the greatest importance is Graham's work according to which the Helmholtz layer consists of external and internal parts. The external part contains hydrated counterions which are retained at the interface by electrostatic forces. The internal part contains ions which are chemisorbed by the solid surface, carry a charge of the same sign as that of the solid and have partially destroyed hydrate shells.

Quantitative treatment of Zeta potential : Consider the double layer to be electrical condenser with parallel plates d cm apart and each carrying a charge e per sq. cm. If ξ is the potential difference between the two plates of the electrical condenser, which is given by :

$$\xi = \frac{4\pi \, de}{D} \qquad ...(1)$$

D = Dielectric constant of the medium. This is the fundamental equation for all types of electro-kinetic phenomena.

Let a liquid be forced by electro-osmosis through the fine capillaries constituting the porous diaphragm. The rate of flow depends upon;

(i) Force of electro-osmosis.

(ii) Frictional force between walls and moving layer.

The electrical force causing osmosis = E.*e.*

Here E is the potential gradient across the membrane, e is the charge at the boundary of movement per sq. cm.

Calculation of frictional force : Assuming that there is no slipping of liquid at surface of the wall. If u is the uniform velocity and d is the thickness of the double layer, then the velocity gradient = u/d.

Frictional force = η x u/d.

where η is viscosity of the liquid. ...(3)

The rate of flow will be uniform when these forces balance each other.

$$Ee = \eta u/d$$

$$d = \eta.u/Ee \qquad ...(4)$$

Putting the value of d in Eq. (1), we get

$$\xi = \frac{4\pi\eta n}{DE} \qquad ...(5)$$

If the thickness of the double layer is negligible as compared with pore diameter, then it follows that

$$u = V/q \qquad ...(6)$$

where q = area of cross section of all the pores in the diaphragm; V is volume of the liquid passed during electro-osmosis per sec.

Equation (5) becomes

$$\xi = \frac{4\pi\eta.V}{E\ Dq} \qquad \text{(Putting V = qu)} \qquad ...(7)$$

If r is the radius of particles and the electro-osmosis takes place in a single capillary, then q = πr^2, we have

$$\xi = \frac{4\pi V}{r^2\ DE} \qquad ...(8)$$

$E = \frac{I}{qkv}$ where I = strength of current,

ky = sp, conductivity of the liquid. Eq. (8) becomes

$$\xi = \frac{4\eta Vk_V}{r^2\ DI} \qquad ...(9)$$

If the liquid is allowed to accumulate on one side of a diaphragm, then a hydrostatic pressure sets up which balances the flow of electro-osmosis.

According to *Polaeuille's* equation

$$V = \frac{\pi r^2 P}{8\eta e} \qquad ...(10)$$

where V = volume of the liquid flowing per second,

l = length of capillary,

η = viscosity of the liquid,

P = pressure difference and

r = radius of capillary.

Substituting the value of V in eq. (8) we get,

$$P = \frac{2\xi DEI}{\pi r^2} \qquad ...(11)$$

From the above equation (11), it is clear that for a given diaphragm material, the difference of potential is proportional to applied voltage, *i.e.*, E, I, D and S do not depend upon the dimensions of the diaphragm, πr^2 is also constant. This is in good agreement with *Widemaan*'s results.

Measurement of zeta potential : Zeta potential is best measured by the following methods.

1. *From streaming potential* : The velocity of a liquid flowing through capillary tube varies with the distance x from the centre of the tube. According to Poiseuilles, the velocity is given by the following relation :

$$u = \frac{P(r^2 + x^2)}{4\eta l} \qquad ...(12)$$

where r is the radius and l is the length of the tube. Consider a moving part of the layer Fig. (8), which is at a distance (r – d) from the centre. The velocity of such a layer will be given by the expression

$$u = \frac{P[r^2 - (r - d)^2]}{4\eta l} \qquad ...(13)$$

Since d^2 is small as compared to 2rd and hence the former can be neglected. Under these conditions the velocity of the double layer is

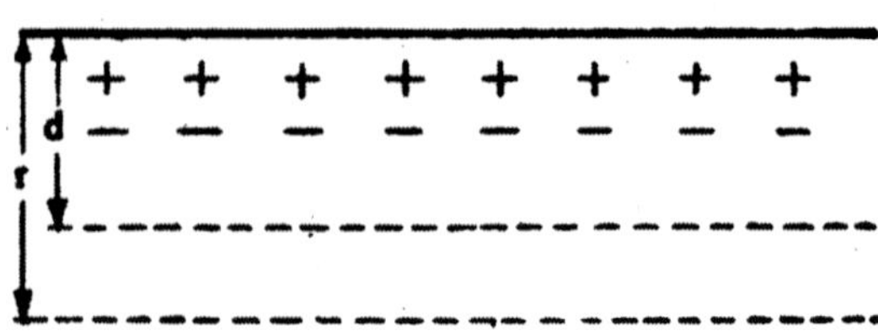

Fig. 1.26

$$u = \frac{Prd}{2\eta l} \qquad ...(14)$$

The intensity of the current produced by this flow of the counterions is the electric charge across the unit tube.

$$I = 2\pi ru\sigma \qquad ...(15)$$

$2\pi r$ = area of the double layer moving in unit time.

σ = charge density.

Putting the value of u from Eq. (14) in (15), we get

$$I = \frac{P\pi r^2 d\sigma}{\eta l} \qquad ...(16)$$

Consider the specific conductivity of the liquid to be uniform then the actual conductivity of the liquid will be $\frac{\pi r^2 k}{l}$

The streaming, potential S may be given by the expression (according to Ohm's law).

$$S = I.R = I = I.\frac{l}{\pi r^2 k} \qquad ...(17)$$

Put the value l from Eq. (16) in (17), we get

$$S = \frac{\sigma d\,P}{\eta k} \qquad ...(18)$$

We know $\xi = \frac{4\pi\sigma d}{D}$ then substituting the value of a from this equation in (18), we obtain

$$S = \frac{\xi D.P}{4\pi\eta k} \qquad ...(19)$$

From the above expression we find that streaming potential is proportional to driving force P. Eq. (19) can be employed in the calculation of zeta potential from streaming potential.

2. *By electrophoresis* : It is also possible to measure the zeta potential from the electrophoretic mobility (u_e) which may be defined *as the velocity with which a solid cylindrical particle is moving through a liquid under the influence of an applied electric field of unit potential gradient. The electrophoretic mobility is given by the following equation:*

$$u_e = \frac{\delta D}{4n\eta}$$

The above relation has been deduced on the basic assumption that the particle is solid in nature. However, this equation may be applied to any particle, *i.e.*, solid, liquid or gas, suspended in a liquid. The main limitation of this equation is that it cannot be applied to the particles of different shapes. According to Smoluchoswki's treatment, the electrophoretic mobility does not depend upon the shape of moving particle. According to Debye and Huckel (1924) if the thickness of-the double layer is very small in comparison with the radius of the particle, then the velocity for electrophoresis may be given by the following equation :

$$u_e = \frac{\xi D}{6n\eta} \qquad ...(21)$$

The above equation is employed for determining zeta potential. The value of u, is for the potential gradient of 1 e.s.u. Also, ξ is expressed in e.s.u. If the electrophoretic mobility is for potential gradient of 1 volt per cm (1 e.s.u. = 300 volts per cm) and the zeta potential is in volts, then Eq. (21) may be put in the following form :

$$u_e = \frac{\xi D}{4\pi\eta} \times \frac{1}{9} \times 10^{-4}$$

or $$\xi = \frac{3\pi\eta u_e}{D} \times 9 \times 10^4 \text{ volt} \qquad ...(22)$$

The value of S comes out to be 0.03 to 0.06 volt if the electrophoretic mobility u, for small colloidal particles is about 2 to 4 $\times$ 10^{-2} cm per second in water and η is 0.01 c.g.s. units and D is nearly 80.

Zeta potential can be measured experimentally with the help of a micro-electrophoretic cell shown in Fig. 1.27. It essentially consists of a flat rectangular glass cell A with a very small but uniform depth throughout. At the two ends, the celi A is prolonged into two cylindrical tubes B and C, which terminate into two vertical tubes K and K' and two outlets D and D'. Each end consists of a double way stop cock F and F'. The function of which is to allow any liquid or sol in A, B and C to be put in communication with either the vertical tubes K and K' or outlets D and D'.

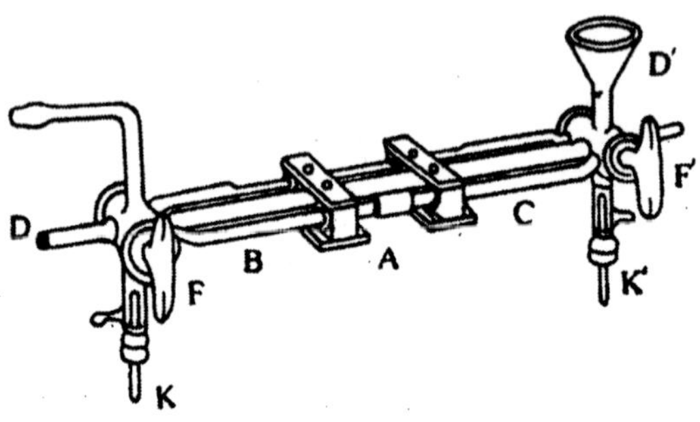

Fig. 1.27

A small amount of gelatin sol is made first conducting by adding a little KCl to it. It is then poured in vertical tube K and K' and allowed to form a gel. The sol is then placed in the microelectrophoretic cell (*i.e.*, B, A and C) by manipulating the stop cocks F and F', and permitting outlet D' as inlet and sucking through E. By turning the stop cocks, the sol is now placed in communication with the gel already present in vertical tubes K and K'. A powerful microscopic cell lies exactly under the objective of the microscope. The sol particles in A at a particular level are focused and a direct current of about 300 volts is applied through two copper wires immersed in a solution of $CuSO_4$ which rests on the gels placed in K and K'. The circuit is completed by inserting a sensitive milliammeter, suitable resistance and a commutator. The sol particles will move towards oppositely charged electrodes and their rate of motion is noted from a previously calibrated scale within the eye piece Now we use Eq. (22).

From Eq. (22) zeta potential can be calculated, provided η, μ, K, q, D and I are known. The measurement of zeta potential thus involves the measurement of the rate of moving colloidal particles under the influence of an electric field ; other factors can be known easily, The (μ) can be best calculated by *Burton*'s method. It should be noted that this method of determining zeta potential is applicable only to emulsions.

3. *By Electro-osmosis :* The apparatus used is shown in Fig. 1.28. The coagulum obtained by the addition of a neutral electrolyte to a sol is kept in a U-tube, each arm of which is provided with a small side tube. The side tubes are joined through pressure

tubing to a rectangular bent glass tube having a uniform cross section. The coagulum is now centrifuged in order to form a compact diaphragm filling the bottom as well as a little way up each of the two arms of the U-tube. Rest of the U-tube (above the diaphragm), the side tubes and the rectangular glass tubes are now completely filled with the liquid obtained during the formation of coagulum form the sol. Two Pt black electrodes are inserted into the two limbs of the U-tube, through which any electric field may be applied and the continuity of the liquid in the rectangular tube is stopped by introducing an air bubble at the middle of the tube. The circuit is completed by introducing a milliammeter, a suitable resistance and a commutator as usual.

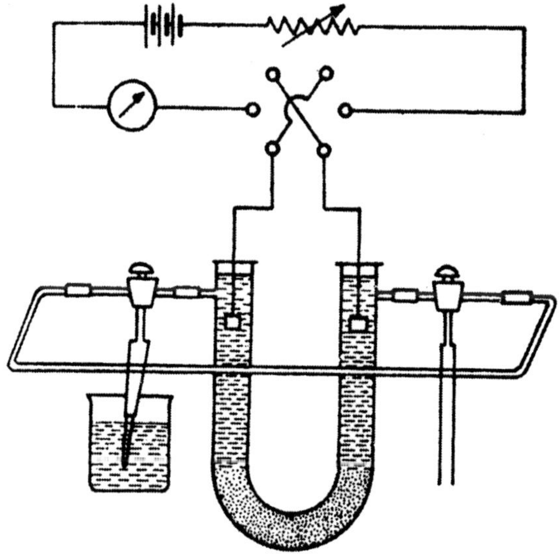

Fig. 1.28

If an electric field is applied, electro-osmotic flow of liquid takes place through the diaphragm and the air bubble introduced in the middle of the rectangular tube will now move slowly in the tube. The volume rate of flow of liquid through the diaphragm can easily be calculated from the rate of flow of the bubble. Now we know from equation (22) that

$$\xi = \frac{4\pi\eta KV}{DI} \cdot (300)^2 \text{ volt}$$

where V is the volume of liquid transported per second by electro-osmotic flow. The direction of flow of the bubble shows the nature of the charge of the diaphragm formed by the coagulum.

Variation of Zeta Potential with Concentration

The complete nature of variation of zeta potential with the concentration of added electrolyte can be only understood by combining two electrokinetic phenomena, viz., Electro-osmosis and electrophoresis. It has been observed that the magnitude of zeta potential gets decreased with the increasing concentration of added electrolyte. It has also been observed that if zeta potential is plotted against log C, where C is the concentration, then straight lines would be obtained. Thus, zeta potential may be represented by an empirical equation, ξ + A–B log C where A and B are constants. It is generally observed that increase of OH^- ion concentration increases the negative zeta potential while H+ ion concentration decreases it.

Zeta potential is also influenced by various other factors, like surface density of the charge on the wall, dielectric constant of the medium and the thickness of the double layer etc. The surface density of the charge on the wall may be markedly increased or decreased by changing the number and kind of adsorbed ions. An increase in the ionic concentration will decrease the effective thickness of the double layer and will also affect the value of dielectric constant of the medium close to the wall layers.

Zeta Potential and Stability of Sols

It is known that the most common method of producing precipitation of colloids is by the addition of electrolytes, which though essential in small concentrations for stability, produce in higher concentration coagulation or flocculation of the dispersed phase. Although the stability of a sol depends upon several factors, but the most important one is the electric charge carried by the particle or zeta potential which results because of the charge in consequence of which there are repulsive forces between the particles.

If an electrolyte is added to the sol, the adsorption of ions occurs and the absolute value of the charge of the particle as well as that of zeta potential is decreased. This has been confirmed by numerous experiments. It should be noted that coagulation takes place only when the charge or zeta potential has been reduced to zero.

Importance of Zeta Potential

Zeta potential is of great importance in gaining some idea about the mechanism of coagulation in which colloidal particles coalesce to form large and large aggregates which finally precipitated out. It appears that when zeta potential is reduced below the critical value, the repulsion between approaching particles is also reduced to such a great extent that those coallescing with certain velocity can join together to form aggregates and ultimately coagulation occurs.

Zeta potential gives us important idea about the mechanism of precipitation. It has been seen that the presence of ion of opposite sign to that of colloidal particles leads to decrease in the value of zeta potential. The decrease of zeta potential leading to precipitation takes place by adsorption of ions of opposite sign to that carried by the colloidal particles.

From the knowledge of zeta potential it is also possible to explain the phenomenon of anomalous precipitation. When an electrolyte is added, in excess of a certain amount, to a sol it will not alter the zeta potential greater than the critical value. The sol is thus stabilized, although further addition of electrolyte will cause the zeta potential to decrease again and allows the coagulation to take place.

Influence of Ions on Electrokinetic Potential

The addition of salts alters the electrokinetic potential, *i.e.,* zeta potential. Also, the higher the valency of the ions of added salts, the greater is the influence of electrokinetic phenomenon.

Freundlich and Fttisch (1923) studied the variation of the electrokinetic potential with different electrolytes in a glass capillary tube. They observed that the zeta potential of glass water (pure) interface is negative. When KCl or $BaCl_2$ is added, there occurs an increase in the negative charge of the glass due to the action of the chloride ions. With the addition of more electrolyte. The zeta potential increases further, then reaches to maximum and finally decreases (Fig 1.29).

When the cations of high valence, (*e.g.,* $La^{3}+$, Th^{4+}, are added, they affect the zeta potential considerably. Further; the influence of the anions is so small that it cannot be observed. It is also evident that at high concentrations of thorium nitrate the zeta potential first becomes positive and then tends to reverse in sign due to the effect of negative ion (Fig. 1.29).

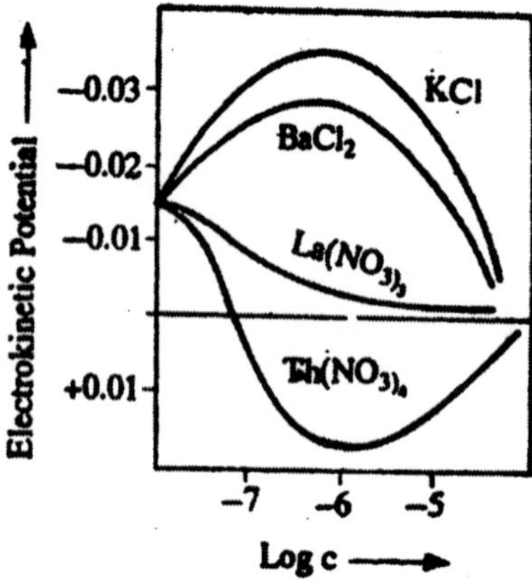

Fig. 1.29

Other factors affecting the zeta potential : Following factors have been found to influence the zeta potential:

(i) Surface density, thickness of the double layer and the dielectric constant of the medium.

(ii) Nature and concentration of the dissolved matter. It also depends upon the nature of the material of the wall.

(iii) The relation between the zeta potential and concentration of electrolytes is complicated. If S, is plotted against log C, where 'C' is the concentration, the straight lines are obtained. The zeta potential is represented as $\xi = A - B \log C$, where A and B are constants. Thus there exists a linear relationship between the zeta potential and the concentration of the sol.

SIZE OF COLLOIDAL PARTICLES

True image is not seen when the colloidal particles are examined under an ultra microscope, but simply the position is indicated. According to Bruyn colloidal particles would not have a diameter less than 5–13mμ since smaller particles do not polarise light.

There are various methods of determining the size of colloidal particles. Few of them have been described :

1. *By means of an ultra microscope* : The sol is diluted to the extent such as to allow its particles to be counted in a known volume

of the cell under the microscope. The number of particles (n) to be seen in unit volume was determined by directly counting the number of particles in a small illuminated volume (Volume containing M gms of the dispersed phase in unit volume) of the sol in the slit ultramicroscope. The volume of the particles (V) can be found out by the relation

$$V = \frac{M}{d}$$

where M represents the total weight of the particles in one ml,

d is the density of the particles. But V is also equal to nv, where v is the volume of a single particle, and n is the number of particles in one ml. Hence,

$$vn = \frac{M}{d} \text{ or } v = \frac{M}{nd}$$

If the particles are assumed to be spherical, then

$$n = (4/3)\, \pi r^3$$

where r is the radius of the particles, then

$$\frac{4}{3}\pi r^3 = \frac{M}{nd}$$

or $$r^3 = \frac{3M}{4\pi nd}$$

or $$r = \left[\frac{3M}{4\pi nd}\right]^{1/3}$$

2. *Rate of diffusion method :* We know that the diffusion coefficient 'D' is given by

$$D = \frac{RT}{NF} \qquad \text{...(1)}$$

where R is solution constant, T is absolute temperature, N is Avogadro's number and F is frictional resistance.

According to *Kirchhoff,* F is given by :

$$F = 6\pi r\eta$$

$$D = \frac{RT}{6\pi r\eta N} \qquad \text{...(2)}$$

where r is radius of the particle and η is the coefficient of viscosity.

Sutherland modified the above equation (2) and gave,

$$D = \frac{RT}{N} \cdot \frac{\left(1 + k\frac{\lambda}{r}\right)}{6\pi r\eta}$$

where K is constant (=0.155), and λ = mean free path of molecule of the dispersion medium. From Eq. (3) the value of 'r' can be calculated. Eq. (3) is only applicable to small particles because for large particles, λ/r vanishes and the equation reduces to Eq. (2).

3. *By involving Electromagnetic theory:* According to the electromagnetic theory of light, the wavelength (λ) which suffers maximum absorption when passed through a sol is related to the radius of colloidal particle by the following expression

$$r = \frac{\sqrt{3}.\lambda}{4\pi\mu}$$

where μ = refractive index of the dispersion medium.

With the help of the above relation, the size of the colloidal particles can be determined by simply knowing the refractive index and the wavelength (λ). The size of the particles of gold sols evaluated by the method lies between 4.9 and 5.2 × 13.6 cm; μ being 1.33.

4. *By involving Tyndall effect :* Rayleigh showed that the intensity of the scattered light (Is) can be expressed by the following relation :

$$I_S = K\frac{v^2}{d^2\lambda^2}$$

where v = volume of the colloidal particle

d = distance between the particle and the observer.

λ = wave-length of scattered light.

K = constant.

The relation is obeyed by insulating materials and not by metallic particles.

The limitation of the method is that the amount of scattered light depends not only on the size but also on the shape of the particle. This method is only suitable for determining the ratio of the size of the colloidal particles. For two sols of equal concentration, consisting of particles of different sizes, we will get

$$\frac{v_1^2}{v_2^2} = \frac{Is_1}{Is_2}, \frac{\lambda_1^4}{l_2^4} = \frac{r_1^6}{r_2^6}$$

5. *Equilibrium distribution :* The diffusion of particles in a sol is because of Brownian movement. In a diffusion cylinder the particles are driven up, in order to distribute evenly throughout the dispersion medium. Against this driving upward, however, acts the gravitational force pulling the particles down. With the result the system will acquire the equilibrium which is called the *sedimentation equilibrium.* This type of equilibrium takes place only when the system remains undisturbed and when the particles are not too large and heavy.

In case of sedimentation equilibrium for colloidal particles of uniform size the following expression has been derived by *Perrin* thermodynamically.

$$\log_e \frac{n_1}{n_2} = \frac{N}{RT} \cdot v\,(d_1 - d_2)\,g\,(h_2 - h_1)$$

where n_1 and n_2 are the average number of particles per unit volume at heights h_1, and h_2, v is volume of particles and d_1, d_2 are the densities of the particle and medium respectively. The equation be employed for determining the particle size by substitution.

$$v = \frac{4}{3} \pi r^3$$

6. *By making use of ultrafiltration :* It is possible to determine the particle size by ultrafiltration of the respective sols through ultrafilters of different pore sizes. For this purpose a series of ultrafilters is prepared, and the pore size can be determined by filtering colloids of known particle sizes.

   Another similar method involves dialysing through porous membranes. It has been observed that if m gms of a substance pass through a membrane in a certain time, then M, the molecular weight of the dialysing substance can be given as follows :

$$m = \frac{k}{\sqrt{(M)}}$$

   where k refers to a constant characteristic of the membrane and can be determined by using a substance of known M.

7. *By making use of altracentrifuge* : The determination of particle size or molecular weight by means of the ultracentrifuge can be done in two ways.

   (1) Sedimentation velocity method,

   (2) Sedimentation equilibrium method.

1. *Sedimentation velocity method* : It is possible to determine the size of colloidal particles, and particularly those of proteins, by the study of their distribution under the influence of an applied field. The method was developed by *Svedberg and his coworkers* (1925). In this method a specially designed high speed instrument called *ultracentrifuge* is used. With the help of this instrument the force exerted on a particle can be as great as millions of gravity or more.

If a particle undergoing sedimentation in a liquid medium is subjected to the gravitational force, then the gravitational force F is given as follows :

$$F = g \text{ (mass of the particle–buoyancy)} \quad ...(1)$$

Now, mass of the particle $= \frac{4}{3}\pi r^3 \rho$

and buoyancy = vol. of particle × density of liquid medium $= = \frac{4}{3}\pi r^3 \rho_1$

where g refers to the acceleration due to gravity, r the radius of the particle and p the density of the particle. Therefore,

$$F = g\left(\frac{4}{3}\pi r^3 \rho - \frac{4}{3}\pi r^3 \rho_1\right)$$

$$= \frac{4}{3}\pi r^3 g\,(\rho - \rho_1)$$

The sedimenting particle will come across frictional resistance R which is proportional to its velocity $\frac{dx}{dt}$, at a given time and is given by

$$R = f\,\frac{dx}{dt} \quad ...(3)$$

where f represents the frictional force. According to Kirchhoff it is given as follows :

$$f - \pi r \eta$$

Hence Eq. (3) becomes $R = 6\pi r\eta \frac{dx}{dt}$

where η refers to the viscosity of the medium.

If the drift velocity -. is constant, these two forces given by equations (2) and (4) will balance each other. Hence.

$$\frac{4}{3}\pi r^3 g(\rho - \rho_1) = 6\pi r\eta \frac{dx}{dt}$$

or $$\frac{dx}{dt} \frac{2r^2(\rho - \rho_1)}{9\eta} dt \qquad ...(5)$$

If w refers to the angular velocity of rotation, then the force exerted on a particle at a point distance x from the axis of rotation would be $w^2x.g$ times gravity. On putting g equal to w^2x in Eq. (5), we get

$$\frac{dx}{dt} \frac{2r^2(\rho - \rho_1)w^2x}{9\eta}$$

or $$\frac{dx}{x} \frac{2r^2(\rho - \rho_1)}{9\eta} dt \qquad ...(6)$$

$\frac{dx}{dt}$ represents the velocity of the particle when it is at a distance x cm from the axis of rotation. As the particle is undergoing sedimentation, the value of x and hence that of $\frac{dx}{dt}$ goes on varying continuously.

Suppose the particles be at a point distance x_1 after time t_1 and at a distance x_2; after lime t_2. Now integrating the above equation between these limits, we have

$$ln \frac{x^2}{x_1} = \frac{2w^2r^2(\rho - \rho_1)}{9\eta}(t_2 - t_1) \qquad ...(7)$$

The factor $ln \frac{x^2}{x_2} \Big/ w^2(t_2 - t_1)$ is termed as sedimentation constant (S) and hence the above equation becomes as follows :

$$S = \frac{2r^2(\rho - \rho_1)}{9\eta} \qquad ...(8)$$

If the sedimentation constant S is known, it is possible to determine the radius of the particle r from Eq. (8). The sedimentation constant can be determined by passing a beam of light through a cell containing the colloidal system under examination and allowing it to fall on a photographic plate. This process is to be repeated at various intervals of time (t) without stopping the ultracentrifuge, and position (x) of the upper boundary of the particles can be noted. The system will be a mono-dispersed system *i.e.,* it will be having particles of uniform size, if the value of S for successive intervals of time is constant. For a polydispersed system two boundaries are usually detected and from their position at various times, it is possible to calculate the respective sedimentation constant. Substituting the value of S in equation (8) the size of the particle can be calculated. This equation can also be used to determine the molecular weight. If m is the actual mass of the particle, then M, the molecular weight of the particle will be mN, where N is the Avogadro's number. If the particle is regarded to be spherical, then,

$$m = \frac{4}{3}\pi r^3 \rho$$

Hence $$M = \frac{4}{3}\pi r^3 \rho N \qquad ...(9)$$

From this equation the value of M can be calculated. If the particles are not spherical, then equilibrium method described below is used to determine the particle size.

Method : The outline of the ultracentrifuge is shown in the Fig. 1.30.

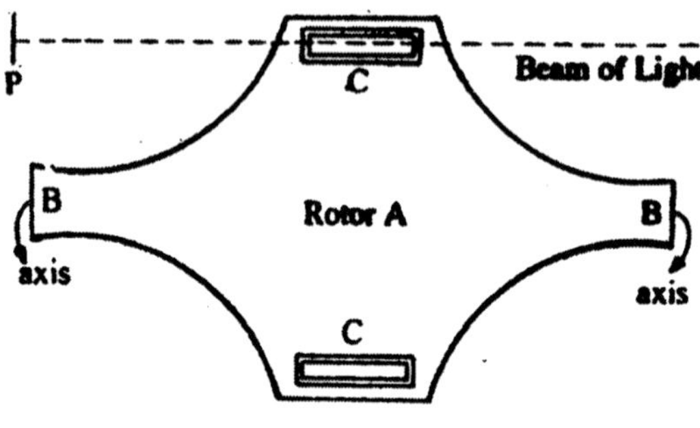

Fig. 1.30

It consists of a rotor A which is moving or rotating at a very high speed along the axis BB. The solution or suspension under examination is kept in a cell C which is provided with quartz window. On rotating the solution very rapidly, the sedimentation of particles takes place and a moving boundary gets produced, behind which only solvent is present. A strong beam of light is allowed to pass through the cell having the system under investigation and then allowed to fall on a photographic plate P. This process is to be repeated after various intervals of the time (t) without stopping the centrifuge and the position x of the boundary is noted. From this, the value of sedimentation constant and hence that of r can be calculated.

Sedimentation equilibrium method : This method has been based on the fact that if the amount of material driven outward by the centrifugal force gets balanced by the amount diffusing in the opposite direction, a state of equilibrium, known as sedimentation equilibrium is reached and the position of the boundary of the particles will remain unchanged.

Let a particle move a small distance dx in an infinitesimally small time dt. Now the amount of substance moving outward (dw) per sec per sq. cm,, under the action of centrifugal force in time dt will be c.dx, where c refers to the concentration of particles. Hence

$$dw = c\ dx \qquad ...(10)$$

Now it is known that the coefficient of diffusion D may be defined as the weight of material diffusing across a plane of one sq. cm. in area in unit time under a concentration gradient $\frac{dc}{dx}$ of unity. Therefore the weight dw diffusing across 1 sq. cm. in time dt may be put as follows :

$$dw = D\left(-\frac{dc}{dx}\right)dt = -D\frac{dc}{dx}dt \qquad ...(11)$$

If the sedimentation equilibrium gets reached both these values of dw given by Eqs. (10) and (11) are equal but of opposite sign. Hence

$$dw = cdx = D\frac{dc}{dx}dt \qquad ...(12)$$

or
$$\frac{dc}{c} = \frac{1}{D}\cdot\frac{dx}{dt}.dx$$

But
$$D = \frac{RT}{6\pi r\eta N} \text{ and } \frac{dx}{dt} = \frac{2w^2 \times r^2(\rho - \rho_l)}{9\eta}$$

On putting these values of D and $\frac{dx}{dt}$ in Eq. (12), we get

$$\frac{dc}{c} = \frac{4}{3}\pi r^2 N \cdot \frac{w^2(\rho - \rho_l)}{RT}x.dx \qquad ...(13)$$

$$= \frac{Mw^2(\rho - \rho_l)}{RT\rho}x.dx \qquad ...(14)$$

Where M denotes the molecular weight of the particle.

Integration of this equation yields:

$$\int_{c_1}^{c_2} \frac{dc}{c} = \int_{x_1}^{x_2} \frac{Mw^2 (\rho - \rho_1)}{RT.\rho} x.dx$$

or $$ln \frac{c_2}{c_1} = \frac{Mw^2 (\rho - \rho_1)}{2RT_\rho} \left(x_2^2 - x_1^2\right) \quad ...(15)$$

From equations (13) and (15) it is possible to calculate the radius and molecular weight of the particle respectively.

In order to calculate the value of M the same process is used as was used in determining the sedimentation constant. If absorption of light is assumed to be proportional to the concentration of the particles, the ratio $\frac{c_2}{c_1}$ at two depths x_1 and x_2 can be calculated photometrically from the blakening of the photographic plate.

(8) *X-ray method*

(9) Electron diffraction method.

METHODS OF DETERMINING PARTICLE SHAPE

Colloidal particles may have very different shapes. The most common types of shapes are depicted in the Fig. 1.31

Fig. 1.31

Although it becomes very difficult to give a classification of colloids based on particles shape, but the following type of classification has been found to be of great value.

Many methods are now available which are used for determination of particle shapes. The most important methods are given as follows :

1. X-ray method
2. Electron Diffraction method

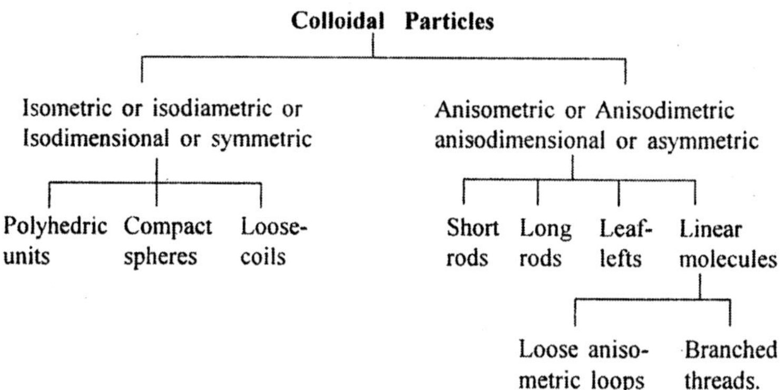

Fig. 1.32

(1) *X-ray method of determining the size and shape of colloidal particles :* As most lyophilic colloids are having crystalline properties, it is possible to use X-ray methods to find the size and shape of the particles. The X-ray method permits particle size determinations down to approximately 10.7 cm, of edge length, assuming the particles to be cubic in shape.

The first determination of this kind were carried out by *Scherrer* who estimated the size of particles of colloidal gold protected by gelatin, equal to 8.62 mμ. The volume V of the particles was 6.4×10^{-19} cm^3 and the mean weight of the particle was $m = V \times \rho = 1.24 \times 10^{-17}$ gm, where p, the density of the solid gold was 19 4g/cm^3. Since the lattice parameter of gold is 4.07Å. the number of lattice planes in each grain in the direction of the cube edge can be calculated as follows:

$$\frac{S}{a} = \frac{\text{Size}}{\text{lattice parameter}} = \frac{86.2}{4.07} = 21.2$$

since 8.62 mμ = 86.2Å

This reveals that there were 21 parallel lattice spacings in a gold particle of average size. Scherrer determined the diameter of nuclear gold at 18Å by means of broadening of X-ray diffraction lines. The same method was also employed by *Brill* (1933). Soot was first regarded to be amorphous, but now it is known that some soots are crystalline in nature but with an extremely small particle size. The powder patterns have been found to be very similar to those of graphite, except that the interference rings are broader. In other words, it can be said that soot consists of very small graphite crystals. The dimensions of these crystals were measured by *Brill* (1933).

The size of the particles can also be determined by small angle scattering but it is found that the latter is advantageous over the broadening of diffraction lines in the respect that small angle of scattering depends upon the density fluctuations in the sample. The small angle diffraction patterns are due to the action of monochromatic X-ray scattered by the separate particles of the sample, the scattering depends solely upon the electron distribution within the sample, which in turn is determined by the boundaries of the grains, their size and shape. In other words, scattering pattern gets formed by the separate electron packets having the size of the grains of the sample. This type of scattering has been studied by *Hosemann (1939), Guinier* (1939) *and Kratky* (1938).

According to *Yodwitch (1951)*, it is possible to determine the size of the particles from peak analysis and from stop analysis of the experimental curves. Using copper and aluminium radiations, he investigated the particle sizes of gold sols by both methods of analysis.

(2) *Electron diffraction method of determining size and shape of particles :* The size of the smallest particles which can be determined by X-ray method is about 10Å and the method cannot be improved even by employing short wavelengths. Electron diffraction method has been found to be most suitable for particle size determination in the range of 100–10A° and even below 10A° *i.e.,* 1 mμ. Hence for the determination of the size of slightly smaller particles electrons can be successfully used, and this method of determination has been found to be more accurate than X-ray methods.

According to *de Broglie (1924)*, the material particles also possess the properties of waves. The wave length of diffracted electrons is given as follows :

$$\lambda = \sqrt{\left(\frac{150}{V}\right)}\ 10^{-8}\ \text{cm}.$$

where V is the applied voltage. From this equation it is evident that the wavelength of the beam of electrons is shorter if the voltage is higher. This property of shorter wavelength is of great importance in the determination of size of very small particles.

Electrons get scattered by crystal lattices in the same manner as are X-rays. However, because of shorter wavelength only front reflection patterns are produced, which could be recorded on a plane photographic film. The electron diffraction method is as accurate as the X-ray method.

The diffraction methods are used for calculation of grain sizes and estimations of shapes. The resolving power of an ordinary microscope may be increased to about 0.17μ when visible light with a minimum wavelength of about 4000Å is used; but for very short radiation, such as X-rays which would enable us to reach a higher resolving power, there are no materials with which to built a convenient optical system.

It means that no material is available for lenses transparent to light of wavelength short enough to give a resolving power much less than 0.1μ. Consequently the idea of an *electron microscope* arised in several minds. An electron microscope, as the name implies, is a device to magnify minute objects (not directly visible in detail to the naked eye) where a beam of electrons is employed instead of light rays as in the ordinary microscope. The electron microscope, therefore, depends upon the following two principles.

(1) *A beam of electrons under a constant voltage is having the properties of a beam of light and at the voltage used the wavelength corresponds to 0.05Å.*

(2) *Electrons can be focused by suitable electric and magnetic fields, very much like light rays get focused by glass lenses.*

Electron microscope is an instrument in which the glass or quartz lenses of a light microscope are replaced by electrostatic or electromagnetic lenses and in which an electron beam instead of a beam of light is used. Also, the voltage and current being closely controlled. The colloid to be studied is kept on a very thin collodian film and a focused beam of electrons is passed through.

The image can be seen using a fluorescent screen, or recorded on a photographic plate. The whole interior of the instrument is evacuated so as to allow the electrons uninterrupted paths.

With the help of the instrument several viruses could be photographed and are found to be rod shaped. The size and shapes of various pigments and paints have also been estimated by electron microscope.

BROWNIAN MOVEMENT

Brownian movement is a phenomenon of fundamental importance for all colloidally dispersed systems. Colloidal particles diffuse from a region of a higher concentration to one of lower concentration. However, their diffusion speeds are much less than that of ordinary solutes which dissolve as small molecules.

It was observed by *Robert Brown*, an English botanist that when pollen grains were suspended in water, they showed a continuous motion in zig-zig way when viewed under an ultra-microscope (Fig. 1.33). He named this phenomenon as *Brownian movement*. Later on, all the colloidal systems were found to show this movement of the particles of the dispersed phase. For example, if a beam of light is allowed to enter a dark room, the dust particles in air (which constitute a colloidal system) are seen dancing in a cone of light. This provides a visible proof of this phenomenon.

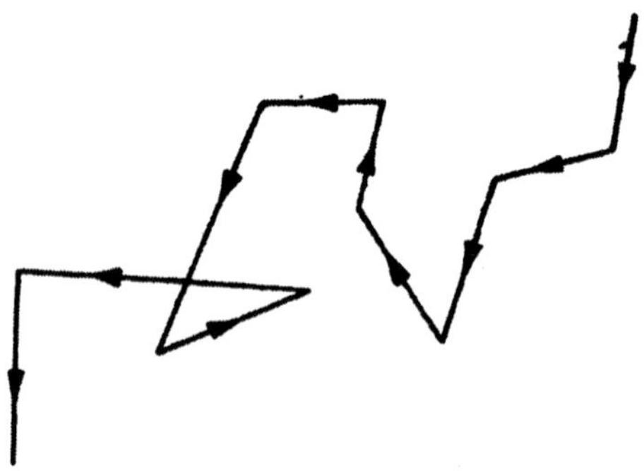

Fig. 1.33

According to *Brown*, smaller particles execute a more rapid and brisk motion than the larger ones, and the motion is not dependent upon (a) the streaming movement of the liquid in which the particles are suspended (b) the mutual attraction or repulsion of the particles, (c) the action of capillary forces and, (d) the evaporation of the liquid under the microscope. The phenomenon of Brownian movement can be best demonstrated in milk. If a sample of diluted milk is observed under an ultra microscope, numerous tiny droplets of fat, swimming and oscillating chaotically in all directions will be found.

Explanation : It is supposed to be due to the unequal bombardment of the colloidal particles by the molecules of this dispersion medium.

With the increase in the size of the particle, the chances of unequal bombardment decrease (Fig. 1.33), the Brownian movement, too disappears. It is due to very fact that the suspensions do not exhibit this phenomenon.

The above explanation involves a large probability in the disparity of impact forces on opposite sides of a particle when it is small rather than when it is large in size and there should be high speeds for smaller particles and only a sluggish movement of heavier ones. It has been proved to be so by several experiments.

Factors Affecting Brownian Movement

Quantitative measurements reveal that the movement, is observable when the particles have a diameter not larger than 0.005 mm. The smaller the particles, the more rapid would be their movement. Hence, the Brownian movement cannot be observed in ordinary suspension because the mass of each particle in this case is so large that the bombardment of the particles gets reduced, the probability of unequal bombardment rapidly increases and consequently the Brownian movement becomes more and more violent.

Important contributions to the understanding of Brownian movement were made possible by the use of ultramicroscope. Zsigmondy in 1905 investigated a large number of colloids by means of ultramicroscope and he came to the following conclusions.

(1) *Small particles execute a more rapid, brisk and vigorous motion than the larger ones.*

(2) *The movement does not change with time and remains the same for months and even for years. For example, this motion persists even when confined for years in underground cellars.*

(3) *The movement is affected by the temperature and it has been observed that the movement gets increased with increasing temperature.*

Importance

1. *Confirmation of kinetic theory :* In fact, the Brownian movement is a direct demonstration of ceaseless motion of molecules shown by kinetic theory.

2. *Stability of colloidal solution* : Brownian movement does not allow the colloidal particles to settle down due to gravity and thus is responsible for their stability.
3. *Determination of Avogadro's number* : We can calculate the Avogadro's number with the help of Brownian movement.

AVOGADRO'S NUMBER

It is an important physical constant and is independent of the chemical nature of the substance and is thus of universal application. However, it is defined as :

"The number of molecules contained in a gram mole or simply a mole of a substance."

The best present value accepted for Avogadro's number is 6.023×10^{23} molecules per gram mole. In the case of solids and liquids one gm mole means the molecular weight expressed in gms. Thus, if the molecular weight of carbon is twelve (12). It means that 12 g of carbon will have 6.023×10^{23} molecules of carbon. Molecular weight of CO_2 is 44 which means that 44 gms of CO_2 will contain 6.023×10^{23} molecules.

In case of gases, it is equal to the number of molecules contained in 22400 c.c. at N.T.P. because 1 gm mole will occupy 22400 c.c. of any gas at N.T.P. In case of gases, it is sometimes expressed as the number of molecules in one c.c. of the gas. In that case it is not Avogadro's number but is called *Lsochmidst's Number.*

Determination of Avogadro's Number

1. *Perrin's First Method* : The method is based on Brownian movement. Perrin considered that the suspended particles behave just like gas molecules. We know that under the influence of gravity, the molecules m a column of gas are not distributed uniformly and consequently there will be more molecules at lower than at higher levels. Therefore, due to difference in the gravitational forces, the difference in potential energy between molecules at heights h_1 and h_2 per molecules of mass m is given as:

$$= ms\,(h_2 - h_1) \qquad ...(1)$$

The difference in potential energy per mole N will be

$$= mNg\,(h_2 - h_1) \qquad ...(2)$$

where N is the Avogadro's number.

According to Boltzmann, the vertical distribution of the gas molecules is given by

$$\frac{n_2}{n_1} = e^{-E/RT} \qquad ...(3)$$

where n_1, n_2 are the number of molecules in a given volume at energy states I and II respectively. E is difference in potential energy per mole in two states. Substituting the value of E from Eq. (2) in Eq. (3); we get

$$\frac{n_2}{n_1} = e^{-Nmg\,(h_2 - h_1)/RT} \qquad ...(4)$$

or $$\log_e \frac{n_1}{n_2} = \frac{Nmg\,(h_2 - h_1)}{RT}$$

If p represents the density of the particle and p that of medium then a factor $\left(\frac{1-\rho'}{\rho}\right)$ has to be introduced in R.H.S. of above equation in order to make account of the buoyancy of the particles in the medium. We have,

$$\log_e \frac{n_1}{n_2} = \frac{Nmg\,(h_2 - h_1)}{RT}\left(1 - \frac{\rho'}{\rho}\right) \qquad ...(6)$$

Eq. (6) can be written as

$$\frac{RT}{N}\log_e \frac{n_1}{n_2} = gv\left(h_2 - h_1\right)(\rho - \rho')$$

where $$v = \frac{m}{\rho} \qquad ...(7)$$

Considering the particles to be spherical, then its volume V = 4/3 πr^3, is the radius of particles. Eq. (7) can be written as

$$\frac{RT}{N}\log_e \frac{n_1}{n_2} = \frac{4}{3}\pi r^3 g\left(h_2 - h_1\right)(\rho - \rho')$$

In the above equation Avogadro's number N can be calculated, if the densities (ρ and ρ'), the radius of particle (r) and the ratio of particles (n_1/n_2) are known.

(a) *Densities of particles* : As the suspension used consists of gamboge and mastic, the particles were subjected to fractional centrifuging to get uniform size. In this method, a known weight w_0 of suspension was evaporated to dryness. It was then heated at 100°C to get a constant weight w_p. Hence the weight of water will be $w_0 - w_p$. It is assumed that there is no volume change when suspension is made, w_0. We have the relation

$$\frac{w_0 - w_p}{\rho_0} + \frac{w_p}{\rho} = \frac{w_0}{\rho'}$$

Volume of water + Volume of particles = Volume of suspension

where ρ_0, ρ and ρ' are densities of water, particles and suspension respectively.

(b) *Volume of the particles* : This is done by counting the number of particles in a given volume of suspension which was then evaporated to dryness. The weight of remaining solid, *i.e.*, residue, is determined. From this weight, the weight of one particle can be calculated which when divided by density of the particle gives the value of v.

(c) Ratio of number of particles (n_1/n_2) : The ratio of number of particles at two different heights is determined as follows :

An opaque disc with a small hole was placed over the focal plane of microscope. The use of this disc is that the field of vision is limited to 5 or 6 particles at the moment. In this way the exact number could be "known by instantaneous observation. Total count for say, 100 readings at each of the two levels at a distance h apart is equivalent to counting the number of granules in an area 100 times as great as that actually under observations. The ratio of the two totals gives the ratio of numbers of particles (n_1/n_2) per unit volume at two heights.

Perrin, with this method, obtained values for N between 5.6×10^{23} and 9.4×10^{23}. These values are quite close to the most accepted value, *i.e.*, 6.023×10^{23}.

2. *Perrin's second method* : Perrin's second method of determining the Avogadro's number made use of diffusion theory of Einstein as applied to Brownian movement.

This discussion consists of two parts :

(i) *Connecting diffusion co-efficient with Avogadro's number.*

(ii) *Connecting diffusion co-efficient with mean replacements.*

Einstein's Equation

Consider a *cylindrical vessel filled* with a dilute solution in which the molecules of the solute are large compared to those of the solvent. The vessel is divided into two parts A and B by means of the semipermeable piston S. If the concentration is greater in A than in B, the piston to move is the difference in osmotic pressure, K, between A and B. K becomes zero (K_0) when the concentrations in the two compartments are equal.

It follows from this that the force of osmotic pressure is responsible *for the equalisation* of concentration caused by diffusion since the diffusion may be prevented by applying an external force, equal to K, the osmotic force.

Consider next, cylinder of a solution of unit cross section without a piston, but with a higher concentration in the left hand than in the right hand portion. The osmotic pressure 'P' acts on the plane E form the left and osmotic pressure E' acts on the plane E' from the light.

Designating the distance from the end of the vessel to E by x and to E' by x + dx, the volume of the solution under consideration is dx. The osmotic pressure per unit volume of the solution is

$$K_0 = \frac{p-p'}{dx} = -\frac{p'-p}{dx} = -\frac{dp}{dx}$$

For dilute solutions $p = n\,RT$

where n is the number of gm moles of the solute per unit volume.

Hence the effective osmotic force K acting on the dissolved substance per unit volume is

$$\frac{dp}{dx} = K_0 = -RT\frac{dn}{dx} \qquad ...(1)$$

The velocity + of the motion of the molecule under the influence of a force, K, is

$$v = \frac{K}{f} \qquad ...(2)$$

where f is the frictional resistance of the molecules.

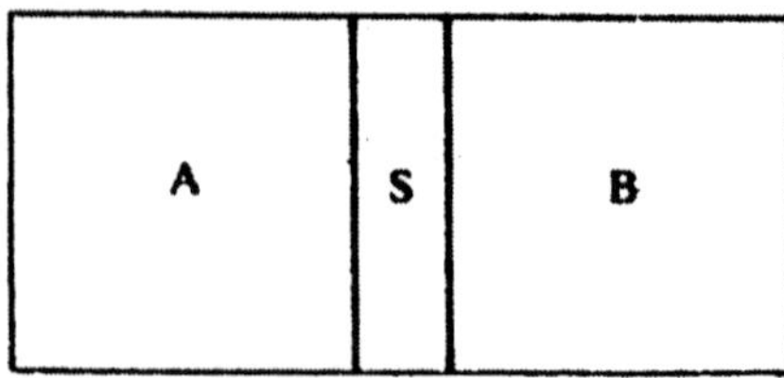

Fig. 1.34

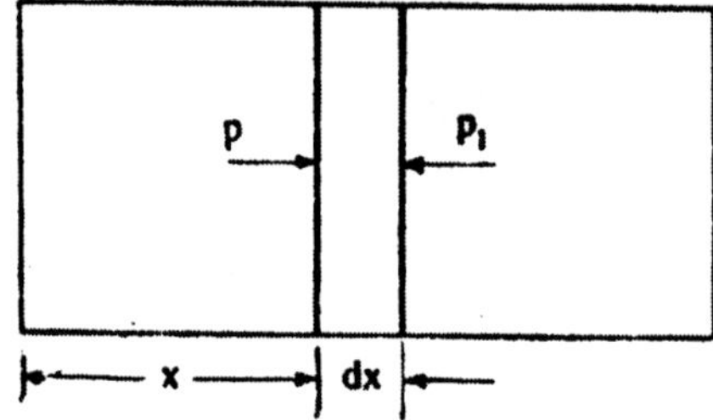

Fig 1.35.

For molecules that are spherical and large compared to the molecules of the solvent Stoke's law applies

$$f = 6\pi\eta r \qquad ...(3)$$

in which η is the viscosity of the solvent and r is the radius of the sphere.

In a unit volume under consideration there are n gm moles of the solute, *i.e.*, nN actual molecules, where N is the Avogadro's number. Since the force K_0 is distributed among the nN molecules it will impart to each a velocity, *i.e.*, 1/nN of the velocity it would impart to a single molecule. Hence, from equations (2) and (3) velocity of the molecule resulting from K_0, is

$$v = \frac{1}{nN}\frac{K_0}{6\pi\eta r} \qquad ...(4)$$

Since K_0 is osmotic force, it follows from equation (1) that

$$vn = \frac{RTdx}{N}.\frac{1}{6\pi\eta r}\frac{dn}{dx} \qquad ...(5)$$

or

$$\frac{vn}{-dn/dx} = \frac{RT}{N.6\pi\eta r} \qquad ...(6)$$

The amount of the solute which will diffuse across unit area at constant rate in unit time under unit concentration gradient is the specific diffusion rate or diffusion constant. According to Pick's law we have

$$- D \frac{dn}{dx} = vn \qquad ...(7)$$

The minus sign signifies that the solute diffuses in the direction of decreasing concentration. On combining Eqs. (6) and (7), we obtain

$$D \frac{RT}{N} \cdot \frac{1}{6\pi\eta r} \qquad ...(8)$$

The above expression relates the diffusion constant in terms of the viscosity of the medium and radius of the particle which are large compared to that of two media.

(ii) *Relationship between diffusion constant and their regular random motion of the molecules resulting from thermal energy which tends to eliminate any equality in concentration,*

Consider the diffusion of dilute solution of Spherical molecules (solute) large compared to those of solvents along the cylindrical cross section represented in Fig. 1.36. Suppose AB is an imaginary plane at a distance Δ from either PQ or RS and suppose n_1 and n_2 be the concentrations of particles in the planes PABQ and ARSB. respectively.

By probability it can safely be assumed that one half of the particles contained in the plane ABQP and ABSR must have crossed the plane AB from left to right and right to left respectively. Thus the number of particles crossing from left to right in time (across AB = $1/2\Delta\ n_{2)}$.

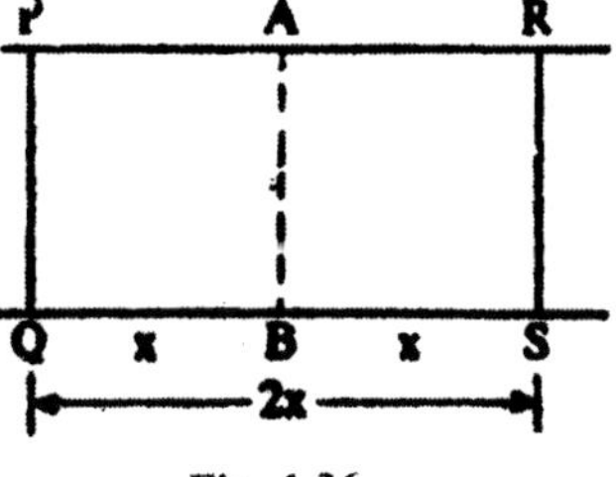

Fig. 1.36

Similarly, the number of particles passing through AB from right to left in the same time t = $1/2\Delta\ n_1$.

Therefore, the net transfer of particles from left to right

$$= 1/2\ \Delta\ (n_1 - n_2).$$

Since the co-ordinate of plane AB is x, equation (9) can be expressed in the differential quotient as follows:

$$\frac{n_1 - n_2}{\Delta} = \frac{dn}{dx} \text{ or } n_1 - n_2 - \Delta \frac{dn}{dx} \qquad ...(10)$$

Hence from equations (9) and (10) the amount in gm. moles of the solution which diffuses through AB in time t is

$$-\frac{1}{2}\Delta^2\frac{dn}{dx} \qquad ...(11)$$

and the amount that diffuses in unit time (second) is

$$-\frac{1}{2}\frac{\Delta^2}{t}\frac{dn}{dx} \qquad ...(12)$$

Dividing Eq. (12) by +dn/dx gives $-\Delta^2/Zt$, the quantity that will diffuse under unit potential gradient. This is again the diffusion co-efficient, Δ *i.e.*,

$$D = \frac{\Delta^2}{2t} \qquad ...(13)$$

Combining Eqs. (13) and (8), we get

$$\Delta^2 = \frac{RT}{N}\frac{t}{3\pi\eta r}$$

or

$$\Delta = \sqrt{t\left(\frac{RT}{N}\cdot\frac{1}{3\pi\eta r}\right)}$$

The value of Avogadro's number N can be calculated from the observed values of Δ, r and η for a sol.

Electrolytic method : From Faraday's experiments, we know that when 96500 coulombs of electricity are passed through a solution of an electrolyte, *e.g.*, an aqueous solution of $AgNO_3$, one gm. equivalent of the ion is discharged at each electrode. Incase of silver each ion is univalent or each atom contains one elementary positive charge when the atom forms a cation. The actual magnitude of the charge is equal to 4.8×10^{-10} e.s.u. This when multiplied by Avogadro's number gives the total amount of electricity required to discharge one gm. atom of silver ions :

$$\therefore \qquad N \times 4.8 \times 10^{-10} = 96500 \times 3 \times 10^9$$

$$N = \frac{96500 \times 3 \times 10^9}{4.3 \times 10^{-19}} = 6.03 \times 10^{22}$$

(One coulomb is equal to 3×10^9 electrostatic units)

4. *Radioactivity method :* On disintegration of radium atoms, α-particles are given out and each α-particle loses its charge and

becomes a helium atom. From the rate of α-particles formed we can find the volume of helium gas resulted. Now it will be possible to determine N_1, the number of helium atoms (or molecules, since helium is monoatomic) in 1 c.c. Hence, N can be evaluated.

On the basis of observation it was found that one gram of radium liberated during its disintegration 0.043 c.c. of helium gas at N.T.P. On the basis of calculation, 1 gm of radium emitted in a year 11.6×10^{17} α-particles. It means that 0.043 c.c will have 11.6×10^{17} helium molecules. One mole of a gas occupies 22,400 c.c. Then, number of molecules in 1 mole, *i.e.*, the Avogadro's number, will be given by

$$\frac{11.6 \times 10^{17} \times 22.400}{0.043} = N = 6.05 \times 10^{23}$$

Furthermore, it was found that when one gram of radium is in equilibrium with its products the number of α-particles produced per second is four times the number (ϕ) emitted from 1 gram of radium per second. The value of ϕ can be measured by the method of Geiger. From the value of ϕ, N can be evaluated.

5. *X-Ray Method :* Now a new and more precise method for the determination of Avogadro's number is available. The method depends upon determining the diffraction pattern of a crystal with the help of X-ray. For example, it is found that in case of sodium chloride crystal, Na^+ ions and Cl^- ions are arranged in a cubic lattice, with spacing d. One mole of NaCI therefore consists of 2N ions and if V is the molar volume of solid NaCl, then

$$2Nd^3 = V = \frac{M}{\phi}$$

where M is the molecular weight and ϕ the density. The value of d can be calculated by means of Bragg's equation which is given by

$$n\lambda = d \sin \theta$$

where n = small integer 1, 2...

λ = wavelength of X-rays employed

θ = glancing angle

d = z distance between successive identical planes

The above equation can also be written as

$$d = \frac{n\lambda}{2 \sin \theta}$$

Hence knowing the value of X and by measuring the glancing angle (θ) by means of ionisation spectrometer, d can be calculated.

Some Interesting Results from Avogadro's Number

1. *Number of molecules in a drop of water :* The volume of 1 gm mole of water at 4°C is 18 c.c. of water, therefore, contains $\frac{6.023 \times 10^{23}}{18}$ molecules. A drop of water whose volume is 0.05 c.c. will contain $\frac{6.023 \times 10^{23}}{18} \times 0.05 = 1.67 \times 10^{23}$ molecules.

If water is allowed to evaporate at such a rate that a million molecules leave it every second, whole drop of water will take about fifty million years to evaporate.

2. *Number of molecules in one human breath :* An average human being breathes out about 400 c.c. of water vapour and carbon dioxide. We know that 22.4 litres of air contains 6.023×10^{23} molecules at N.T.P. and approximately the same number will be present at ordinary temperature. 400 c.c. of air will, therefore, contain about 10^{23} molecules.

Stability of Sole

Stability of lyophobic sols : The stability in case of lyophobic sols is due to the presence of electric charge on them. In order to precipitate a colloidal sol, the particles must coalesce so as to form aggregates. In the case of lyophobic sols such agglomeration is avoided by the presence of electric charges of the same sign. Due to electrostatic attaction such sols are stable. For example, a AgBr sol (if there is present an excess of $AgVSO_3$) contains +vely charged bromide particles, on account of the adsorption of Ag+ ions on the surface of the particles.

Stability of lyophilic sols : In case of lyophilic colloids the stability is due to :

(1) *Electric charges* of same sign.

(2) *Salvation, i.e.,* the tendency to bind with the solvent. The layer of the molecules of the solvent form an envelop which prevents the aggregation of the colloidal particles.

The lyophobic colloids can be readily coagulated by removing the electric charge, while in lyophilic colloids, the charge removal may decrease stability but does not lead to coagulation.

This fact may be illustrated by considering proteins at their isoelectric point, where the particles are uncharged. Although, under these conditions the protein dispersions show a minimum of stability, they do not always precipitate as they are still protected by a surrounding layer of water. If, however, this layer is also removed by any means, coagulation occurs.

Isoelectric point : The colloidal particles will be positively charged in the presence of acid due to the adsorption of H^+ ions and negatively charges in the presence of alkali because of the adsorption of OH^- on the surface. This clearly reveals that there must be some intermediate hydrogen ion concentration at which the particles are neither positive nor negative. Thus, the isoelectric point in the case of colloids may be defined as :

"The concentration at which the colloidal particles have no charge is known as the isoelectric point".

Hardy observed that a colloidal solution of albumin is negatively charged. If a small amount of acid is added, the negative charge decreases. On adding more of acid a stage is reached when the charge changes to positive. At certain concentration of acid, the electric charge is neither positive nor negative. So isoelectric point may also be defined as the pH at *which the mobility is zero.*

Gelatin has the isoelectric point at pH 4.7 indicating that it has no electrophoretic motion at this pH. At a pH below 4.7, it moves towards the cathode while at pH above 4.7 it moves towards the anode.

The stability of lyophilic colloids at isoelectric point will be minimum. But there are exceptions to this rule, *e.g.,* silicic acid, which has been found to possess maximum stability at isoelectric point.

The isoelectric point plays an important role in the stability of the sol. The stability of the sol is due to the electro kinetic potential or zeta potential ξ. The zeta potential of the sol is plotted against the concentration of the electrolyte added, as shown in Fig. 1.37. It can be seen from the graph that at the point IP the curve I intersects the abscissa. This is the *isoelectric point* because at this point the zeta potential is zero. If the particles have not yet coagulated they will not move under the influence of an electromotive force. According to Hardy at the isoelectric point the colloidal particles are electrophoretically inert. Most of the colloids coagulate at this point. In case if they do not coagulate they are least stable, for example they can be readily coagulated with alcohol.

The isoelectric point is also found in protein sol. A protein molecule is amphoteric in nature since it is the combination of a number of amino acids. In acid medium this sol becomes positively charged at an intermediate hydrogen ion concentration the particles will have equal positive and negative charges. The isoelectric point which was found to be different in different proteins. Hence the gelatin sol is having an isoelectric point at pH 4.7. It indicates that in this pH range it is not having electro-phoretic motion. It moves towards the cathode when pH is below 4.7 and moves towards anode when the pH is more than 4.7.

Factors Affecting the Stability : In addition to charge and solvation. There are various other factors governing the stability of a colloidal system.

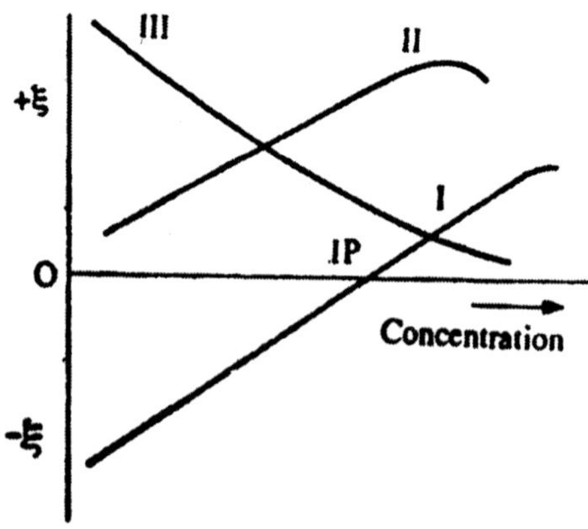

Fig 1.37 : The change of zeta potential with the addition of electrolyte. (The isoelectric point)

1. *Brownian movement :* Because of Brownian movement, the particles are in constant rapid zig-zag motion. With the result, the aggregation of particles is prevented. The sol becomes unstable as soon as this movement ceases.

2. *Addition of electrolyte :* The stability of a sol is closely related to the potential difference between the particles of dispersed phase and dispersion medium. It was observed by *Hardy, Schultz, Picton* and *Linder* that the suitable quantity of the electrolyte

added affects the potential difference between the dispersed phase and dispersion medium. Therefore, the stability of a sol largely depends upon the concentration of added electrolyte.

3. *Rate of addition of electrolytes :* The rate of addition of electrolyte is responsible for the stability of sols. According to Dhar, the phenomenon is known as acclimatization of sols. There are two types of acclimatization.

   (a) *Positive acclimatization :* In this case small quantities of electrolyte are added slowly, in order to coagulate the sol.

   (b) *Negative acclimatization :* Here the amount of electrolyte required to coagulate a sol is more.

   According to *Kresteinskaja* and *Moltschanova*, the phenomenon of acclimatization is due to slow chemical changes produced by the interaction of the colloid and the added electrolyte.

4. *Effect of dilation :* According to *Chaudhary and other* (1928) the dilution of a sol also affects its stability in two opposite ways.

   (i) Decrease of charge as well as the total surface of sol particles with dilution makes the sol unstable if the potential at which the coagulation of sol takes place and the relative adsorption of all ions on the surface remain unchanged.

   (ii) The greater distances between the particles of a diluted sol favour stability. *Mukerjee (1930)* tried to relate the coagulation of the sol with the stability by the measurement of migration velocity. The effect was found to be complicated and it is not possible to give a relation between the two quantities.

5. *Effect of temperature :* The increase of temperature generally decreases the stability of colloids. This may be due to the dissolution of precipitating electrolytes from the walls of the containing vessel at the higher temperature. But Reid and Burton (1928) have shown that in many cases, *e.g.,* copper hydro sol, heat alone is sufficient to bring about the coagulation.

   According to *Dhar and Parkash* (1930), a larges number of sols, like sols of ferric, aluminium, zirconium, and stannic hydroxide, vanadium pentoxide and copper ferrocyanide need small amount of electrolyte at 60°, for coagulation than at 30°.

6. *Ultraviolet radiation and X-rays :* Many hydrophobic sols may be coagulated by ultraviolet radiations. X-rays and even by radiations furnished by radium, but the knowledge in this direction is at present very little developed.

   Lal and Ganguly (1930) studied the coagulating influence of ultraviolet light and have observed that hydro sols of silver iodide, arsenate and thiocyanate, silver, vanadium pentoxide, gold, thorium hydroxide, and arsenious sulphide may be coagulated by exposure to ultraviolet light, whether the sol were positively or negatively charged.

7. *S:abilising by protecting films :* If two sols of hydrous ferrous oxide are prepared by two different methods, say, first by adding a few ml of conc. $FeCl_3$ drop by drop to pure boiling distilled water and the second by adding few ml of 1% gelatin to the $FeCl_3$ before using, then the relative stability of the sols will be indicated by the greater amount of electrolyte required to coagulate the second sol. This is due to the fact that a protective layer around each particle is formed by the gelatin and this protective layer thus prevents the sol to some extent from being coagulated. Various other sols could also be protected by such type of films and generally lyophilic colloids, such as gelatin, agar, starch etc. are used for this purpose.

8. *Ultrasonic irradiation :* Ultrasonic irradiation can either promote or arrest coagulation depending on the properties of the colloid and on the frequency and intensity of radiation. It was shown by *S. Prakash* and *A. K. Ghosh (1959)* that the dissolution of the colloidal silver particles takes place upon irradiation of certain silver sols.

9. *Boiling or freezing :* Certain sols are very sensitive towards boiling or freezing. *Lottermoser (1944) A. Buzagh and S. Rohrsetzer (1959)* have shown that arsenious sulphide sols purified by electrodecamation coagulate not only upon boiling, but also on freezing to –9°C and subsequent thawing.

10. *Mechanical Agitation :* According to *Freundlich and Loebmann* (1929), the mechanical agitation of the sol generally decreases the stability. They observed that the sols of cupric oxide and

chromium hydroxide can be readily coagulated by mechanical agitation. If f is the rate of coagulation and s is the rate of stirring then according to *Pauli* (1938),

$$f \propto \sqrt{s}.$$

Theories of Stability of Sols : Various theories have been proposed to explain the stability of sols. Some of them have been noted below :

1. *Powis theory* : According to Powis, coagulation is caused only when charge on the colloidal particles is removed. The charge on the colloidal particles need not remove completely. A lowering of potential of the interface to a certain value called the *critical potential* is essential to bring about the *coagulation.*

2. *Perrin and Rice theory* : *Perrin* and *Rice* suggested that the stability of the sol is related to the negative free surface energy of the double layer. They found a certain degree of dispersion for which there is an equilibrium between normal positive free energy and negative free energy of the double layer. It means that a colloidal suspension could be stable in thermodynamic sense. No lyophobic system is ever stable in thermodynamic sense. This theory is not applicable to practical cases.

3. *Salvation theory* : *Kruyt* showed that hydration of solvation of colloids is related to the stability of the sols. Every lyophilic sol has a thin-layer of solvent around its colloid particles which do not allow the sol to coagulate easily. It can be coagulated by the addition of dehydrating agent which removes the solvent layer (generally water), then it is possible to coagulate the sol.

4. *Gyemant-Lewis theory* : According to *Gyemant* and *Lewis*, the stability of sol is responsible for the surface tension and charge of the sol.

Thermodynamic Stability of Colloidal Solutions : The *lyophilic sols* as those of gelatin, gum arabic, agar-agar etc. are molecularly dispersed sols and their colloidal behaviour is due to the large-size of the molecule. The substances which form lyophilic sols are spontaneously soluble in the solvent and form stable systems. For example, gelatin will dissolve off itself in water like cane sugar or any other truly soluble substance. Since lyophilic sols are naturally soluble, they form stable sols or solutions and are not precipitated unless a very large excess of an electrolyte is added to them. *These sols are, therefore, thermodynamically stable.*

The *lyophobic sols* as those of AS_2S_3, $Fe(OH)_3$ etc., are neither molecularly dispersed, nor are formed by just bringing the material in contact with the solvent. For example, a sample of arsenious sulphide will not dissolve as such in water, because it is not spontaneously soluble in water and for preparing them special methods are required. In the colloidal state their particles remain suspended in water because the electrical charge of the particles prevents them from coming together and forming larger particles. *Thermodynamically, therefore, lyophobic sols* are unstable and as soon as the charge on the particles is neutralised, they form a precipitate.

Now, an unstable system has a greater potential energy than the stable system. The potential energy of a lyophobic sol is due to the work done against the surface tension of the liquid. If the number of particles is N, the potential energy of the colloidal system is :

$$A = 4\pi r^2 g\ K$$

where r is the radius of each particle and γ the surface tension of the liquid. It is this potential energy which makes the lyophobic systems unstable.

PROTECTION OF COLLOIDS

Protection : It is observed that when certain hydrophilic colloids such as gum, gelatin, agar-agar etc. are added to a hydrophobic colloid, the stability of the latter is markedly increased. Now the addition of the small amounts of electrolytes does not cause the precipitation of the hydrophobic colloid. This action of the hydrophilic colloids to prevent precipitation of the hydrophobic colloid by the electrolytes is called protection and the hydrophilic colloid is called protective colloid.

It is further observed that the protective colloid not only increases the stability of the hydrophobic colloid but the latter can be evaporated to dryness and the dry mass peptised by simply shaking with water. Thus the protective colloid converts an irreversible (hydrophobic) colloid into a reversible colloid.

Some examples of protective colloids are :

(i) Soluble substances like $Ca_3(PO_4)_2$ are held as colloids in blood due to protective action of protein in blood.

(ii) To prevent clogging in pens; superior pen inks contain some protective colloids.

(iii) Casein in human milk is better protected than in the cow's milk. It is because of this that cow's milk is more easily coagulated.

(iv) Protargol and Argyrol powders are the protected forms of colloidal silver.

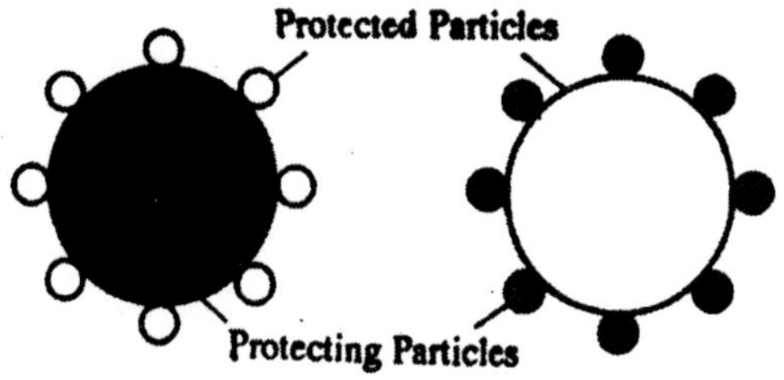

Fig 1.38 : Formation of protective colloids.

Explanation : The particles of the protected colloid get adsorbed on the particles of the hydrophobic colloid, thereby forming a protective layer around it (Fig. 1.38). The protective layer prevents the precipitating ions from coming in contact with the colloidal particles. According to a recent view the increase in stability of the hydrophobic and colloid is due to the mutual adsorption of the hydrophilic and hydrophobic colloids. It is immaterial which is adsorbed on which. In fact the smaller particles whether of the protective colloid or of the hydrophobic colloid are adsorbed on the bigger particles.

Preparation of protected colloid : It is possible to get a protected sol by means of protective one. By means of protection, colloid in higher concentration can be obtained. Furthermore, it is important in preventing the growth of primarily precipitated particles.

Protected sol of platinum : Paal and *Amberger* prepared protected sols of several metals. The platinum protected sol was obtained as follows :

Chloroplatinic acid is added to sodium lysalbate (decomposition product of protein) taken in excess of water. To this NaOH is added till we get reddish-brown liquid which is then treated with hydrazine where N_2 gas is evolved. This solution is kept for 6 to 7 hours and then dialysed. The black brittle and lustrous residue so obtained was dissolved in water which gives a protected platinum sol.

Gold Number : The power of the hydrophilic colloid to prevent the precipitation of a lyophobic colloid by addition of an electrolyte depends upon the nature of the substance. The protective character of various hydrophilic substances can be expressed quantitatively by gold number. The *gold number* according to *Zsigmondy* may be defined as :

"The number of milligrams of the protective colloid which must be added to 10 c.c. of a given gold sol so as to just prevent its precipitation by addition of 1 c.c. of 10% NaCl solution."

Smaller the gold number, higher the protective power of a colloid. Gold numbers of some protective colloids are given below:

Protective colloid	***Gold number***
Haemoglobin	0.03 to 0.07
Starch	15 to 25
Gum Arabic	15 to 25
Gelatin	0.005 to 0.1

The protective power was also measured by *Ostwald* in terms of *Congo Robin number. It is the amount of a protective colloid in mg which prevents colour change in 100 ml. of 0.01% Cango robin dye solution to which 0' 16 gm equivalent of KCl is added when observed after 10.15 minutes.*

Theories of Protective Action

Many theories have been put forward to explain the phenomenon of protective action. Some of the theories are given below.

1. *Bechhold Theory* : According to *Bechhold* (1905), the protective action is attributed to homogeneous encircling of the suspended particle by the particles of the protective colloids, thereby forming a protective sheath to each particle of the hydrophobic colloid. But according to *Zsigmondy*, Bechhold's view can only be accepted in the case of coarse suspensions, but not in case of particles of size not far removed from molecular dimensions, as are encountered in the highly dispersed red gold sols. Hence there are objections to the Bechhold's theory.

2. *Billitzer's Theory* : According to *Billitzer* (1905), protective action may be ascribed to the adsorption of any precipitating electrolytes by the protective colloid, and not by the union of the particles of the two colloids. This was again disproved by

Zsigmondy who suggested the mutual adsorption of the particles of the two sols. His view was confirmed by the fact that protection is not complete immediately after mixing the two sols and a definite time is needed for the union of particles. *Zsigmondy's* view of mutual adsorption proposes a union of the particles of the two colloids to form a complex which is having the stability of the protective colloid.

3. *Williams and Chang theory* : According to *Williams and Chang* (1957) the protective action may be ascribed to the fact that negatively charged hydrophilic colloids will often form a coating around the negatively charged hydrophobic colloid. Similarly, positively charged hydrophilic colloids will often form a coating around the positively charged hydrophobic colloids. The hydrophobic colloid then serves as hydrophilic colloid and is less readily precipitated by the properly charged ions. In other words, *Williams and Chang* assumed that the hydrophilic substance envelopes the particles of the hydrophobic sol. In order to do this, the number of hydrophilic molecules must be larger than those of the hydrophobic particles. Approximate calculations have revealed that in most cases the hydrophobic particle gels covered by a monomolecular layer of the molecules of .the prorective colloid. The multilayer adsorption on the surface of the hydrophobic particles, assumed by various authors, was rejected by Williams and Chang.

Applications of Protective Action

The action of protective colloids plays a significant role in many biological systems. It finds use in a number of biological preparations and pharmaceutical products. Silver sols find use in medicines for intravenous injection, or as an ointment. Argyrol is a colloidal silver and silver oxide sol having pronounced antibacterial properties. The stabilization of colloidal suspension is carried out by certain protective colloids.

HOFMEISTER SERIES OR SALTING OUT OR LYOTROPIC SERIES OF THE SOLS

It is known that the lyophilic electrolytes need larger amount of electrolyte in order to bring about coagulation. Hence, the addition of large amount of electrolyte to precipitate the dispersed substance of lyophilic sol is termed as *"salting out"*. In order to coagulate the lyophobic sols comparatively smaller amount of electrolyte is required. The salting out effect depends upon the nature of ions and the salt of a given metal. Thus a series is set up in the order of their effectiveness. This series is sometimes termed as Hofmeister series or the lyotropic series.

Hofmeister (1890) obtained the concentration of different salts required to coagulate the egg albumin. The results obtained for various anions and cations arc given here. Hence the lyotropic series for cations is as follows :

$$Mg^{+2} > Ca^{+2} > Sr^{+2} > Ba^{+2} > Li^{+} > Na^{+} > K^{+} > Rb^{+} > Cs^{+}.$$

The series for anions is as follows :

Citrate > tartrate > $SO_4^{-2} > Cl^- > NO_3^- > ClO_3^- > I^- > CNS^-$

The above effect has been found to be very similar to the order of salting out of aniline, ethyl acetate and ether from aqueous solutions of various ions.

It was shown by *Michaellis* (1925) that for Kobalyi (a colloidal jelly with neutral charge) the effect of different electrolytes follows the lyotropic series. It has been observed that with lithium ion the shrinkage would be maximum and with Na^+ and K^+ ions, would be minimum.

The Hofmeister series provides an idea of the order of temperature to which gel must be heated before transforming into the sol in the presence of anions.

It has been estimated that the SO_4^{-2}, tartrate and citrate ion inhibit the swelling of gelatin and some other gels. The Cl^-, ClO_3^-, Br^- and I^- favour the imbibition of water. In the solutions of the iodide the gel formed gets dispersed at room temperatures.

COACERVATION

According to *Bungenberg*, *Jong* and *Kruyt* (1930) the formation of a liquid precipitate by the mutual coagulation of hydrophilic colloid is termed as *coacervation.* For example if a warm and concentrated solution of gelatin is mixed with the solution of gum arabic liquid coacervates, then droplets or coacervates are obtained. The coacervates appear generally in the form of viscous drops instead of continuous liquid phase. The literary meaning of coacervation (Latin word) implies *heap together*.

It is assumed that the cells of lightly bound water molecules surrounding the particles prevent them from coalescing. The electrostatic attration of their opposite charges hold a number of particles together in the form of a droplet. The mixture of the two hydrophilic sols separated into two layers only when the concentration of the colloids is high and when the sign of the particle charge is the same. Both of the layers are having both electrolytes. If the two oppositely charged hydrophilic sols are mixed then droplets also get formed. The coacervates can again be dispersed by the addition of iodide or thiocynate ions or by the ions of high valency which are able to reduce the zeta potential of each sol.

SENSITISATION

If a small amount of a hydrophilic colloid is added to hydrophobic colloid, then it becomes more sensitive to the coagulating influence of electrolyte as compared with the original sol. This phenomenon is termed as *sensitization* of the hydrophobic sol. An example is that if a little gelatin is added to ferric hydroxide sol, in amount insufficient to produce coagulation, then the sol remains still positively charged and becomes more sensitive for coagulation. Similarly if a little amount of gelatin is added to red gold sol, the further addition of the electrolyte would bring about coagulation more readily. The non-electrolytes such as alcohol is also able to sensitize the colloid.

Zsigmondy (1916) was the first scientist who attempted to make a systematic study of the sensitizing action of certain hydrophilic colloids on gold sol by introducing a new term called *U-numbers*, which may be defined as follows :

"*The number of mgs. of a hydrophilic sol which is sufficient to produce the colour change from red to blue* in 10 c.c. of gold sol. In the following table the U-numbers of certain hydrophilic sols have been included :

Table 1.3

S.N.	*Colloidal sol*	*U-numbers*
1.	Gelatin	0002–00.04
2.	Casein	0.002–00.04
3.	Histidine	0.1–0.2

Theories of sensitization : It has been a matter of much uncertainty to give a definite cause of sensitization.

(i) *Adsorption theory :* The cause of sensitisation is ascribed to the adsorption of oppositely charged hydrophilic sol by the hydrophobic one, if the hydrophilic colloid is ionic in nature. More correctly it can be said that the ions of the hydrophilic colloids get adsorbed by the hydrophobic one and the potential of the hydrophobic sol is reduced very closely to its critical coagulation potential. This view has been confirmed by the fact that sensitizing influence of a sol would depend upon the hydrogen ion concentration of the hydrophilic sol.

(ii) *Ghash and Dhar's theory' :* Another view regarding the sensitization of positively or negatively charged sol was provided by *Ghosh* and *Dhar* who suggested that the sensitizating effect of a positively charged hydrophobic sol *e.g.,* ferric hydroxide by the negatively charged hydrophilic colloid like gelatin, and albumin is attributed to the neutralisation of charge of each. On the other hand a negatively charged sol such as arsenious sulphide by some hydro philic sols is due to the repression of the hydrolysis of the sols because a little amount of H^+ ions are already present in gelatin-albumin or tannin.

The non-electrolytes such as alcohol, ether and acetone are also found to sensitize a sol. It has been found that if a hydrosol is shaken with a liquid which is immiscible in water, the sensitization as well as coagulation would take place simultaneously. For example the substances like chloroform, CCl_4 or benzene may sensitize the silver sol. The water soluble substances such as alcohol, ether, acetone are having more sensitizing effect.

THEORIES OF ORIGIN OF CHARGE

The presence of electric charge on the colloidal particles was established by *Linder* and *Picton.* Colloidal systems are stable because of the presence of this charge. Hence the problem of origin of charge is of great applicability. There are several mechanisms and various factors by which the charge can come into existence. Following are some of the views which have been proposed.

Older Views : According to older views, the charge on the colloidal particles was supposed to be due to the following :

(a) *Charge due to ions* : It has been shown that in electrolytic colloids, *e.g.*, sodium palmitate, ferric hydroxide, etc. it is assumed that such colloids dissociate into ions and the charge is ionic in nature. But the view is not valid in case of non-electrolytic colloids such as clay, smoke etc., which carry a charge.

(b) *Charge being frictional* : According to this view charge results by the friction between the molecules of the dispersion medium and colloidal particles. It was also proposed that the charge depends upon the difference of the dielectric constants of the medium and the dispersion material

It was also found that when the dielectric constant of the medium is greater than that of the dispersed material, the particles become –vely charged relative to the medium and vice-versa. This explains why most of the hydro sols are negatively charged. The reason is that water has high dielectric constant (about 80).

(c) *Dissociation of dispersion medium* : This view was put forward by *Hardy*, according to which the dispersion medium is dissociated into ions and these ions are adsorbed by colloidal particles. It was also suggested that it was possible to change the sign of the sol by placing it in an environment of ions of opposite charge to which it has. This theory fails in case of non-ionisable dispersion medium such as organic solvents.

(d) *Association of electrolytes* : It has been found that small quantities of electrolytes are associated with colloidal systems and if they are removed by any means, the sol becomes unstable. The charge on the colloidal particles is due to the association of such electrolytes.

Modern Views

(a) *Preferential adsorption of ions* : According to this view, the charge on the colloidal particles is due to preferential adsorption of either positive or negative ions on the surface of colloidal particles. The existence of the ions in solution can be shown with the help of following examples (Fig. 1.39).

1. Colloidal gold formed by the Bredig's arc method is more stable if water in which electrodes are dipped contains traces of alkali. Negative OH^- ions get adsorbed on the colloidal gold particles.

2. If dilute solutions of $AgNO_3$ is added to excess of KI, negative AgI sol is formed, but if dilute KI solution is added to excess of dil. $AgNO_3$ positively charged sol of AgI is obtained. This is because in the first case I^- ions being present in excess are adsorbed by AgI to give [AgI) I^-)], while in the latter case Ag^+ ions which are present in excess are adsorbed to form [AgI) Ag^+].

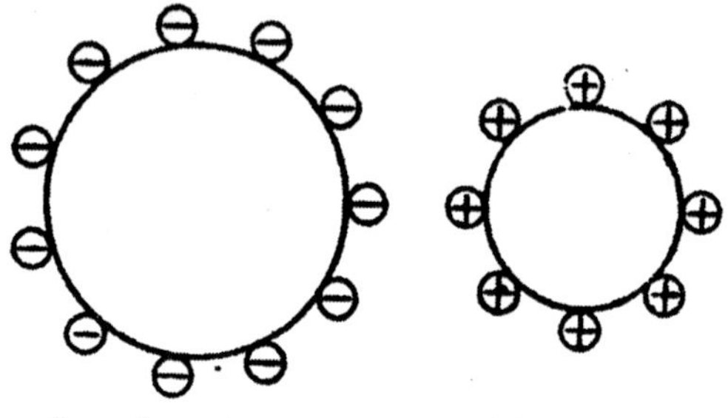

Adsorption of –ve ions Adsorption of +ve ions

Fig. 1.39

The above example thus conclusively proves that charge on the colloidal particles is due to the adsorption of ions present in the dispersion medium.

It may be pointed out that as a general rule. The ions preferred by the colloidal particles for adsorption are those which are common to them.

(b) *Electrical double layer theory :* The concept of the existence of electrical double layer of the positive and negative charges at solid liquid boundary was given by *Helmholtz*. He assumed that electrical double layer is virtually an electrical condenser, with parallel plates, no more than a molecule distance apart.

The double layer consists of two parts, one part known as *fixed* part of the layer and is fixed on the surface of solid. The second part consists of a diffused or mobile layer of ions which extends into the liquid phase. It contains ions of both signs.

The ions which are preferentially adsorbed by colloidal particles are retained in the fixed part of the d- :ble layer. The actual charge is due to the presence of these ions. The oppositely charged ions are present throughout the diffused portion of the double layer.

Zsigmondy explained the formation of both +vely and –vely charged sols of stannic oxide.

When freshly precipitated SnO_2 was peptised by KOH, a –vely charged sol is obtained. KOH reacts at the surface of SnO_2 particles to form potassium stannate, which dissociates to give K^+ ions and stannate ions (SnO_2^{--}) and form the two components of the double layer.

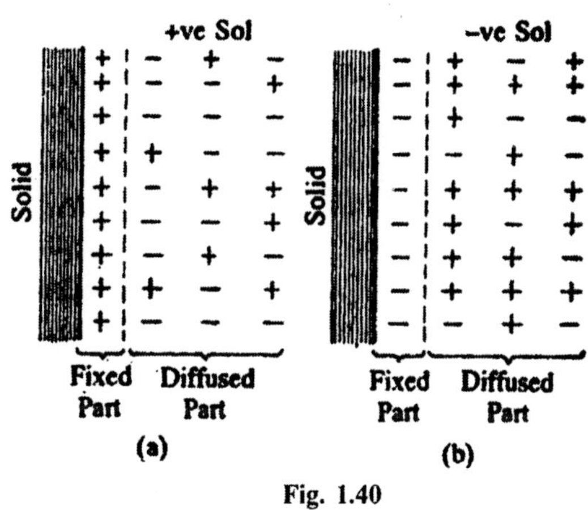

Fig. 1.40

$$2KOH + SnO_2 \rightarrow K_2SnO_3 + H_2O$$

$$K_2SnO_3 \rightleftharpoons 2K^+ + SnO_3^{--}$$

Here SnO_3^{--} are retained on the fixed part of the double layer and it is negatively charged while the K+ ions go in the dispersion medium [Fig. (1.41)].

$$[SnO_2.xH_2O]\ SnO_3^{--} : 2K^+$$

A positively charged SnO_2 sol prepared by peptization with HCl, a little of $SnCl_4$ is obtained which dissociates into Sn^{+4} and Cl^- ions. The stannic ions are adsorbed on the particle side and Cl^- ions go in the dispersion medium.

$$[SnO_2.xH_2O]\ Sn^{+4} : 4Cl^-$$

(c) *Orientation theory :* According to this theory, molecules with polar groups are oriented when adsorbed at the solid-liquid interface. Thus, if the positive ends of adsorbed molecules are turned outward the suspension will have characteristics of a positively charged body. When the charge is reversed in another fluid, it can be made by a 180° reversal of the oriented molecules at the interface.

Significance of charge on the colloid : The existence of electrical charge on the colloidal particles plays a significant role. It accounts for :

(a) *Stability of the colloidal systems :* Because of similar charge on the colloidal particals they mutually repel one another and do not come close enough to coelasce and form bigger particles to settle down. This is provided by the fact that coagulation takes place immediately if charge on the particles is neutralised by the addition of electrolytes or persistent dialysis.

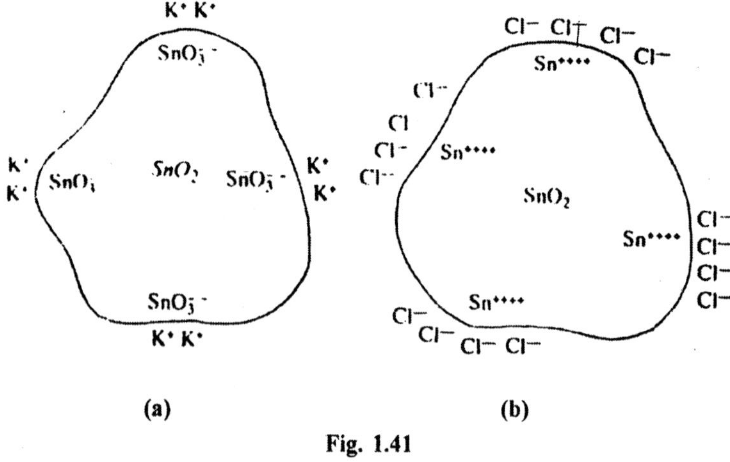

Fig. 1.41

(b) *Industrial applications :* A large number of industrial applications of colloids are based on the existence of electrical charges on them.

DETERMINATION OF CHARGE ON A COLLOIDAL PARTICLE

The magnitude of the charge can be determined by various methods; two of which have been described here.

(1) *Barton's first method using Stake's formula :* According of Lamb, the velocity (v) of an isolated particle in an electric field is given by the expression

$$v = \frac{X\rho}{\beta}, \text{ where } \beta = \frac{\lambda}{l} \qquad ...(1)$$

Here X = potential gradient applied

η = coefficient of viscosity

l = facility of slip against the immediate layer of the molecules of the medium in which it is surrounded.

p = charge density of the particles.

A colloidal particle can be considered to be a spherical condenser, having two layers. The charge density of a spherical condenser is given by actual charge

$$\rho = \frac{\text{actual charge}}{\text{sufface volume}} = \frac{e}{4\pi r^2} \qquad ...(2)$$

where r is the radius of the particle.

From Eqs. (1) and (2), we have

$$v = \frac{Xe.l}{4\pi r^2 \eta}$$

or

$$X.e = \frac{4pr^2.v.n}{l} \qquad ...(3)$$

The capacity C of a spherical condenser is given as

$$C = \frac{r^2}{d}.K \qquad ...(4)$$

K = dielectric constant of the medium

d = distance between the two layers.

If P be the potential difference between the particles and the medium, then the capacity is given by

$$C = \frac{e}{P} = \frac{r^2}{d}.K \text{ [From eq. (4)]}$$

$$e = \frac{Pr^2}{d}.K \qquad ...(5)$$

Substituting the value of e in Eq. (3), we get

$$\frac{XPr^2K}{d} = \frac{4\pi r^2 v.\eta}{KX}$$

$$P\frac{l}{d} = \frac{4\pi\eta v}{KX} \qquad ...(6)$$

Burton observed that both l and d arc of same magnitude, *i.e.*, 10^{-8} cm., l/d therefore, be taken as unity. We, therefore, have

$$P\frac{10^{-8}}{10^{-8}} = \frac{4\pi\eta v}{KX}$$

or
$$P = \frac{4\pi\eta v}{KX} \qquad ...(7)$$

Substituting this value of P in Eq. (5), we get

$$e = \frac{4\pi r^2 \eta . v}{X} \qquad ...(8)$$

The charge on colloidal particles can be determined from the expression (8), provided the value of r, η, d, v and X are known. These can be evaluated as follows :

(i) The viscosity coefficient η can be seen from the standard table at a particular temperature.

(ii) The radius of particles (r) can be determined by any of methods described earlier.

(iii) The value of d as shown by Gouy and Helmholtz is approximately equal to 10^{-8}cm.

(iv) The value of v and X can be determined by Burton's method.

A part of clean and dry U tube is filled with the dispersion medium and some of the sol is run slowly till it reaches equal heights in the limbs. To one arm of a positive and to the other arm a negative electrode was inserted and definite potential was applied in one direction. As a result of the potential, colloidal particles move towards the oppositely charged electrodes and their movement can be seen by ultramicroscope. Let h_1 be the increase in the height of the boundary in t_1 seconds. Reverse the current, whereby the particles will now begin to move in opposite direction. Let the corresponding increase in height be A, in time t_2 seconds. The velocity of migration (v) can be evaluated by the expression,

$$v = \frac{h_1 + h_2}{t_1 + t_2}$$

The potential gradient (K) can be determined by the expression

$$X = \frac{Z}{L}$$

where Z = applied voltage in volts

and L = distance between two electrodes in cms.

(2) *Barton's second method :* This method is based upon the fact that at isoelectric point the migration velocity is zero. Isoelectric point is defined as the point where the stability of a colloid becomes minimum and coagulation may take place. In order to calculate the charge on a single particle :

(a) the number of colloidal particles in the volume of suspension required for precipitation, and

(b) the quantity of precipitating ion required to cause precipitation are determined.

THE SPONTANEOUS AGEING OF COLLOIDS

The term ageing is used for *the spontaneous changes in colloids sol when kept for regular time intervals*. The most common change taking place in colloids is the increase in particle size. The ageing is different from coagulation. The process of coagulation takes place by certain means. The effect of container in which the sol is contained and the effect of radiations like ultrasonic, and cosmic radiations influence the stability of the sol.

The 'life curve' of the sol : Feitknecht, Signer and *Berger* (1942) attempted to investigate the changes in size and shape of nickel hydroxide sol. The particles arc of laminar type.

It can be prepared by the peptisation of Ni $(OH)_2$ precipitate. If the degree of dispersion is plotted against time the curve as shown in Fig. 1.42 is obtained. As soon as the peptisation gets completed the maxima of the curve is obtained. The I phase refers to the birth and childhood of the sol.

In phase II the degree of dispersion becomes maximum and in III the particles aggregate and the curve decreases. But the sols prepared by condensation methods exhibit the different behaviour. In this case the degree of dispersion first decreases, remains almost constant and finally decreases again, as revealed by the Fig. 1.43.

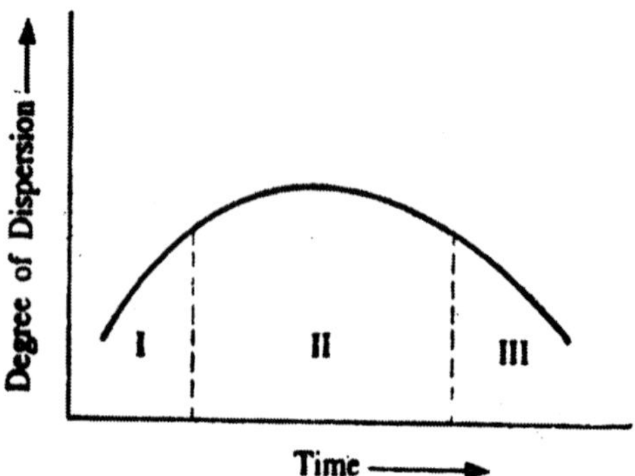

Fig. 1.42 : The life curve of $Ni(OH)_2$.

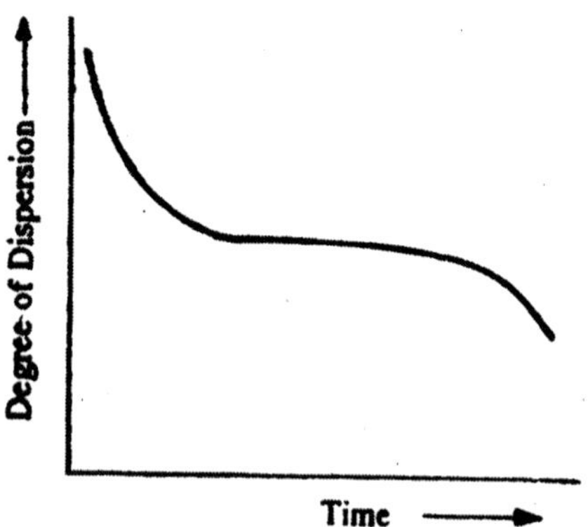

Fig. 1.43 : The life curve of a sol prepared by condensation method.

Factors Which Affect the Spontaneous Ageing

The factors affecting the process of spontaneous ageing are as follows :

(i) *Changes in particle size and shape : Mohlethaler* (1948) proved that there occurs some change either in size or in shape of colloidal sol during ageing. Me found that, in the freshly prepared sol of vanadic acid, the particles are not symmetric and arranged by a number of smaller particles, if kept for some time the particle grows in size and would become asymmetric. The viscosity and streaming potential increase with time.

(ii) *Chemical changes :* In ageing some times, some chemical changes occur in the sol. *Malfltano* (1904) and *Heymann* (1928) found that the chemical composition of particles of ferric hydroxide changes with time. During ageing the amount of chlorine gets decreased.

The hydrolysis of ferric chloride takes place in stages which are given as follows :

$$FeCl_3 + H_2O \rightarrow Fe(OH)Cl_2 + HCl$$

$$Fe(OH)Cl_2 + H_2O \rightarrow Fe(OH)_2\,Cl + HCl$$

$$Fe(OH)_2Cl + H_2O \rightarrow Fe(OH)_3 + HCl$$

The freshly prepared sol consists of the particles rich in Fe(OH) Cl_2 but consists of $Fe(OH)_2Cl$ when kept for some time. The hydroxy chloride are converted into ferric hydroxide and hydrochloric acid. It provides the clear explanation of the chemical changes taking place during ageing.

(iii) *Brownian movement :* As Brownian movement is due to unequal bombardment of solute molecules by the particles of dispersion medium, it is the chief factor responsible for ageing. It has been confirmed in the case of various colloids that the rate of ageing gets increased with the temperature of the colloidal sol.

(iv) *Solubility of dispersed phase in dispersion medium :* The ageing is also affected by the solubility of dispersed phase in the dispersion medium. It was reported by *Kolthoff* and Bowers (1954) that in the sol of AgCl, AgBr and Agl, the Agl is most stable than AgBr and AgCl. The order of ageing is given as follows :

$$AgCl > AgBr > Agl$$

This is attributed to the fact that the AgCl particles are having relatively greater solubility and thus dissolve, becomes more coarser and finally crystallises out on the larger particles. This is termed as recrystallisation ageing. In case of calcium oxalate sol it may very easily be crystallised out than AgCl, AgBr or Agl.

(v) *Impurities in the sol :* The factor governing the process of spontaneous ageing is the impurities present in sol. In every hydrosol the small amount of electrolyte is always present which is able to promote the process of ageing. For example the colloidal gold may be having the impurities of KOH (if prepared by process). In this case the pink gold sol becomes violet or blue when kept for regular time interval These impurities are generally present in sol and in chemicals taken for preparation. *Theories of spontaneous Ageing* :

Very little work has been done on the spontaneous ageing. However, the following possibilities may be regarded to be taken place in connection with the phenomenon of spontaneous ageing.

(i) *End to end aggregation of particles* : According to this concept in ageing there is end to end aggregation of the short rods in fresh sol. In the case of vanadium pentaoxide sol the particles aggregate as rod shaped. The sols of linear or laminar type may set upon ageing in this way.

(ii) *Growth of recrystallisation* : In the case of vanadic acid sol there occurs the growth in particle size during ageing. In case of silver halide sol there occurs a change due to the solubility of the dispersed phase in the medium.

(iii) *Partial coagulation* : It was reported by *Jonker* (1942) that stantaneous ageing is ascribed to the partial coagulation which is probably caused by the insufficient repulsion between the small particles. This type of coagulation has been found to depend upon temperature, size and shape of the sol particles.

COAGULATION OR FLOCCULATION

The colloidal sols are stable by the presence of electric charges on the colloidal particles. Because of the electric repulsion the particles do not come close to one another and coalesce. The removal of charge by any means will lead to the aggregation of particles and hence precipitation immediately. *The process by means of which the particles of the dispersed phase 19 a sol w precipitated is known as coagulation or flocculation.*

Electric charges on lyophobic particles can be removed by the application of an electric field as is used in electrophoresis. But a common method of producing precipitation is by the addition of electrolytes.

The precipitate after being coagulated is known as coagulum.

Methods for coagulating a sol : There are several methods employed for coagulating a sol. Some of them are noted below:

1. *By the addition of electrolytes* : In this method large amount of electrolytes are added to the sols which cause precipitation. This is due to the fact the colloidal particles take up the ion whose charges are opposite to that on colloidal particles. With the result that the charge on colloidal panicles is neutralised and coagulation takes place. In case of arsenious sulphide sol (negatively charged) coagulation takes place by adding $BaCl_2$. It is due to the fact that negatively charged particles of the arsenious sulphide sol take up barium ions, resulting neutralisation of the charge on the colloidal particles and hence lowering the stability of the sol.

It has been observed that generally the greater the velency of the added ion, the greater is its power to cause coagulation. An ion having an opposite charge to that of the particles of the sol is responsible for coagulation. This ion is generally called as *active ion*. For example, calcium chloride is approximately 100 times more active than Nad in the coagulation of a silver sol. The particles of a silver sol &TS stabilised by negative charges and for the coagulation the valency of the cation is effective. Similarly for the coagulation of positive sols the valency of the anion is decisive. Further, the precipiting power of an electrolyte increases very rapidly with increase in the valency of the cation or anion, the ratios being approximately 1 : 40 : 90 for the ferric hydroxide sol and 1: 70:500 for the arsenious sulphide sol. Thus in the coagulation of ferric hydroxide sol, the coagulating power increases in the order of $Cl^- > SO_4^{--}$ PO_4^{--} $Fe(CN)_6^{--}$, while in the coagulation of arsenious sulphide sol, the coagulating power increases in the order of $Na^+ > Ba^{++} > Al^{+++}$. This importance of velency was first recognised by Schaize (1882) and more data were obtained latter by *Linder, Picton, Hardy* and *Freundlich*. The coagulation values of NaCl, $BaCl_2$ and Lat $(NO_3)_3$, for the silver sol prepared by reduction of silver carbonate with tannin are 30, 0.5 and 0.003 millimol per litre respectively. The *coagulation or flocculation power* can then be expressed as the reciprocal of these flocculation values *i.e.*,

$\frac{1}{30}:\frac{1}{0.5}:\frac{1}{0.003}$ *i.e.*, NaCl : $BaCl_2$; $La(NO_3)_3$ as 0.033 2 : 33.3 or 1 : 60 : 10000.

2. *Physical methods* : The coagulation of some sols can be carried out by :

(a) mechanical treatment,

(b) beating or cooling,

(c) irradiation,

(d) vigorous shaking,

(e) treatment with electric current, etc.

3. *By continuous dialysis :* We know that traces of electrolytes are present in the colloidal system which are necessary for the stability. If the sol is subjected to continuous dialyser the colloidal system becomes unstable.

4. *Salting out :* Coagulation of lyophilic sol can be made by the addition of sufficient high concentrations of certain ions. Thus salting out of lyophilic colloids is due to the tendency of ions to become solvated, causing the removal of adsorbed water from the dispersed particles.

5. *By hydrated ions :* Since ions can also differ in the degree of hydration or solvation, this factor also plays an important role in the precipitation of sols.

6. *By removal of electric charge :* Removal of electric charge on lyophobic particles by means of the application of electric field results precipitation. This is accomplished by electrophoresis.

The coagulating effect of electrolytes on 'hydrophobia sols was studied by *Schulze, Hardy, Linder* and *Picton.*

1. *Hardy-Schulze law :* According to them, the greater the valency of the active ion, greater is the power to cause coagulation. Active ion is responsible for coagulation.

Thus, in the case of positively charged sol the coagulating power of anions is in the order of

$$[Fe(CN)_4]^{-4} > [PO_4]^{-3} > [SO_4]^{-2} > [Cl]^-$$

In the case of negatively charged sols, the coagulating power is in order of

$$Al^{+++} > Br^{++} > Na+$$

The coagulation values of NaCl. $BaCl_2$ and $La(NO_3)_3$ for silver sol are 30, 0.5 and 0.003 millimoles/litre.

The reciprocal of coagulation value is regarded as the coagulating power or flocculating power *i.e.,*

$$\frac{1}{30} : \frac{1}{0.5} : \frac{1}{0.003} \quad \textit{i.e.,}\ 1 : 60 : 1000$$

Table 1.4

	Arsenious sulphide sol		*Ferric hydroxide sol*	
S.N.	*Electrolyte*	*Coagulation value milli moles/litre*	*Electrolyte*	*Coagulation values (milli moles/litre)*
1.	NaCl	52	KCl	132
2.	KCl	51	K_2CrO_2	0.225
3.	$BaCl_2$	0.69	K_2SO_4	0.210
4.	$MgSO_4$	0.22	$K_3Fe(CN)_6$	0.096
5.	$AlCl_3$	0.093	$K_4Fe(CN)_6$	0.085

2. *Whentham rule :* The coagulating power of a series of ions of the same sign is related to a constant raised to the power representing the valency of each ion.

$$P_1 : P_2 : P_3 = x : x^2 : x^3$$

P_1, P_2. P_3 represent the coagulating powers of mono, bi and trivalent ions, and x is a constant greater than unity.

Factors Affecting Coagulation

1. Nature of the sol.
2. Nature of charge on sol particles.
3. Valency of the precipitating ions.
4. The mode of defining coagulation.
5. Time allowed for coagulating; concentrations of electrolytes.
6. The ratio of surface to mass of dispersed phase. Large quantities of precipitants are needed for more concentrated and more finely dispersed sols.
7. Rise in temperature favours coagulation.

8. Addition of dehydrants like alcohol induces coagulation of emulsoids.

9. Centrifugal forces greatly enhance the setting out of particles due to difference in the densities of the medium and the dispersed phase.

10. Agitation with agitators. The surface of the agitator provides a slowing down of Brownian motion and a resulting aggregation of particles. Separation of butter by churning is an example of this type.

Coagulation with similarly charged ions : It was observed by *Wannow* that the coagulation values of similar ions like Li^+, Na^+ and K^+ do not differ very much for coagulation of negatively charged sols. The Table 1 is having the data obtained by Wannow.

From the Table 1.4 it can be seen that there is a little difference between the coagulating powers of Na^+ and K^+ ions or between Ba^{+2} and Mg^{+2}. The coagulation value increases with increase of atomic weight or ionic radii in the series, Na, K. It is observed from the table that the coagulation value of HCl is lower than sodium or potassium chloride. The H^+ ions are thus very active in causing coagulation. *Boutaric* and *Perrean* (1928) reported that in the coagulation of arsenious sulphide with HCl, HNO_3 or H_2SO_4 the coagulation always occurred at pH 1.2.

Effect of salt mixture in coagulation : The effect of mixture of salt like NaCl + KCl or $Bacl_2$ + $CaCl_2$ was also tried by *Dumaoski* and *Vinnikowa* (1934) on a colloidal sol and it was found to be additive. For example 0.02 M BaCl, is having the same coagulating power on a colloid as 0.02 M $CaCl_2$, a mixture of $BaCl_2$ and $CaCl_2$ of molarity 0-04 M is having the same coagulation value as 0.04 M Bad, and 0.04 M $CaCl_2$.

Theories of coagulation : Several views have been proposed to explain the mechanism involved during coagulation. Some important views are cited below:

1. Adsorption theory : Freundlich developed a theory according to which coagulation occurs by the adsorption of ions of opposite charge to that of colloidal particles.

During the coagulation of negatively charged As_2S_3 sol by $BaCl_2$ solution, it is assumed that negative charge on the colloidal particles thus becomes unstable, aggregate and settle down in the form of a precipitate.

If we take the case of different electrolytes to coagulate the same sol, the equivalent amounts of ions of different valencies should be adsorbed. For example the quantities of uni-bi– and trivalent ions adsorbed should be in the ratio

$$3 : 1.5 : 1$$

The classical Freundlich adsorption isotherm has been found to applicable *i.e.*, $x/m = kc^{1/n}$. If c_1, c_2 and c_3 refer to the concentrations of uni-, bi- and trivalent electrolytes then we have

$$kc_1^{1/n} = 3,\ kc_2^{1/n} = 1.5 \text{ and } kc_3^{1/n} = 1.0$$

where k and n are the same for each electrolyte, then the relative amounts are

$$c_1 : c_2 : C_3 = 3^n : 1.5^n : 1^n$$

If it is regarded that the n is having reasonable value as 6 then the ratio is

$$729 : 11.4 : 1$$

These results have been found to be in fair agreement with experiment as in Table 1 for uni-, bi- and trivalent electrolytes.

2. *Critical potential theory* : The phenomenon of coagulation was explained by *Gouy, Barton, Debye* and *Hackel* on the basis of *electrical double layer*. According to them the ions which are responsible in precipitation are adsorbed on the fixed part of the double layer. For example in the case of As_2S_3 sol when peptised by H_2S can be shown as

$$[As_2\ S_3]\ S^{--} : 2H^+$$

On the addition of $BaCl_2$, it is represented as

$$[As_2S_3]S^{--} : Ba^{++}$$

Because Ba^{++} ions are responsible in precipitation and replace H^+ ions which are then passed into the bulk of the solution making it acidic. Zeta potential measurements have revealed that during coagulation the zeta potential values are decreased. There is a critical value of zeta potential at which coagulation occurs.

3. *Verwey and Overbeek's theory* : According to these workers the coagulation takes place when the particles of sol approach each

other. Verway, Overbeek and others calculated the forces and energies of repulsion and direction between two particles surrounded by ionic atmospheres. The attractive forces are van der Waal forces acting at the surface while van der Waal London potential forces also operate between the colloidal particles. The attraction is opposed by electrostatic forces of repulsion.

Verwey calculated the relative concentration of uni-univalent electrolyte (NaCl) and bi-bivalent electrolyte ($MgSO_4$) and trivalent ion on negative arsenious sulphide sol. The concentration ratio is

$$1 : \left(\frac{1}{2}\right)^6 : \left(\frac{1}{3}\right)^6 \text{ or } 100 : 1.6 : 0.13.$$

This theory has been found to be quite successful in explaining the Hardy-Schultz law.

In the case of ferric hydroxide sol, there occurs a considerable discrepancy.

This theory presents a complete explanation of coagulation except in case of ferric hydroxide sol, there occurs some discrepancy.

4. *Ostwald theory : Ostwald* (1948) put forward a new concept about the phenomenon of coagulation. According to his view the coagulation does not depend upon the discharge of particles but due to some physical properties of the dispersion of medium. The coagulation takes place when a definite value of activity coefficient is reached. According to him the particles are pushed out of the dispersion medium. It was shown by *Vonck* (1952) that the activity view of Ostwald regarding coagulation has been found to be in good agreement with the Verwey theory.

KINETICS OF COAGULATION

We know that the stability of hydroyhobic colloids is judged by their rate of coagulation. It means that valuable information about the coagulation process can also be obtained from studies of kinetics of coagulation. Coagulation in different instances may proceed at different rates. The course of coagulation with time is known as kinetics of coagulation and is generally determined by two factors.

(1) *Brownian movement of the particles,* and

(2) *Their interaction when they are close together.* The rate of coagulation can be best measured by determining the decrease of the number of particles after certain intervals of time. *Zsigmondy* made the first quantitative study of the rate of coagulation. He measured the length of time required for a gold sol which changes its colour from red to violet when an electrolyte is added to it.

He found that as the amount of electrolyte was increased, the time required for the colour change decreased rapidly first,, then slowly and finally reached a constant value which was independent of the valency and the nature of coagulating ion. *Zsigmondy*, therefore, concluded that the explanation of the kinetics involves two types of coagulation :

(a) *Rapid coagulation :* This type of coagulation results at a high concentration of electrolyte. It takes place as a result of complete or nearly complete neutralisation of an electric charge on the surface of colloidal particles. It is independent of valency and nature of coagulating ion.

(b) *Slow coagulation :* In this type, the coagulation takes place as a result of only partial neutralisation of the charge on the surface of colloidal particles. It depends upon the concentration of the electrolyte. *Kinetics of rapid coagulation.*

The rate of rapid coagulation is completely determined only by Brownian movement.

Smoluchowski derived the mathematical expression relating the velocity of rapid coagulation with the law of probability governing the colloidal and adhesion of the colloidal particles. Each collision results in aggregation only when the distance between the colliding particles becomes smaller than R, the radius of sphere of attraction. In other words, R is the distance between the centres of two particles at which a lasting contact is formed. In the simplest case R = 2r, where r is radius of the particles, when particles just touch. Such a two double particle can take up another single particle or also another double particle or the particle.

If R is much greater than 2r, there are no chances of collisions to take place, hence particles will not collide and the sol would be less stable.

First of all, Smoluchowski calculated this decrease in the number of single particles. These particles get diffused in the sphere of attraction of a stationary particle. Now the treatment was extended to the larger number of particles in motion. He attempted to calculate the number of double and decrease in triple particles. He found that the number first increases. It can be seen from the graph (Fig. 1.45) that the velocity curve represents a maxima showing the decrease in number of single, double of triple particles. It is evident that the number of collisions produced depends upon the concentration of particles and the temperature.

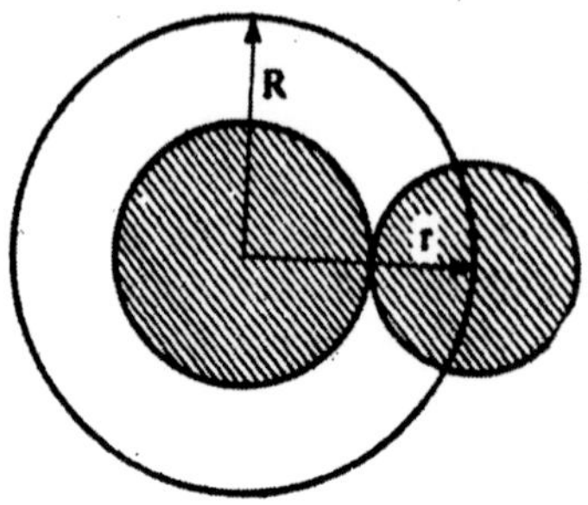

Fig. 1.44 : Aggregation of particles.

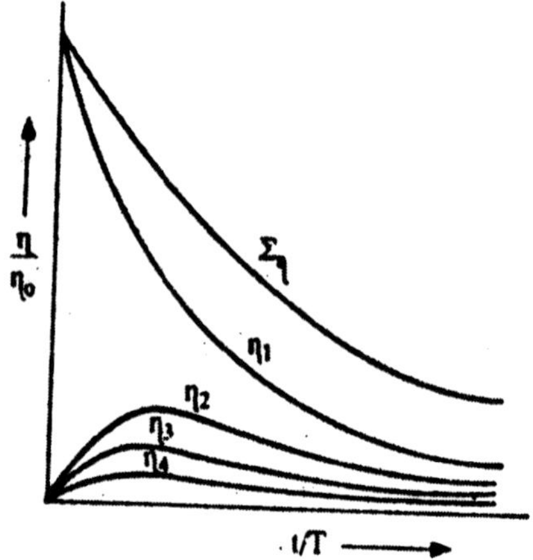

Fig. 1.45 : Coagulation as a function of time.

Smoluchowski showed that the number of single particles decreases with time according to a bimolecular equation, given by

$$n = \frac{n_0}{1 + 4\pi DRn_0 t} = \frac{n_0}{1 + \frac{t}{T}} \quad ...(1)$$

n = number of particles per ml at time t.

n_0 = original number of particles per ml.

T = time of coagulation.

D = diffusion coefficient.

R = radius of attraction.

The time $t_{1/2}$ from the beginning of coagulation to that at which the number of particles diminished by one-half is called the *coagulation time*. For fast coagulation, according *to Smoluchowski*,

$$\frac{n_0}{2} = \frac{n_0}{1 + 4\pi \, D R n_0 \, t_{1/2}} \qquad \text{...(2)}$$

Equation (1) can now be solved for single, double or triple particles at a given time. Thus,

$$n_1 = \frac{n_0}{(1 + t/T)^2}$$

$$n_2 = \frac{n_0 . t/T}{(1 + t/T)^3}$$

where n_1, its and n_k represent the number of single particles, double particles and of multiple particles made of k single ones respectively. It should be noted that the term $4\pi DRn_0$ in equation (1) was assumed to be constant, because the values R and D are inversely proportional to each other. In the course of coagulation the particles grow larger and hence R must also increase, the diffusion coefficient must, however^ decrease, since large particles move more slowly than small ones.

Tests of Smoluchowski equation. *Smoluchowsiki*'s theory is generally tested by the following facts :

(i) As represented in the above equation, the coagulation time should be inversely proportional to the number of particles originally taken, *i.e.*, n_0

(ii) The coagulation should follow the course of a bimolecular equation as given above.

(iii) There should exist a region of concentration of electrolyte where the coagulation is independent of this concentration.

On taking the above facts into consideration the Smoluchowski's theory was tested. The decrease in the number of particles was calculated by counting them in ultramicroscope. A definite quantity of the electrolyte was added to the gold sol. After a definite time interval the sample was taken and stabilised by adding gelatin to prevent the change in the number of particles. The particles were then counted in usual way in determining the size of particle.

From diffusion coefficient D, n and n_0, the value of R can be calculated. From ultramicroscope countings, the radius of particles (r) were determined and then compared with R/r. As the value of R/r is merely constant, it means that coagulation occurs if R is twice as larger as r which suggests that the particles can coalesce only when just touching.

Wintgen and Ehringhous (1923) observed that thc theory may be applied even to the coagulation of gold sol in molten borax at 1000°C. The value of R/r obtained was 2.3. It may be regarded that the aggregation process are approximately same at lower and higher temperatures. It is independent of the medium.

Kinetics of Slow Coagulation

The slow coagulation takes place by the partial neutralisation of electric charges. In slow coagulation the lesser amount of electrolyte is added which then leaves the residual electric charge. So the kinetics of slow coagulation would be different from rapid one. Here each collision will not always give rise to aggregation due to the small concentration of electrolyte added.

Smoluchowski suggested that the slow coagulation has been found to depend upon the following factors:

(i) collision of the particles,

(ii) charge on particles,

(iii) the rate of moving particle, and

(iv) the repelling force in zeta potential zone.

In slow coagulation, the lesser amount of electrolyte has the partial neutralisation of electric charges. Hence the kinetic Study of slow coagulation involves the coagulation of electrically charged particles. In slow coagulation the probability of aggregation is not dependent upon collision but on the viscosity of particles and mutual repelling force.

In coagulation, it is to be noted that it becomes essential to include the extent to which the electric charge on particles is removed by adding the electrolyte.

Smoluchowski in his treatment of slow coagulation kinetics, suggested the introduction of a new factor S which denotes the fraction of the collision which is effective in producing aggregation. Hence the total number of particles present in a sol at the time t is given by the expression,

$$Sn = \frac{n_0}{1 + \xi\,(t/T)} \qquad ...(3)$$

It was found that the above equation is not applicable in many cases. The reason of deviation is that the electric charge is not uniformly distributed at the surface of particle. Therefore the two particles when approach each other at uncharged spots, will form the aggregate like the bunch of grapes.

On taking the above view of Smoluchowski as true, it is evident that the sol coagulated by various amounts of electrolyte would pass exactly through same stages only after different times. This stage is characterised by the adsorption and time after which the sol reaches the same *adsorption*. Hence there should be a constant relation- ship between these values.

Experimental Proof : This view was confirmed by Mukherjee, the experimental outlines of the experiment are given as follows:

(1) He investigated the effect of adding an electrolyte (*e.g.*, KCl) on the ruby gold sol which turns blue as soon as electrolyte is added. He investigated the rate of adsorption spectrophotometrically in the region of 6400Å. By adding varying amount of N.KCl *i.e.*, 1, 2, 3, 4 c.c. the change of colour from red to blue was observed. The plot of adsorption coefficient against time (vide Fig. 1.46) reveals that the adsorption coefficient increased with the time.

(2) According to *Mukherjee* the sol of same colour and same adsorption coefficient must possess same number of particle size. The reason for this is that the different amounts of added KCl have different values of adsorption coefficients. Thus, for slow coagulation the total number of particles can be calculated at different time intervals respectively. Thus at time

$$t_1, \qquad \Sigma n = \frac{n_0}{1 + \zeta_1\, k n_0\, t_1} \qquad ...(4)$$

$$t_2, \qquad \Sigma n_1 = \frac{n_0}{1 + \xi_2\, k n_0\, t_2} \qquad ...(5)$$

and $\quad t_3, \quad \Sigma n_2 = \dfrac{n_0}{1 + \xi_3 \, kn_0 \, t_3}$ ...(6)

where Σn, Σn_1, Σn_2 denote the total number of particles at time t_1, t_2 and t_3 respectively. As the solutions are having the same adsorption coefficients so the number of particles should also be the same, then

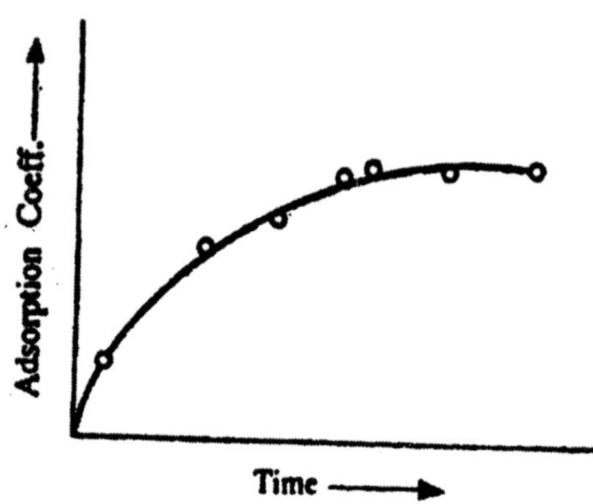

Fig. 1.46 : Adsorption coefficient and time.

$$\Sigma n = \Sigma n_1 = \Sigma n_2$$

or $\quad \dfrac{1}{t_1} : \dfrac{1}{t_2} \; \dfrac{1}{t_3} = \xi_1 : \xi_2 : \xi_3.$ ...(7)

This equation has been found to be satisfied for every value of adsorption coefficient.

Similarly, it is possible to test the original Smoluchowski's equation. For all adsorption coefficients the ratio of $\frac{1}{t_1} : \frac{1}{t_2} : \frac{1}{t_3}$ was never found to be constant, which reveals that the Smoluchowski's equation does not hold good.

According to Smoluchowski the time $t_{1/2}$ (*i.e.*, the time of beginning of coagulation to that which the number of particles has diminished by one half) will be longer. It is attributed to the fact that all collisions causes the decrease of number of particles. Thus, the equation (1) of rapid coagulation changes into the form

$$\frac{n_0}{2} = \frac{n_0}{1 + \xi \, 4\pi \, RDn_0 \, t^{\delta_{1/2}}}$$

where $t^{\delta_{1/2}}$ represents the time in slow coagulation.

The values of ξ, *i.e.*, the fraction of collision in aggregation as observed by *Kruyt* and *Arkel* (1920) lies between 0 0036 to 0-65. If higher the concentration of added electrolyte *e.g.*, Bad; the value of ξ would approach to one. *Tuorila* (1928) investigated the dispersion of paraffin in water. He found that the charge number of effective collisions gets increased with decreasing particle size.

Modern theory of slow coagulation : According to modern theory of slow coagulation, the coagulation takes place via two processes a and b. Thus the mechanism may be put as follows :

$$\text{charged particles} \xrightarrow{a} \text{discharged particles} \xrightarrow{b} \text{coagulation}$$

The mechanism may be explained by taking the following example :

Suppose KC1 (where K^+ ions are active ions) is added to a –vely charged sol. In this case every charged particle is surrounded by a double layer and only those ions will penetrate which possess a minimum critical value of kinetic energy. Since discharge process may be brought about only by a very small fraction, the speed of process "a" is finite. This shows that in slow coagulation neutralization of electric charge occurs with a measurable rate, while in rapid coagulation this process is instantaneous. In short it can be said that in slow coagulation the processes "a" and "b" take place alongwith the electrolytic pull of coagulating and chemical interaction of adsorbed and coagulating ion. All these processes give the rate of reaction which were not considered by Smoluchowski.

GELS

Gels : According to Ostwald a gel is a colloidal *semi solid system rich in liquid phase*, Thus a gel is having two components one is solid and other is a liquid. For example, silicic acid gel in water, sodium oleate gel in water or gelatin in water. The curdled milk or a piece of meat is also an example of more complex gels. When a sol *e.g.*, gelatin, agar-agar is cooled, it sets to a semi-solid gel. This phenomenon is termed as *gelation.*

Jelly: Jelly is the common name of a gel. According to McBain a gel may be defined as a colloidal system consisting of the both liquid phases rich and dry system, but a jelly is an elastic, coherent system which is very rich in liquid phase. Table jellies are generally prepared from gelation ; fruit jellies (*e.g.*, jams, preserves). *McBain* employed the same terms for gels and jellies.

Xerogels : There is a similar term designated by xerogel which implies dry gels *e.g.*, sheet of gelatin, a foil of cellophane, or a piece of dry gum arable.

It has been estimated that xerogels usually have less liquid than solid. The jellies have 95% or more liquid. Gelatin jellies have 95 to 99% of water. Similarly agaragar jellies have 99.8% water and only 0.2% of the solid polysaccharide. Fibrin clots contain the very small amount of water.

Classification of Gels

The gels may be classified as—

(a) *On the basis of dispersion medium* : It is possible to classify the gels on the basis of dispersion medium. If gel is either in alcohol, benzene or water then it is termed as *alcogel*, *bemogel* and *hydrogel* respectively.

(b) *On the basis of size of particles* : This classification of gels is done on the Size of particles which form the frame work. Hence a gel may be *colloidal* or *coarse*. The common gelatin jellies pectin, agar-agar are the colloidal gels.

(c) *On the basis of chemical composition* : On the basis of chemical composition gels can be classified as *inorganic* and *organic*. The gels containing an inorganic solvent are termed as inorganic gels while the gels having an organic solvent are termed as *organogels*.

(d) *On the basis of their properties* : The gels can be classified on the basis of some of their mechanical properties. *Thus it may be rigid gels*, elastic jellies, non-elastic jellies and *thixo-tropic* gels.

   (i) *Elastic or heat reversible gels* : Examples of these gels are the gels of gelatin, agaragar. *These gels are reversible to heat.* Such gels may be prepared by dissolving the substance in warm water, it is then cooled till it sets. If these gels are dehydrated they convert into the elastic solid form. The gel is again regenerated by adding water.

The rate of settling of such gels has been found to depend upon the setting time. The greater the rate of under cooling, the smaller its settling rate. *Hattiangdi* (1948) represented, on the basis of quantitative experiments, the time of setting by the equation.

$$t = ac^{-E/RT}$$

where t refers to the time of setting, *a* a constant, a the concentration, E the energy of activation and T the absolute temperature. The substances like sulphate and acetate salts generally increase the setting point whereas the nitrides, iodides and thiocynates lower it.

In elastic gels the fibrils composing the gel are flexible. On hydration these fibrils expand again to form the gel again.

(ii) *Non-elastic or irreversible type gels :* Examples of such gels are the sols of silicic acid or alumina. These gels may be prepared by adding water to dry solid. When dried they lose their acidity and becomes glassy. The rigidity of these gels increases with time. In the case of silicic acid gel the volume decreases on drying. This decrease is having a particular limit, beyond which the air enters in the cavities of the gel.

(e) *Thixotropic gels :* These are the gels which when shaken form a sol and on standing are converted into the form of the gel. The gels of this nature are termed as thixotropic gels. The frame-work of such gels is so weak that it is broken on standing. This phenomenon of sol-gel transformation is termed as *thixotropy.* The gels which are prepared on cooling are known to exhibit such a property. These are liquified on cooling.

Methods for the Preparation of Gels

(i) *By coagulation or decrease in solubility :* Many colloidal solutions could be transformed in the form of a gel by coagulation. Such transformation has been characterised by:

(a) the shape of particle

(b) the concentration of the sol

(c) degree of salvation.

The sol of pectin sets upon by the addition of alcohol or sugar. In the same manner the sols of aluminium hydroxide or ferric hydroxide set upon by the addition of coagulating substance, provided the concentration of the sol is high.

(ii) *By chemical reaction :* The gels can also be prepared by chemical reactions of concentrated solutions if one of the products of the reaction is insoluble and the particles are having the tendency to aggregate in the form of linear colloid. The gelation is also promoted by solvation. An example of this type includes the formation of $BaSO_4$ gel, by shaking a saturated solution of barium thiocyanate with manganous sulphate. The gels of aluminium hydroxide can be prepared by mixing a concentrated solution of aluminium salt and ammonium hydroxide.

(iii) *By cooling of colloidal solutions :* The gels of some sols can be prepared by cooling their hot solutions. For example the agar-agar; gelatin, soap are soluble in hot solvents, when they are cooled they form a gel. The gelation of this type has been found to depend upon:

(a) the temperature of gelation

(b) the time of gelation,

(c) viscosity of the medium, and

(d) the minimum concentration of the colloid.

(iv) *By double decomposition :* The gels of some sols can be prepared by the process of double decomposition. Thus silicic acid gel can be prepared by adding water to sodium silicate. During this double decomposition free silicic acid gets formed. Due to this acid liberation, it reapidly sets to a solid gel.

(v) *By exchange of solvents :* In the preparation of some of the gels, the gelation takes place by the exchange of solvent in which the sol is insoluble. Thus in the case of calcium acetate gel, pure alcohol can be added to the aqueous, solution of calcium acetate, in this way whole of the calcium acetate goes into alcohol which then sets .in the form of a gel having the liquid. This method of exchange of solvent has been found to be very similar as the preparation of colloidal sols by this method.

Properties of Gels

(i) *Electrical conductivity* : According to *Kistler* (1931) no change of electrical conductivity takes place when a sol converts into a gel. This property reveals that there is simply the change of state during this conversion.

(ii) *Optical properties* : Leick suggested that the gels exhibit the property of double refraction. The existence of double refraction can be ascribed either because of the presence of some property from the start in gel or formed due to internal tension. For example dry gel of gum showed the phenomenon of double refraction similar to the glass under the compression or tension In the case of swollen gel it was observed that a quick pressure at first produces a double refraction but rapidly changes into double refraction of reverse sign.

On the other hand the gels exhibit the property of polarised light. *Leick* determined the refractive indices of different gelatin gels.

(iii) *Elastic properties* : The gels exhibit the elastic properties as in the case of a solid.

For example, the properties like compressibility, flexibility, tensile strength can be followed by gels.

(iv) *Swelling or Imbibition of gels* : The hydrophilic or plastic gels are having the property of adsorbing definite amount of water or other liquid, causing the volume of the gel to increase. The swelling is somewhat similar to *imbibition* which implies the uptake of a liquid by porous material without any observed increase in volume. For instance if a sheet of gelatin is immersed in water it absorbs water and swells up. Similarly a piece of unvulcanised rubber swells up when placed in chloroform or in benzene. According to *Muller* (1949) the process of swelling is not entirely reversible. After the removal of liquid it returns to its initial state. The extent of swelling has been found to depend upon—

(a) temperature of the system

(b) pH of the solution

(c) individual nature of gel.

(v) *Syneresis :* According to *Graham* (1864) the syneresis may be defined as *the spontaneous liberation of the liquid from a gel.* Elastic and non-elastic gels both shrink when they are kept standing.

Examples—

(1) Perspiring of cheese is an example of syneresis.

(2) In the case of silicic acid sol it has been found that when a suitable electrolyte is added, it sets to form a jelly and then shrinks in size. The syneresis may be considered to be the *reversal of swelling.*

Syneresis also occurs in the ageing of unstable gels. It has been observed in a fresh jelly with unstable framework that when the frameworks becomes coarser the liquid gets expelled out from it. The rate of draining of the liquid depends upon the unstability of the framework. The degree of syneresis has been found to depend upon :

(i) temperature

(ii) pressure of electrolyte

(iii) dispersion medium

(iv) state of the gel

According to *Prasad and coworkers* (1952) the syneresis may take place by the mutual precipitation of sols possessing oppositely charged particles. For example if a positively charged sol of copper hydroxide is added to negatively charged nickel hydroxide sol, the syneresis occurs. For the gelation the small amount of the electrolyte becomes necessary.

If the concentration of the electrolyte is very small the syneresis occurs easily. The coarse opaque gels always synerised more readily than transparent colloid disperse jellies of the same material.

Importance of Syneresis : The phenomenon of syneresis finds much importance in the case of some biological problems. For example in the ageing of tissues the water content of a young body is greater than the water contents of an aged .body and secretion of liquids from glands.

Structure of Gels

There is much controversy regarding the structure of gels. However, various views have been given from time to time. A few of them are given as follows :

(a) *Honey comb theory : Butschli* (1897-1900) made first attempt regarding the internal structure of gels, on the basis of microscopic observations. He suggested that many gels are having the fine honeycomb structure which can be made visible by special methods. On the microscopic examination it was found that gelatin gel after hardening with formaldehyde exhibit this type of structure. He considered that the honey comb wall is having thickness of about 0.2μ – 0.3μ. This theory explains only the rigidity of the gel but fails to explain the phenomenon of syneresis of gel.

(b) *Solvent theory :* Mortin-Fisher regarded a gel as a system of two components, the first is liquid solute while the other as the solid solvent. *Kartz* on his investigation of X-ray studied the swelling gels and found that during swelling the X-ray spectrum do not get changed. This shows that a gel takes up the liquid intermicellarly and by intra-molecularly. By X-ray spectra it was found that the dimensions of the crystal lattice remain constant. This reveals that the internal structure of a gel is miceller in nature.

(c) *Granular fine structure theory : Zsigmondy* showed by ultramicroscopical observations that gelatin does not conform to the structure of honey comb as pointed out by *Butschli* (1900). Thus, *Zsigmondy* and *Backmann* suggested that gelatin gels are having granular fine structure. According to the above view the phenomenon of sol gel transformation may be considered as a transfer of water from the continuous to the discontinuous phase.

The *Zsigmondy* theory receives the support from the results of diffusion of electrolyte into gels. The rate of diffusion of dilute electrolyte in gel is much similar to the rate of diffusion into water. This similarity in diffusion reveals that in gels there are open wall channels in the structure of gels. If the rate of diffusion is slow in concentrated gels it is regarded that these channels being so thin to resist the entry of water in gels.

(d) *Fibrilar theory :* According to this theory the particles in gels are arranged in the form of fibre which are then intervened. These fibres are responsible for the great cohesion and elasticity of gels. The experiments carried out by *McBain* and *Bavert* confirm the existence of such fibres in the case of soap gel and barium malonate gel. This theory is further modified by considering that the gel consists of dimensional net chains. The elasticity of gels and their properties of sol gel transformation have been also accepted by this view.

(e) *Von Weimarn's theory* : According to this theory the tissue of the gel is composed of bundles of thin crystalline needles. In each bundle the needles are spread readily from the centre and the ends of the different bundles get interwoven giving a continuous structure. *Von Weimarn* postulated that the gels according to the degree of dispersion of their primary structural elements may be classified into *macro crystalline micro crystalline, ultra micro crystalline and subultramicro crystalline.*

(f) *Thomas Sibi theory* : According to this theory the structure of gel is in agreement as proposed by Von Weimarn. He concluded that when a sol is cooled, it forms a cluster of needles which interpenetrate to yield a firm mass. He also proposed that units of structure may be interwening hair and not the straight needle.

Structural units of gels : It is possible to classify into three structural units as given below :

1. *Gels with unstable frame-work* : This type of gels occurs in the ferric hydroxide, aluminium hydroxide and graphite gels. This frame-work also occurs in gels in which the structural elements are little symmetric. The particles get joined in such cases by very weak cohesive forces or van der Waal's attraction. Such gels are generally thixotropic, *i.e.*, their frame-work is so weak that it gets destroyed even on shaking. The gels of linear macromolecules such as unvulcanised rubber or polystyrene belong to this group. When polystyrene is kept in a liquid it first swells and then slowly dissolves into the solution.

2. *Gels with metastable frame-work* : Metastable frame-work generally occurs in protein gels, *e.g.*, gelatin gels. These gels are characterised by the fact that these gels are rich in liquid elastic and get transformed into liquids by heating. The structural elements are joined into a net work by secondary bonds. Cohesive forces or primary valencies also exert some importance in promoting the stability of these gels. A partial orientation of the chains may take place if two or more chains can be aligned into

bundles and linked by hydrogen bonds or other linkages. The net work of the linear macro molecule has been shown in the Fig. 1.4.7.

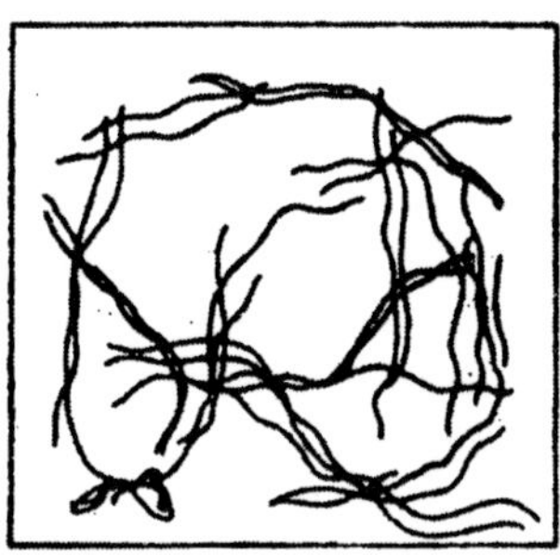

Fig. 1.47 : Gels with metastable frame-work.

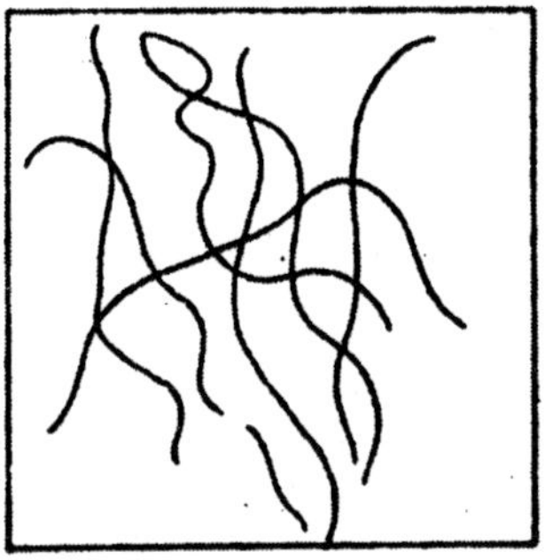

Fig. 1.48.

3. *Gels with stable framework :* The gels of stable framework are the gels of concentrated silicic acid. These gels can be obtained by adding acid to sodium silicate solution. In the reaction first mono-silica acid gel gets formed which then undergoes polymerisation to form space polymers too. According to *A.E. Alexander* and *P. Johnson* (1949) the resulting gel would be rigid and stable and cannot be transformed into a liquid system.

Finally a three dimensional net work of Si–O is obtained. Vulcanised rubber is an example of cross linked xerogel. The long and twisted molecules of natural rubber do not get linked by primary valencies into a spatial frame work, but they are essentially free. Because of this reason unvulcanised rubber swells readily and the dissolves more or less in a number of solvents.

When heated with sulphur, the sulphur atoms react with the chains and linking them together. The structure of non-vulcanised rubber has been given in the Fig. 1.48.

Applications of gels : The process of gel formation is used in the following ways :

(1) Gelatin and agar-agar gels are employed in laboratories for making liquid junctions.

(2) Silica gel is used in laboratory as a dehydrating agent in desiccator.

(3) Silica gel is employed in industry and also used to support the platinum catalyst. It also acts as a resistance to catalytic poisoning during the manufacture of H_2SO_4 from contact process.

(4) Solidified alcohol, a gel, is used as fuel in picnic stoves and is made from alcohol and calcium acetate.

(5) Boot-polishes and animal tissues have the gel structures.

Sol-gel Transformation

It is the phenomenon in which a sol is converted into a gel. Fernau and *Pauli* showed that a sol of CeO_2 having about 10 gm. per litre was converted in the form of gel by addition of an electrolyte. This electrolyte added brings about coagulation. If the sol of CeO_2 is kept for 200–300 days then it losses its power and form a precipitate in the place of a gel by adding electrolyte.

Examples of sol-gel transformation : The phenomenon of sol-gel transformation can be seen in some lyophilic sols such as :

(1) Gelatin in water and agar-agar in water. In such cases the phenomenon is limited at a particular temperature. It has been observed that if we prepare a gelatin sol by heating gelatin in water at 70°C. it is seen that it sets to form a gel at lower temperature. If this gel is again heated, it loses water and liquifies to form a sol. The process has to be repeated many times.

(2) *Arisz* gave the formation of a very unique example of sol-gel transformation. He observed that if a 10% gelatin sol in 32% glycerol was cooled from 77°C to 44°C, the mixture sets to form a gel at this temperature. If the same mixture is cooled at 35°C a gel is formed. When it is heated to 44°C, it first liquifies and then sets into a gel again.

Cause of sol gel transformation : The reason of sol gel transformation is that a certain degree of undercooling is responsible to convert the process of solidification to crystallisation. It is regarded that the size of crystals gets increased after the solidification. Since the meeting point of the crystals depends upon the size of particles, it is clear that when reheating occurs, the junction point of the sol and gel gets loosened and the conversion of sol into gel takes place.

Physical changes in sol gel transformation : The various changes take place in sol-gel transformation. A few of them are given below :

(i) *Thermal changes* : In certain cases the heat is evolved when a sol is transformed into the gel. This heat evolution may be due to the latent heat of crystallisation. In the cooling curve of gelatin a small break can be seen. This shows that the sol-gel transformation is followed by the evolution of heat. In the case of soap sol conversion into gel, the same behaviour can be seen. In the case of cooling curves of solutions of low molecular weights some breaks or arrest points have been observed if the solute starts to separate.

(ii) *Optical change* : Similar to the ordinary colloidal solutions the sol-gel transformation shows the *scattering of light* or the phenomenon of *Tyndall effect*. This *Tyndall effect*.increases during the conversion. It reveals that there occurs an increase in the size of particles during the change. In the case of a gelatin sol in water it was observed that at about 70°C the Tyndall cone is very weak. As soon as the sol gets cooled, this Tyndall cone becomes stronger. Again on heating it becomes weaker. In other words, it may be said that Tyndall light depends upon the viscosity of the solution. For the study of cone in gelatin gel *Donnan* and *Krishnamurti* suggested that the original anisometric particles in the sol formed were spherical in shape. It was further shown by *Kramer* and *Dextin* that gelatin gel at isoelectric point does not show change in Tyndall effect during the sol-gel transformation. *Derksen* has explained the Tyndall effect on the pH of the solution.

(iii) *Electrical conductivity* : No change in electrical conductivity occurs during the sol-gel transformation. The conductivity was found to increase if the solution is having the trace of electrolytes in it. It may be said that the conductivity has the secondary effect in solgel transformations.

(iv) *Diffusion* : No satisfactory work has been done regarding the change of rate of diffusion during the sol-gel transformation. However it is known that the rate of diffusion of the substances having low molecular weight has been found to be similar to the rate of diffusion of pure solvents.

(v) *Volume changes* : The change of volume occurs during the sol-gel transformation. In the case of sol-gel transformation of gelatin and water a contraction in volume and in the case of methyl cellulose and water the expansion in volume takes place. The reason being the process of crystallisation and dissolution which is a very common behaviour of this process For example in the gelatin of 9% sol of ferric hydroxide, no volume change has been observed since the process is not accompanied by the processes of crystallisation, whereas in the formation of silicic acid gel a considerable increase in volume has been found out.

(vi) *Viscosity :* The viscosity changes during sol-gel transformation. *Aritz* attempted to study the relationship in the case of gelatin sol. A sol of gelatin is first prepared by heating below the temperature 70°C. This is then cooled from a temperature t_1 to t_2.

As depicted in Fig. 1.49, the viscosity of the solution follows the curve a–a'. If the sol is now allowed to stand at a temperature t_1 the viscosity gets increased gradually following the curve a' – b'. Finally it reaches the equilibrium state b.

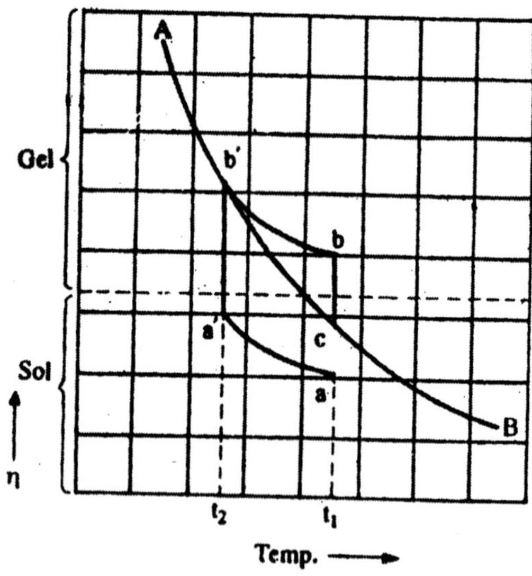

Fig. 1.49.

If the gel is maintained at a temperature t_2 and is rapidly heated to t_1 then the change of viscosity follows the curve b' – b. Then the viscosity gets decreased from b to equilibrium state c. The critical viscosity at which the transformation has occurred is represented by the dotted curve.

The total change of viscosity lies along the curve A–B where the change occurs mainly due to temperature.

In other properties like *vapour pressure* and refractive index the gel behaves exactly like that of sol.

Effect of Dissolved Substance on Sol-gel Transformation

The sol-gel transformation is greatly affected by the dissolved substances. *Pascheles* (1900) attempted to study the changes in the solidifying point of gelatin sol in presence of substances. The effect of anions has been found to be in the order :

SO_4^{-2} > tartarate > acetate > halogens > thiocyanate.

The anions like sulphate, tartarate and acetate group favour the gelation and raise the solidifying point. The rest of the groups were found to lower the solidifying point and increases the time of gelation.

The non-electrolytes such as sugars or polyhydric alcohols also effect the sol-gel transformation. It was reported by Lewites that the above mentioned substances decreases the time of gelation whereas the substances like urea, thiourea and urethane lengthen the time of gelation.

Theories of Sol-gel Transformation

The most satisfactory explanation has been based on the reversal of phase as in the case of emulsions. For example if a liquid X is mixed with another immiscible liquid Y, an emulsion is obtained. If this emulsion is coagulated by adding any electrolyte then the two phases, get reversed *i.e.*, Y becomes the dispersed phase and X the dispersion medium. But this theory appears to be unsatisfactory because if this transformation is taking place then in the reversal of phase there should be a little change in physical properties but it has been found that during sol-gel transformation there occurs sharp change in physical properties such as refractive index, conductivity and viscosity etc.

Rigidity of Gels

Regidity may be defined as the ratio of shearing stress to strain. The shearing stress is the force per unit area while the strain is the angular deformation. Thus the rigidity may be put as follows :

$$\text{Rigidity} = \frac{\text{shearing stress}}{\text{strain}} = \frac{\text{force per unit area}}{\text{angular deformation}}$$

From the above relationship it can be seen the greater is the stress and smaller be the deformation, the greater will be the rigidity.

Many experimental techniques are available for the mechanical testing of rigidity. The compression of jelly under a definite weight or the sagging under its own weight is also possible to measure. The rigidity and other elastic properties can be measured by such tests.

The rigidity of jellies gets increased with concentration and decreased with increasing temperature. In case of gelatin jellies the rigidity gets increased with square of concentration and it also increases with the average molecular weight of the gelatin sample.

Marked difference exists between concentrated, rigid gels, and dilute jellies of low rigidity. The gels of low rigidity are usually unstable and may be thixotropic *i.e.*, they can be liquified on shaking and sets again on standing. According to *Hoffmann* (1952) many unstable gels can be liquefied by ultrasonic radiations. The stable gels of 3ù5% gelatin or of vulcanised rubber are not thixotropic because in this case the frame work is bound together by strong primary valencies or by a large number of secondary bonds. Hence, this strong frame-work may be disrupted by shaking or by ultrasonic vibrations. As investigated by Muller (1949) the very dilute metastable jellies of linear colloids or the gels of ferric hydroxide having weak framework are found to give the property of thixotropy.

Thixotropy

Szegvari and Schalek (1923) showed that if a suitable small quantity of electrolyte is added to a concentrated ferric oxide sol, a partly gel is obtained which has a peculiar property of being liquefied when shaken. When it is kept standing for some time it again sets to the gel. This property can be seen in other colloidal systems *e.g.*, alumina, V_2O_5, zirconium dioxide, stannic oxide and in certain gelatin sols. According to *Peterfi* (1927) the thioxotropy implies the *change of touch*. Thixotropy is also termed as the isothermal reversible sol-gel transformation. A thixotropic gel can be liquefied by ultrasonic *i.e.*, high frequency sound waves. The property of thixotropy has been found to depend upon the concentration of electrolyte. At higher concentration the sol is precipitated which does not liquefy on heating.

The quantitative treatment of thixotropy was first of all given by *Goodeve* and *Withfleld* (1938) by assuming the equilibrium between the spontaneous building of an internal structure and its break down. The apparent viscosity of a thixotropic system was determined at various rates of shear. If η_a denotes the apparent viscosity, n, the extrapolated residual viscosity and 1/S is reciprocal of shear, then these quantities are related by the following relation :

$$\eta_0 = \eta_r + \theta/S$$

where θ refers to the coefficient of thixotropy, which can be calculated by the slope of the straight lines given by the above equation. If apparent viscosity is plotted against the reciprocal of shear 1/S then curves as depicted in Fig, 1.50 are obtained.

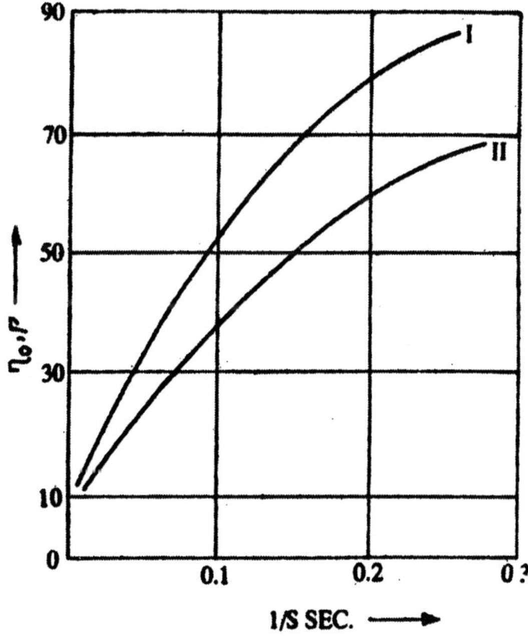

Fig. 1.50 : Plot of η_a vs 1/S.

The residual, viscosity (η_r) which is the intercept of the above straight line refers to the *resistance of the flow at extremely high shear ratio.* If 3% linoleic acid is added to carbon black dispersion, the thixotropy gets reduced. In the above figure the curve I represents the behaviour of thixotropic carbon black dispersion without added linoleic acid whereas the curve II in presence of linoleic add.

Measurement of thixotropy : A comprehensive method is available for the measurement of thixotropy but the feature of the property of thixotropy can be measured by the, time of solidification. However, by the equation,

$$\eta_a = \eta_s + \theta/S$$

if the apparent viscosity η_a, residual viscosity η_t and shearing force S is known, then the coefficient of viscosity θ may be measured. In order to measure the viscosity various types of viscometers are used. But the *rotating cylinder viscometer of Couette* is generally employed. (Fig. 1.51).

It consists of a cyiinder shape container Z_1 in which a colloidal solution has been taken. A small cylinder Z_2 hangs into it. In the upper part a mirror is present which reflects the light beam on a scale. The outer vessel Z_1 is rotated by means of an electric motor, with the result the colloidal solution will move and through it the movement will be transferred to the smaller cylinder Z_2. At the constant speed of Z_1 the viscosity has been proportional to the angle φ through which Z_2 has rotated

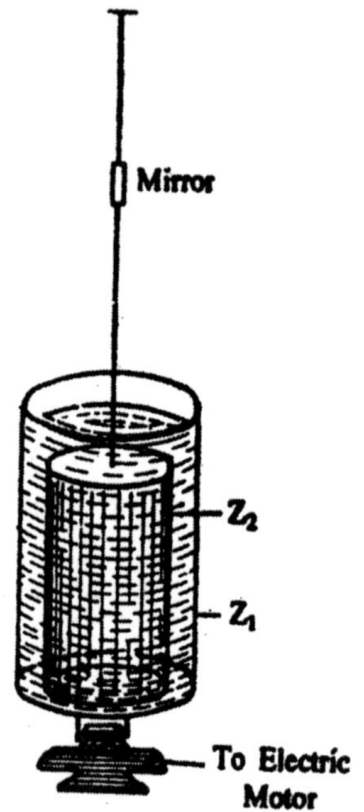

Fig. 1.51 : Viscometer of Couette.

$$\eta < k_1\phi/\omega$$

$$\eta = k_2/\phi$$

The value of k_1 has been found to depend upon the radii and the height of the cylinder.

In this method the angular velocity ω of Z, gives rise to a shear stress which corresponds to the pressure p of the capillary viscometer.

The main advantage of Couette viscometer is that it conveniently permits the shear to get varied within the wide limits.

EMULSIONS

Emulsion is a colloidal system consisting of immiscible liquids, *e.g.*, milk is an emulsion in which particles of liquid fat are dispersed in water.

In common occurrence, however, one of the liquids is water and the other, an oily substance insoluble in it. The suspended droplets are larger than the particles of the sols; it is because of the density differences between the phases being small. Emulsion droplets can be observed under an ordinary microscope and sometimes even with a magnifying lens.

An emulsion is a heterogeneous system consisting of more than one immiscible liquids dispersed in one another inform of droplets whose diameter, in general, exceeds 0.1μ. Such systems possess an extremely small stability which is made by the addition of surface active agents, finely divided solids, etc.

Types of Emulsions : Emulsions are of two types :

(1) *Oil in water (o/w) type:* In these emulsions oil forms the dispersed phase and water, the dispersion medium. For example, milk, vanishing cream, etc. These are also called *aqueous emulsions* (Fig. 1.52a)

(2) *Water in oil (w/o) type :* In these emulsions water is in the dispersed phase and oil in the dispersion medium. For example, butter, cold cream etc. are also called oil *emulsions.* (Fig. 1.5b).

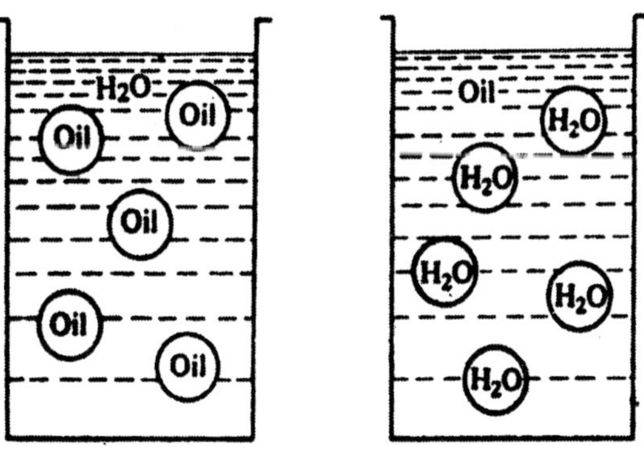

Fig. 1.52 (a) **Fig. 1.52(b)**

In addition to above there is one usual type known as multiple emulsion. As the name indicates, a multiple emulsion is one in which both types of emulsion exist simultaneously. It can be denoted as w/o/w emulsion.

Factors Determining the Type of Emulsions

When two liquids, say oil in water are shaken to form an emulsion, the type of emulsion formed, (*i.e.*, oil in water or water in oil) depends upon the following factors.

(a) *Relative proportion of the two liquids* : As a general rule the liquid present in excess forms the dispersion medium. For example to obtain an emulsion of oil water, water is taken in excess and to obtain water in oil emulsion, oil is taken in excess.

(b) *Surface tension of the two liquids* : The liquid with greater tendency to form spherical drops and hence the dispersed phase. Thus, if the surface tension of an oil in is greater than that of water, it will form an oil in water type emulsion.

Preparation of emulsions : Emulsions are usually obtained by spraying mixtures of phases through narrow nozzles or in counter-rotary agitators. The nature of the emulsifying agent determines the type of the emulsion obtained. According to Bancroft Rule :

"The phase in which the stabilizer is more insoluble becomes the external phase."

Neutral soaps which are insoluble in hydrocarbons but soluble in water give oil-in water emulsions while acid soaps which arc more soluble in hydrocarbons yield water-in-oil emulsions.

Emulsions can also be prepared by using ultrasonic waves.

Emulsion is, however, also possible with stabilizers insoluble in both phases. In these cases *the phase which wets the emulsifier better becomes the outer phase.* Clay, glass powder, calcium carbonate, and pyrites are easily wetted by water and give rise to aqueous emulsion while lamp black, which is more easily wetted by an oil, gives oily emulsion.

A condensation method given by Summer has been employed in preparation of concentrated o/w emulsions.

Characteristic of Emulsions

(i) *Concentration and particles size* : In the case of emulsions the amount of one liquid dispersed in another is relatively much greater as compared to the soles. The maximum amount of one liquid which can dispersed in another cannot exceed 74% of the total volume available. Emulsions more concentrated than 74% have also been found. The diameter of droplets in case of emulsions is of the order of -001-.05 mm. Recently stable emulsions having diameter of 0-0001 mm have also been reported.

(ii) *Optical properties* : A relationship between optical properties and particle size and also between light scattering with the properties of suspensions have been reported. The interfacial areas in emulsions by optical measurements have been determined by Langlois and others in 1954.

(iii) *Viscosity* : The property of viscosity (resistance to flow) is quite important both for practical and theoretical purposes. It provides some information about the structure of emulsions.

(iv) *Electrical Conductivity* : This property is useful in distinguishing between o/w and w/o types of emulsions. The emulsion in which water is the dispersion medium possesses high conductivity than the emulsion having oil as dispersion medium.

Distinction between two types of emulsions : Following tests are used to identify the type of an emulsion,

(1) *Dye Solubility method* : This method of identifying the two types was given by *Robertson* (1910). In this method a small amount of coloured dye soluble in one phase, but insoluble in other is added to the emulsion under examination. If the colour spreads through the whole emulsion, the phase in which the dye is soluble is the continuous phase. For example, if an oil soluble dye, *e.g.*, red Sudan III, is added to .the emulsion under examination, then the emulsion will take up readily the colour of the dye when oil is the continuous phase *i.e.*, the emulsion is of W/O type. If the emulsion is of O/W type it will remain colourless.

(2) *Dilution method* : The basis of this method is that an emulsion is readily dilatable by the liquid which constitutes the continuous phase. A small amount, say two drops of the emulsion to be tested is kept on a microscopic slide or in the depression of a spot plate. A drop of each component is added to each drop and the mixture is slightly stirred. The component which causes uniform mixing with the emulsion when stirred, is regarded to

be the continuous phase. For example, suppose a drop of water is added to a drop of the emulsion under test. If it results in uniform mixing when stirred, then water is the continuous phase *i.e.*, , the emulsion is O/W type. If the two do not uniformly mix, it is of W/O type. Similarly, if a drop of oil is added to the emulsion, it will readily mix with the W/O type emulsion (*i.e.*, oil is the continuous phase). Observation of the dilution under a microscope may often be useful.

(3) *Fluorescence method* : This method is based on the fact that many oils fluoresce under ultraviolet light. Thus the type of an emulsion can be readily identified by examining a drop of the emulsion under fluorescent light microscope. The emulsion is of W/O type, *i.e.*, oil is the continuous medium if the whole field fluoresces, while if a few fluorescing dots are seen; the emulsion is O/W type, *i.e.*, water is continuous medium.

(4) *Wetting of filter paper method* : This method is carried out by placing a drop of the emulsion to be tested on the small piece of filter paper. The emulsion will be of O/W type, if the liquid spreads rapidly, leaving a small drop at the centre. If there occurs no spreading, the emulsion is of W/O type. This test is not valid with oils, which spread on filter paper.

(5) *Conductivity method* : This .method is based on the simple fact that if the conductivity of the emulsion is high, the emulsion is of O/W type, as these emulsions are more conducting. In these emulsions water is the continuous medium and aqueous systems are in main good conductors. On the other hand, if the conductivity of the emulsion is low (or (nil), the emulsion is of

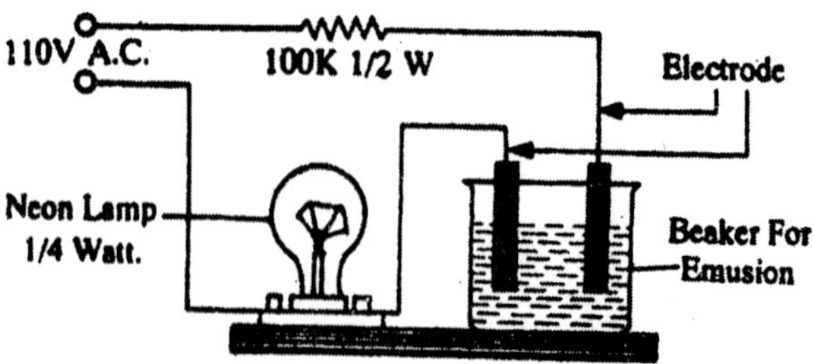

Fig. 1.53

W/0 type, as these emulsions are less conducting (In these emulsions oil is the dispersion medium, and oils are poor conductors). A simple device which may be used for this purpose is shown in Fig. 1.53.

In the case of O/W type emulsion, the neon lamp will glow when the electrodes are dipped into the liquid, whereas the lamp will not glow when the emulsion is of W/O

Emolsifiers

In order to prepare stable emulsions, it is important to add a third component known as emulsifier or emulsifying agent in suitable amounts. Several types of emulsifiers are known.

(1) Long chain compounds with polar groups such as soap, sulphonic acid sulphates etc.

(2) Most of the lyophilic colloids also act as emulsifiers such as glue, gelatin etc.

(3) Certain insoluble powders as clay, lamp black, etc.

(4) Soluble substances like iodine also act as emulsifiers.

Role of an emulsifier : An emulsifier may act in two ways :

1. *It may be more soluble in one liquid than in the other* : In this case L will form a sort of protective film around the drops of this liquid in which it is less soluble and thus prevents them from coming together (Fig. 1.54a). For example neutral soaps which ace more soluble in water than in olive oil is water type emulsion

Emulsification of oil and water by an acid soap

(a)

Emulsification of oil and water by neutral soap

(b)

Emulsification of water and kerosene by soot par-

(c)

Fig. 1.54

(Fig. 1.54b). Acidic soaps which are more soluble in oils than in water, give water in oil type emulsion (Fig. 1.54c).

2. *The emulsifiers may be insoluble in both the liquids but not unequally wetted the by two.*

For example lamp black or soot particles are wetted more by kerosene oil than water. In this case, the surface of the solid is more in contact with the liquid by which it is wetted most (Fig. 1.54c). For example lamp black or soot particles will stabilise an emulsion of water in kerosene oil.

Theories of emulsification : The stability of an emulsion depends upon the following factors :

1. Electric charge in the film or droplets.
2. Thickness and compactness of the protecting film or interfacial layer.
3. Viscosity of the dispersion medium.
4. Density difference between the two liquids.

Various theories have been proposed in order to explain the stability on the basis of above factors.

1. *Quincke and Donnan's theory or surface tension theory of emulsions : Qnincke* and Donnan proposed that during the formation of an emulsion, emulsifying agent lowers the interfacial tension between the dispersed phase and dispersion medium. The emulsifying agent concentrates at the interface. It has also been shown that the interface between a mineral oil and water is lowered by the sodium salts of fatty acids.

2. *Bancroft double interfacial tension theory :* The formation of *Quincke Donnan* theory was made by Bancroft. The formation of various emulsions is explained on the basis of this theory. According to Bancroft, the emulsifying agent forms a film between the two phases which is three molecules in thickness. Out of three molecules, one molecule of the film is of emulsifying agent, one of the dispersed phase, and one of the dispersion medium. Thus, the emulsifying agent forms a film between oil and water. The film has two surfaces one towards water and other towards the oil. If the surface tension between water and emulsifier is less than that of oil and emulsifier, then the film

will tend to bend so as to become convex on the water side and emulsion of oil in water will result. In the reverse case w/o type emulsion will be produced.

The two types of formation of emulsions have been shown in Fig. 5.

3. *Oriented molecular wedge hypothesis* : This theory due to Harkins is based upon the theory of the relative dimensions of the head and tail ends of the soap molecule. According to this hypothesis, a soap of similar molecule consists of a long hydrocarbon chain and a polar end group. At an oil water interface, according to *Harkins* these polar end groups will be thoroughly oriented towards the water and the hydrocarbon chain towards the oil. If the cross section of the polar end becomes larger than hydrocarbon chain, then a stable emulsion will be formed in which the area of water side of the soap film becomes greater than that of oil side, *i.e.,* water will be the dispersion medium. An o/w emulsion will thus An o/w emulsion will thus salts of fatty acids.

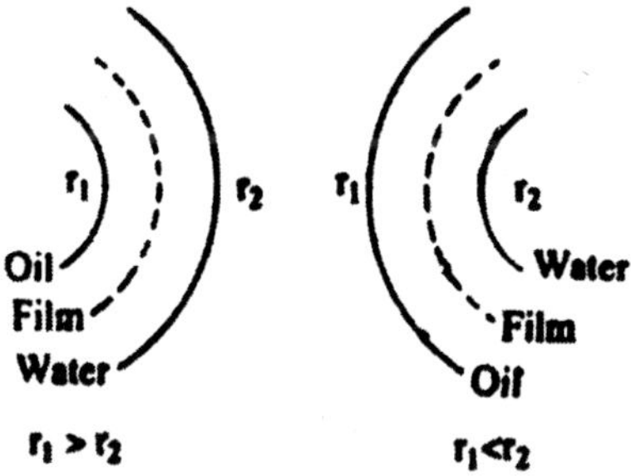

Fig. 1.55 : Formation of emulsions.

On the other hand if hydrocarbon chain is having a larger cross section than the polar group then o/w emulsion would be stabilised.

Unfortunately, exceptions to this theory are known. For example, silver soaps which should, by this theory, yield o/w emulsions actually stabilise the opposite type.

4. *Adsorbed film theory* : This theory was given by *Clowes* who takes into consideration the system of oil and soap solutions. The soap tends to concentrate between oil and water to form a coherent film.

Soaps of monovalent cations are easily dispersed in water but not m oils farm a film which is readily wetted by water than in oil. It will result o/w type emulsion.

On the other hand, a film composed of soaps, of bivalent or trivalent cations gets readily dispersed in oil and wetted by oil. This will result w/o type emulsion.

According to *Bhatnagar "all emulsifying agents which are having excess of negative ions in them and wetted by water and will give o/w type emulsion while all emulsifying agent having an excess of positive ions in them and wetted by oil will result w/o type emulsion".*

Inversion of emulsions or Phase : The change of o/w emulsion into the w/o type or *vice-versa* is called the *inversion of phase* or *inversion of emulsions and the emulsion* is said to have inverted. An o/w emulsion say olive oil in water containing sodium or potassium soap as emulsifying agent can be readily converted into w/o type by adding salt like $CaCl_2$ or $AlCl_3$, etc.

The *mechanism* of the inversion of phase is not yet clearly understood.

Clowes (1913) and *S.S. Bhatnagar* (1920–21) have shown that hydrocarbon or olive oil in water stabilised by a sodium or potassium soap, may be converted into a water-oil system on addition of salts of bi and tervalent cations, *e.g.,* Al, Fe, Cr, Ni, Pb, Ba, Sr and Ca.

Inversion is apparently a function of temperature. It was also observed that the temperature at which the inversion takes place was sensitive to the concentration of emulsifying agent.

Although inversion of emulsions is the subject of considerable investigation, yet it is not well understood is many respects. An explanation of the mechanism of inversion of O/W of systems stabilised with sodium ceatyl sulphate and cholesterol has been given by *Schalman and Cockbain.* This type of emulsion can be further stabilized by the existence of a negative charge resulting from the ionisation of the sodium ceatyl sulphate in the stabilizing film.

If a multivalent cation *e.g.,* Ba^{++} or Ca^{++} is now added to the system, the charge becomes ineffective because of the close association of the ions with the film. This makes the coagulation of the droplets possible. The added ions would be able to convert the liquid film into a solid one, possessing rigidity, so that when the emulsion is shaken, the films round the oil drops break, leaving the water enclosed in irregular "sacks". This distorted appearance has been common in W/O emulsions.

Soaps are capable of stabilizing emulsions. It has been observed that water soluble sodium and potassium soaps, yield O/W type emulsions. On the other hand, calcium soaps, which are oil soluble give W/O emulsions.

Demulsification : The process of separating the two constituent liquids of the emulsion into separate layers is called *demulsification.* It can be brought about by any of following methods :

(1) By addition of an electrolyte to destroy the charge on the dispersed phase.

(2) By chemically destroying the emulsifier.

(3) By distilling off one component.

(4) By freezing one component.

(5) By extraction of one component by solvent extraction or by other suitable means.

(6) By applying -large centrifugal force.

(7) Addition of a dehydrating agent. The method is especially useful for water in oil type.

(8) By addition of large excess of the dispersed phase.

(9) By heating under high pressure.

(10) Irradiation with ultrasonic waves.

The function of heating and high frequency sound waves is to increase the intensity of motion of droplets and hence the coalescence in producing emulsification.

Agents producing O/W emulsions will break W/O emulsions and *vice-versa.* Centrifugation usually brings about creaming and the concentrated emulsion must then be treated chemically in order to induce coalescence.

Importance of Emulsions : Emulsions find manifold applications in various fields.

(1) *Medicine :* Numerous medicines and pharmaceutical preparations are emulsions. In such forms they have been found to be more

effective. Codliver oil, caster oil, petroleum oil are used in medicines, and are emulsions.

(2) *Articles of daily use :* Milk is an emulsion of fat dispersed in water stabilised by case in. Ice-cream is an emulsion. Butter, coffee, fruit jellies, etc., are all emulsions in nature.

(3) *Cosmetics :* The skin penetrating vanishing creams o/w type emulsions and hair creams, cold creams are w/o type emulsions. The lotions, creams and ointments are stabilized by lanoline.

(4) *Industry :* The latex obtained from the sap of certain trees is an emulsion of negatively charged rubber particles dispersed in water.

During the concentration of sulphide ores, froth flotation process is employed. In the process, oil emulsion is added to the finely divided ore and foam produced by passing ore contains most of the particles of the ore.

Emulsion of oils and fats have been employed in leather industry for making soft leather and water proof.

Asphalt emulsified in water is used for building roads, without the necessity of melting the Asphalt.

Emulsions are also employed in oil and fat industry, paints and varnishes, cellulose and paper industry etc. Furthermore, spraying liquids in the form of emulsions are used in agriculture.

MICROEMULSIONS OR MICELLAR EMULSIONS

The term *microemulsion* was coined in 1958 to describe a fairly specific class of colloidal systems. Historically, the term microemulsion was applied to systems prepared by emulsifying an oil in aqueous surfactant and then adding a fourth component called a cosurfactant, generally an alcohol of intermediate chain length. Benzene, water, potassium oleate and hexanol might be the components of a typical microemulsion formulation. What is observed experimentally is that the usual milky emulsion becomes transparent upon addition of the alcohol. Light scattering and an assortment of other techniques reveal that the resulting system consist of either O/W or W/O dispersions with particles having diameters in the 10– to 100–nm size range. Whether oil or water is continuous, the extent of uptake of the other component may be appreciable. In summary, the following differences between micro-emulsions and coarse emulsions should be noted:

1. Microemulsions contain particles of size 10.60 nm, at least an order of magnitude smaller than those in coarse emulsions.
2. Microemulsions are isotropic and optically transparent and coarse emulsions cloudy.
3. Microemulsions form spontaneously; coarse emulsions Ordinarily require vigorous stirring.
4. Microemulsions are stable with respect to separation into their components; coarse emulsions may have a degree of kinetic stability but ultimately separate.

The micro- or micellar emulsions are isotropic, optically transparent (with a particle size of 10-60 nm), and thermodynamically stable. The arbitrary formation of these systems ($\Delta G < O$) is associated either with the presence of a negative interphase tension (due to the high pressure in the film, formed by the mixture of the surfactant and the addition, at the oil-water interface), or with the contributions of the entropy component and also of the energy of repulsion of the electric double layer. Whereas conventional emulsions are thermodynamically unstable systems whose kinetic stability is determined by the forces of repulsion of the electric double layer on the surface of the globules and by the van der Waals forces of attraction (in accordance with the DLVO theory), the thermodynamic stability of micellar emulsions is determined by the frcc energy of double layer formation, the entropy effect (for $R < 20$ nm), and by the forces of the electric double layer; the van der Wals forces of attraction play a secondary role. Micellar emulsions can be considered as swollen micelles.

The term microemulsion seems quite firmly established as the name for the sort of system described above. Still, there has been and continues to be a great deal of controversy as to the exact nature of these systems and the suitability of this vocabulary. The word *emulsion* implies the presence of two phases with an interfacial free energy associated with the phase boundary. Their small size makes the specific area large for microemulsions with a large free energy contribution from γ. Both the spontaneous formation of microemulsions and their stability with respect to separation are hard to reconcile with these considerations. It has even been suggested that the mixed film of surfactant and cosurfactant make the interfacial free energy negative. Alternatively, the increase in overall free energy with decreasing particle size may be onset by a favourable and hence negative $T\Delta S$ term in which ΔS describes the entropy of mixing the microemulsion particles with molecules of the dispersion medium. Since the number of microemulsion particles increases with decreasing particle size, the $T\Delta S$ term becomes more favourable with decreasing size. This idea is hard to apply quantitatively because of uncertainty as to the value—or meaning—of γ in these systems.

A totally different way of looking at microemulsions is to view them as complicated examples of micellar solubilization. From this perspective, there is no problem with spontaneous formation or stability with respect to separation. Furthermore, ordinary and reverse micelles provide the basis for both O/W and W/O microemulsions. From the micellar point of view, it is the phase diagram for the four-component system rather than γ that holds the key to understanding microemulsions.

The difference in perspective between the emulsion and micellar points of view is suggested by Fig. 1.56. The small shaded circle on the left represents a micelle with little or no solubilization. From left to right, the "particles" increase in size owing to increasing solubilization. The circle on the right represents an emulsion particle: an oil drop with a monolayer of surfactant on the surface. An actual continuum of states such as that suggested in Fig. 1.56 is not physically attainable—most emulsions are prepared by comminution rather than condensation—but microemulsions do lie between the extremes. The conflicting schools of thought concerning microemulsions arise from the difference in perspective: One side looks from the emulsion point of view, and the other from the micellar point of view. Swollen micelles is another term for microemulsions: this terminology clearly reflects the different perspective of its originators.

Fig 1 .56: Schematic progression from micelle (at left) to emulsion droplet (at right). Various degrees of solubilization, including microemulsions, lie between the two extremes.

We know that phase diagrams are an effective way of representing the complex behaviour of surfactant systems. Let us take a look at microemulsions in terms of phase diagrams. It turns out that non-ionic surfactants form microemulsions at certain temperatures without requiring cosurfactants. Since only three components are present, these have somewhat simple phase diagrams. This kind of system offers a convenient place to begin. Fig. 1.57 is a composite of both experimental and schematic, interpretive portions.

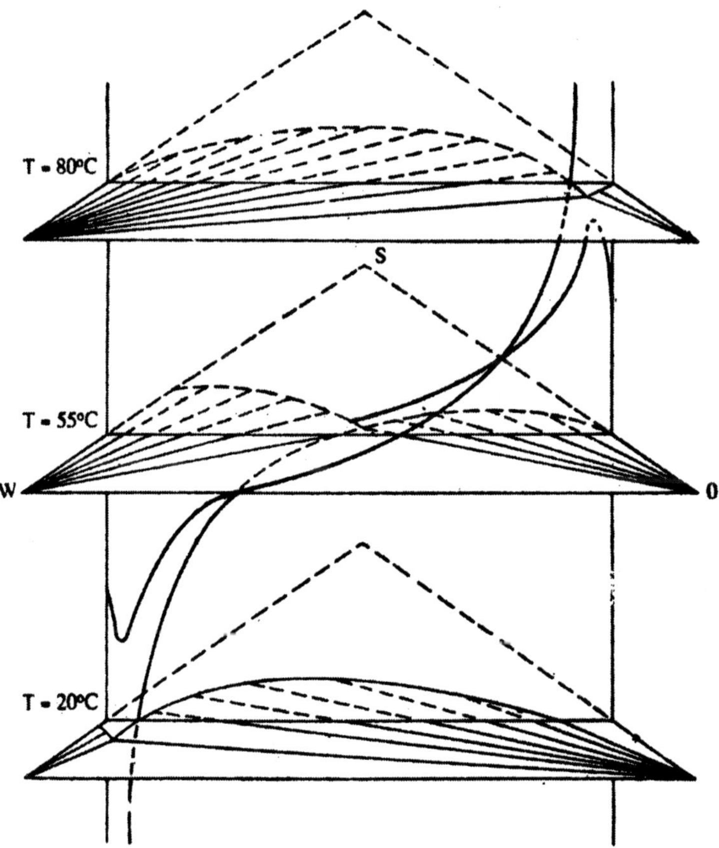

Fig. 1.57 : Rectangular figure shows the phase diagram for water-cyclohexane (at 5% surfactant) versus temperature. Superimposed ternary phase diagrams offer an interpretation of the phases present.

The rectangular diagram shows the experimental behaviour of the system water-cyclohexane-polyoxyethylene-(8.6)-nonyphenol ether. All systems studied contained 5 % surfactant with variable proportions of cyclohexane and water. Temperature was the experimental variable and the nature of the phases present was recorded as the temperature was changed. The presence of different equilibrium phases is represented by different areas on the rectangular diagram. The latter is divided diagonally by a pair of lines that trace a ribbonlike pattern. This pair of lines cross twice so the ribbon is divided into three areas. These, plus the regions above and below the ribbon, define five different phase situations-encountered at various temperatures and compositions.

The idealized ternary phase diagrams that have been superimposed on the experimental plot in Fig. 1.57 help us understand the five different regions on the rectangular diagram. The stacked triangles represent the oil-water-surfactant (abbreviation S) phase diagrams at different (constant) temperatures. Since the experimental data were collected at 5% surfactant the rectangular plot slices through the triangles 5% of the distance toward S from the O-W base of the triangle.

Since the micellar phase change from water continuous to oil continuous with increasing temperature, it is an intriguing question how to describe the micelles in the three-phase region that exists at intermediate temperatures.

In this study it is the homogeneous micellar phases that comprise the microemulsions For this 5% surfactant system it is only over a relatively narrow range of temperatures that is possible to have fairly extensive solubilization of either oil or water in the micellar solutions.

It could be argued that the system described in Fig. 1.58 is so different from those to which the name *microemulsion* was first applied as to make the figure irrelevant to the present discussion. However, one of the ways to represent a four-component phase diagram (p and T constant) is to use a trigonal prism in which one of the components—rather than T—varies along the rectangular faces. Fig. 1.58 (a) is partial phase diagram for the system water-benzene-potassium oleate-pentanol. In this figure homogeneous micellar solutions are represented by unshaded areas in the individual triangles. These micellar systems are equivalent to the homogeneous areas in the ternary phase diagrams in Fig. 58. At equilibrium up to 50 wt% benzene can be incorporated into the system with only minor variations in the maximum water uptake. The resulting four-component microemulsion is the same as that produced by first emulsifying the oil and water with potassium oleate and then adding pentanol. This is, true because we are considering equilibrium phase diagrams. As with all thermodynamic conclusions, the diagram tells nothing about the rate at which equilibrium is achieved.

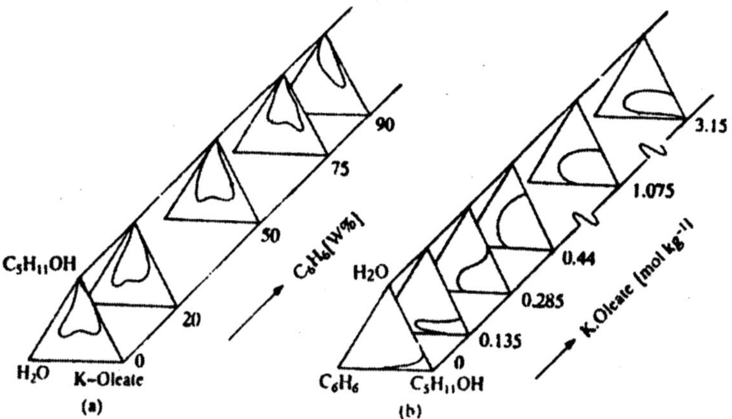

Fig. 1.58 : Two representations of a portion of the phase diagram for the water-benzene-potassium oleate-pentanol system. The unshaded regions represent homogeneous solutions.

Fig. 58(b) shows the same data as Fig. 58(a), replotted with the variables interchanged. Note how the homogeneous area expands in size with increasing potassium oleate concentration up to 0.44 mole kg^{-1}. At still higher concentrations of potassium oleate the homogeneous area shrinks as other surfactant phases compete for the components.

Applications of Microemulsions

Systems in which one liquid phase is finely dispersed in another under the stabilizing influence of one or more additional components find applications in countless areas. As consumers, we encounter many of these every day. Floor waxes, shaving lotions, beverage concentrates, pesticide preparations, cold creams, and pharmaceutical products are a few of the more common examples. In recent years a great deal of research in this area has been directed toward the problem of tertiary oil recovery.

1. *Recovery of oil* : First of all the recovery of oil from natural reservoirs occurs in three stages. During the *primary recovery* stage the pressure of natural gases in the reservoir pushes the

oil out. When the gas pressure is no longer adequate, water is pumped into the reservoir to force the oil out. This is called water *flooding* and represents the second stage of *oil recovery.* Primary and secondary oil recovery leave about 70% of the total oil in place, much of it trapped in the pore structure of the reservoir by capillary and viscous forces.

Numerous methods have been explored to recover at least some of this vast resource. Injection of oil-miscible fluids, gases under high pressure and steam—either separately or in combination—have all been tried with various degrees of success. This is where microemulsions enter the picture. Under optimum conditions an aqueous surfactant solution—which may also contain cosurfactants, electrolytes, polymers, and so on—injected into an oil reservoir has the potential to solubilize the oil, effectively dispersing it as a microemulsion.

Any attempt to represent the trapped oil by a manageable model is bound to be an oversimplification. To see qualitatively, however, how capillary forces trap the oil and how surfactant solutions offer a potential for freeing it, imagine a cylindrical pore containing a slug of oil. Furthermore, assume the oil is in contact with water and that the interface is hemispherical. This assumption about the shape of the interface makes the water-oil-rock contact angle zero and is equivalent to neglecting 6 ($\cos 0° = 1.0$).

Although a primitive picture of oil in a rockey reservoir, this model can be described by a single size parameter r, the radius of the pore and the radius of curvature of the interface. The Laplace equation may then be used to describe the pressure across the oil-water interface, a pressure that must be exceeded to displace the oil. According to the Laplace equation, $\Delta p \propto \gamma/r$ and since r is small. Δp will be large unless r is offset by a small value of γ. Using r values that are sensible for geological structures, it has been estimated that $\Delta P \simeq 500$ psi/ft if γ is on the order of 10 mN m^{-1}. This kind of pressure drop is unattainable under field conditions. Working backward and taking Δp as an attainable 1-2 psi/ft. The Laplace equation shows that γ must be less than about 0.1 mN m^{-1} preferably closer to 10^{-3} mN m^{-1} for effective oil displacement.

Any surfactant adsorption will lower the oil-water interfacial tension, but these calculations show that effective oil recovery depends on virtually eliminating γ. That microemulsion formulations are pertinent to this may be seen by re-examining Fig. 1.56. Whether we look at microemulsions from the emulsion or the micellar perspective, we conclude that the oil-water interfacial free energy must be very low in these systems. From the emulsion perspective, we are led to this coclusion from the spontaneous formation and stability of microemulsion. From a micellar point of view, a "pseudophase" is close to an embryo phase and, as such, has no meaningful γ value.

It is apparent the extrapolating laboratory studies on microemulsions to oil recovery is a formidable task. While laboratory research is conducted with pure solutes, distilled water, and at constant temperature, these are meaningless in the field where the following applies.

1. The oil itself is a complex mixture containing surface-active components
2. Commercial petroleum sulfonates, a mixture of compounds, are the most widely used surfactants.
3. Groundwater contains dissolved minerals and in practice, brine is used as the aqueous component.

In addition, viscosity considerations may be as important or more important than capillarity fortunately, microemulsions also have relatively low viscosities!

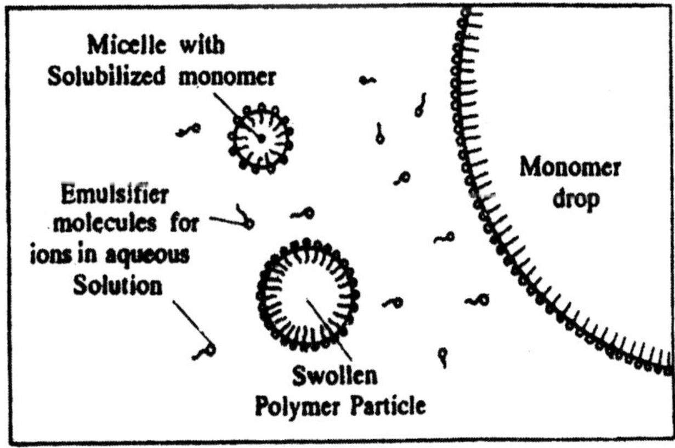

Fig. 1.59 : Schematic representation of the distribution of surfactant and monomer in an emulsion polymerization.

Despite the obvious difficulty of the tertiary oil recovery problem, this is a major area of surfactant research, since the potential rewards for success are very great.

2. *Synthesis of certain polymers :* A second area of microemulsion application is in the synthesis of certain polymers. The process is called emulsion polymerization, a misnomer, since micelles rather than emulsion drops are the site of the polymerization reaction. Because of the commercial importance of polymers, this process has been extensively researched and is quite well understood. We shall only consider some highlights of the process.

Emulsion polymerization is applicable only to monomers which are relatively insoluble in water, such as styrene. A coarse emulsion of monomer in aqueous surfactant is prepared with a water-soluble initiator say, H_2O_2 in the solution. The surfactant concentration is above the CMC, so surfactant molecules are present as monomers, micelles, and emulsifiers at the oil-water interface. Even an insoluble liquid like styrene dissolves in water to some extent. Therefore the monomer is present in coarse emulsion drops, solubilized in micelles, and as dissolved molecules in water. A schematic illustration of the distribution of surfactant monomer and polymer in an emulsion polymerization process is shown in Fig. 1.59.

The H_2O_2 molecules undergo thermal decomposition to form hydroxyl free radicals –OH which initiate the polymerization. The overall reaction for the polymerization of styrene can be represented as :

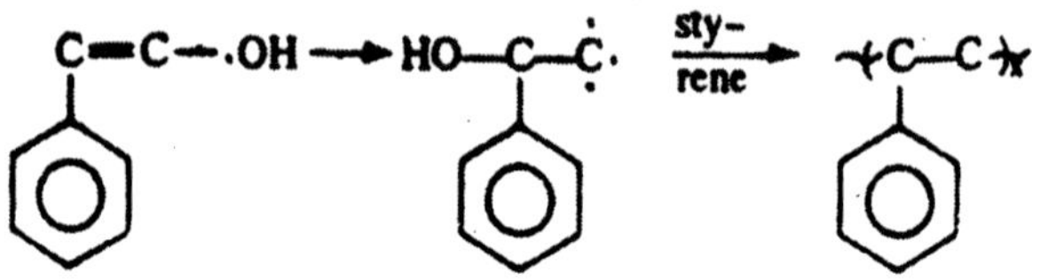

C=C ← ·OH ⟶ HO—C—Ċ· $\xrightarrow{\text{sty-rene}}$ ⟶ ⟮C—C⟯ₓ

Fig. 1.60

When x is large, the uniqueness of the end group(s) can be ignored; familiar polystyrene is the product.

In emulsion polymerization the first step in Fig. 1.60 takes places in water between dissolved monomer and initiator fragments. The resulting free radical is solubilized in micelles, where it quickly reacts with solubilized monomer to form polymer. The low concentration of monomer in the aqueous phase prevents this from occurring to any appreciable extent in the water, although, by diffusion, there continues to be a flux of monomer from emulsion drops into micelles. Likewise, any polymerization that occurs in the coarse emulsion drops themselves is insignificant because of the much greater numerical abundance of micelles than the emulsified styrene droplets. The polymer chains grow in micelles until the process is terminated by reaction with another radical. Thus, polymer growth is either propagating or terminating in micelles at any time; therefore half the micelles in a reaction mixture contain growing chains under stationary state conditions. Both the rate of polymerization and the average molecular weight of the polymer depend on the surfactant concentration—via the concentration of micelles—in emulsion polymerization, while this has no effect on polymerizations conducted in nonmicellar solutions.

The aqueous polymer dispersion that results from emulsion polymerization is called a latex. In applications the polymer may be separated or the latex may be used directly as in paints and floor coatings.

As the conversion to polymer proceeds, the micelles become progressively more and more swollen by the polymer-monomer mixture. As with other microemulsions, it eventually becomes problematic as to whether the resulting dispersed particles should be called micelles or swollen polymer particles with adsorbed surfactant.

FOAMS

Disperse systems in which a gas is the dispersed phase and ½>a liquid is the dispersing medium are called *liquid foams*, while those with a gaseous dispersing medium are called *solid foams*. Since solid foams are produced by the freezing of liquid ones, *i.e.*, they are a "continuation" of the latter, we shall stop to consider only liquid foams. And since dilute liquid foams are kinetically unstable and are therefore of no practical importance, we shall consider only concentrated foams.

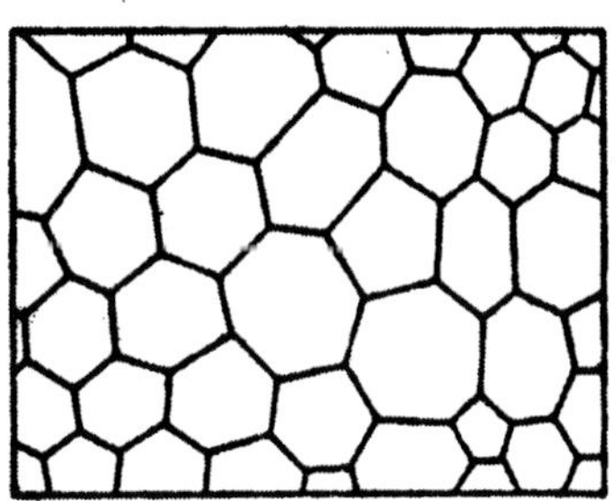

Fig. 1.61 : Structure of a foam (magnified)

Foams are coarsely dispersed systems. The gas bubbles in them are large in size (they can be seen by the naked eye). They are firmly pressed against one another and are separated only by thin intelayers of the dispersing medium—a foam film (Fig. 1.61).

Foams are most frequently produced by dispersion methods such as forcing the gases through narrow orifices into the liquid or intensive agitation of the liquids in the presence of the gases. This is how a foam is formed when soapy water is blown through a pipe, when whipping creams preparing cocktails pouring viscous liquids in the presence of gases into another vessel, etc.

Foams are sometimes obtained by the condensation method as a result of the formation of gaseous products or the evolution of a gas dissolved in a liquid. Examples are the foaming or rising of dough when yeast ferments or sodium or ammonium bicarbonate decomposes in it, the formation of a foam in the production of plastic foams as a result of the decomposition or evaporation of pore-forming agents, and the foaming of concrete by the hydrogen formed by the reaction of an aluminium powder with alkaline substances present in the concrete in the production of foam concrete.

Foams are the most unstable variety of disperse systems. Depending on the viscocity of the dispersing medium, the nature of the solutes and their concentration, the lifetime of a foam (the lifetime of a foam column of a definite height) varies from several seconds to several days. It has been established that absolutely pure liquids do not practically form foams, *i.e.,* the lifetime of the foams in such cases is negligible small. To obtain a foam with a certain lifetime, the liquid must contain a dissolved surfactant.

Substances in whose presence the lifetime of foams is prolonged are called *foaming agents* or *foam stabilisers.*

Two types of foaming agents are distinguished. The first one has a low effectiveness; when the concentration of a foaming agent in a system grows, the lifetime of the foam rapidly reaches a maximum of about several scores of seconds, while with a further increase in the concentration the lifetime begins to shorten. This type of foaming agents includes the lower members of the homologous series of fatty acids and alcohols, and also alkalies. By Traube's rule, the surface activity of the compounds indicated above increases with growth of the carbon chain. Naturally, the concentration at which the highest stability of a foam is observed diminishes.

The second type of foaming agents includes substances that continuously improve the stability of a foam when their concentration in a system is increased. Such substances have a detergent effect and are called *detergents*. They include salts of the higher fatty acids (soaps), alkyl sulphates, alkyl sulphonates, alkyl-aryl sulphonates, and some high-molecular compounds (gelatin, potential, etc.). By using them, the British scientist J. Dewar (1842-1923) succeeded in prolonging the lifetime of a soap bubble up to three years.

When the concentration of detergents in a system reaches a certain value, the stability of a foam grows stepwise, while a further increase in the concentration no longer changes the stability. This phenomenon is explained by the appearance at this concentration of superthin films 5-10 mm thick, which are called "black" because of their absolute light transmission. Such films are very strong and do not become thinner. For every detergent, black films form at a definite concentration that is typical only of this detergent and is its physicochemical property.

The nature of foam stability has not yet been established finally, and opinions on this matter diverge. For intance, J. Gibbs explains the stability of a foam by the "elasticity" of the foam film due to the increase in its surface tension in extension, which facilitates its reverse contraction. P. Rehbinder explains the stability of a foam by the structurization of thin films. B. Deryagin considers the main cause of foam stability to be the wedge effect in thin liquid films that prevent their thinning.

In addition to their stability, foams in chemical technology are often characterized by indices such as their dispersion (the mean diameter of the gas bubbles), the height of a foam column (determined by blowing the gas through the liquid in a cylinder of a standard size), the lifetime of an individual foam bubble, the foam multiplicity, etc. Most frequently, the concept of the *foam multiplicity* β is used to describe foams. It is defined as the ratio of the foam volume V_f to volume of the liquid V_{lq} from which it was formed, *i.e.*, $\beta = V_f/V_{tq}$. Depending on the value of β, particularly, we distinguish "wet" ($\beta < 10$) and "dry" ($\beta > 1000$) foams.

Foams have diverse applications. They are used in the concentration of minerals by flotation, in laundering and washing, in fire extinguishing, in the production of highly porous building and insulating materials (foam concrete, foam glass), in the production of plastic foams (flexible polyvinyl chloride, polyester, polystyrene, elastomeric, epoxy, phenolic foams), etc.

Notwithstanding the diverse applications of foams, there are no less cases when their formation is not desirable. For example, the formation of foams is not tolerated in mixing, distillation, evaporation, and other similar technological operations. Especially great harm is caused by foams in sewage. They sometimes completely cover the surface of water basins and prevent the dissolving of oxygen in the water. This kills the flora and fauna in the basins.

Foams are most frequently destroyed by surfactants—*antifoaming agents*. Being more active, they displace foaming agents. Examples of antifoaming agents are alcohols, esters and silicone oils. Foams are also destroyed by thermal or mechanical action on them.

Theory of Foam Stability

Foam is a term reserved for dispersions of gas in liquid (or solid) in which the volume fraction of the disperse phase is so high that the "colloid" in the system might well be regarded as the network of interconnected films. Examination of creamed emulsions or high phase Volume emulsions as a network would seem to be an insufficiently explored model. The understanding of foam stability and foam breaking thus requires understanding the mechanism of rupture of individual films. A useful approach is closely related to that followed in discussing sol stability. The potential energy (Gibbs free energy) of a suitable quantity of film (unit area or area to contain one mole of surfactant) is plotted against film thickness. Such a diagram is thought to have the qualitative features shown in Fig. 1.62.

A substantial minimum occurs at thickness C, the common black film. The simple model (for an anionic surfactant film) in this region would be double-layer repulsion combined with dispersion attraction. The second minimum at much smaller thickness corresponds to the Newton black film. Its thickness, experimentally of the order of twice that of the largest dimension of the surfactant molecule, is such that the "heads" must be very close to each other at the midpoint of the film. The structure may not differ greatly from that describable as a two-dimensional crystalline hydrate.

Mode of Action of Anti-Foamers

If the objective is to reduce foaming, a tremendous industrial challenge important in cases ranging from the sugar beet industry to emulsion polymerizations to soap boiling itself, the opposite desideratum prevails. We ask what factor or factors weaken the foam film? Putting the problem in this way suggests immediately the verified fact that antifoams can be highly individual and also suggests a general principle. An antifoamer for use with a foaming surfactant should itself be surface-active and also be geometrically incompatible with it in the interfacial film. Alcohols of appropriate chain structure are prime examples of the operation of this principle.

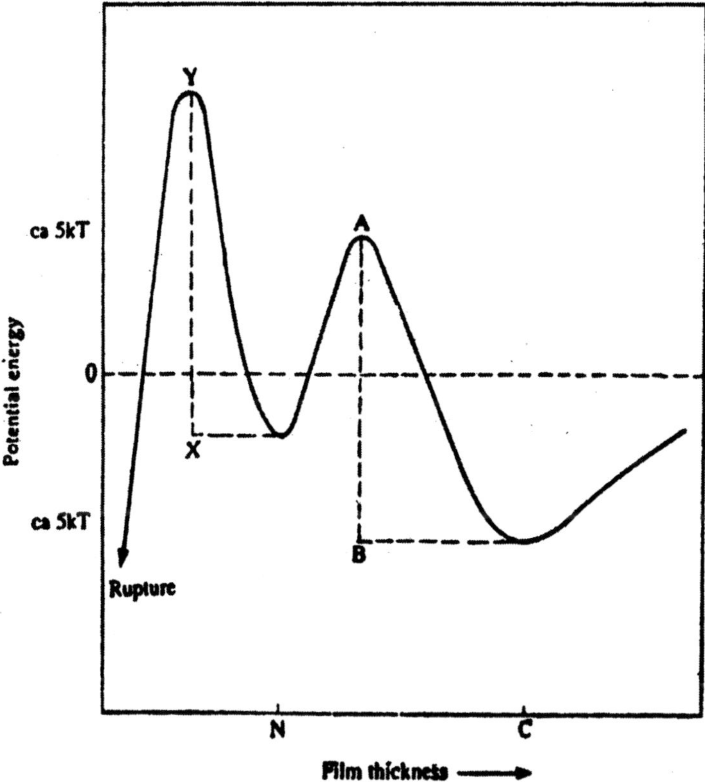

Fig. 1.62 : Schematic energy diagram useful in describing aim stability. C represents the common black film. N represents the Newton black film that can persist only if the minimum is low enough and the barrier to its rupture, XY, is high enough.

A second type of antifoaming principle can be expressed following the observation that in many systems films of an equilibrium thickness will last "forever"—years, if left undisturbed so the principle is to cause such disturbance. If the hypothesis is valid that local fluctuations in surface tension (Marangoni effects) can instigate film rupture, adding substances that produce such disturbances should "kill" foam. Insoluble liquids dispersed as colloidal drops may act in this way by taking some of the surfactant from the foam film. Soluble defoamers may administer their stock to foam films by diffusion through them, and foam-killing gases act similarly. An interesting reported application is the use of ethanol vapour to relieve paroxysmal pulmonary edema (foaming in the lungs). One of the most effective classes of antifoamers consists of polysiloxanes with about 3% solid colloidal filler. In aqueous systems a hydrophobic silica is used as filler. It is believed to function through adsorbing surfactant from the foam film where it is carried by the silicone oil, which wets it the mechanism of action of the particles is not yet precisely understood but may consist of forming an adsorbed autophobic film by adsorption of the erstwhile foam former. Solid particles whose wetting properties hold them at an interface stabilize foams just as they do emulsions.

In closing this section we note that solid foam, sometimes thought to be a product of the modem age of "plastics", is in fact, one of the oldest applications of foaming : *leavened bread*!

AEROSOLS

Aerosols are disperse systems in which liquid or solid (crystalline) particles of the dispersed phase are distributed in the volume of a gas. The size of the dispersed phase particles in aerosols varies within broad limits, from 100 to 1000 nm, including region of colloidal and coarsely dispersed systems.

Two types of aerosols are distinguished, namely, a liquid in a gas—L/G, and a solid in a gas—S/G. In chemical technology, L/G aerosols are generally called fogs (mists), and S/G aerosols are called smoke or dust. Mixed aerosols are often encountered that are called smogs—S. L/G.

Aerosols, like the other disperse systems, are produced by the condensation and dispersion methods. For example, the condensation of water vapour from the air is attended by the appearance of natural fogs and clouds, while the condensation of combustion products (incompletely burned carbon and water vapour) produces industrial smoke. Aerosols can also appear as a result of the chemical reaction of substances (for example, of ammonia and hydrogen chloride vapours, or of sulphur trioxide and water vapour).

In various production processes associated with the mechanical treatment of materials, aerosols form by the dispersion method. Examples are the appearance of dust in mechanical handling operations, in blasting and agricultural work, when grinding cement clinker in the production of cement, and the grinding of grain. Special pulverizing equipment makes it possible to produce aerosols employed in medicine or for controlling agricultural pests.

Aerosols are customarily obtained without surfactants. This is why aerosols are aggregatively unstable, but owing to the features of the gaseous dispersing medium, they have a high sedimentation stability.

Aerosols are similar to colloidal solutions in a number of properties. They arc characterized by thermodynamic instability, Brownian motion, diffusion, sedimentation, the Tyndall effect, selective light scattering, electrophoresis, etc. But the gaseous dispersing medium introduces some novel features; light scattering in aerosols is much greater than in colloidal solutions. Brownian motion and diffusion are more intensive, the electric charge of dispersed particles of aerosols is negligibly Small, and air is a poor conductor of an electric current, therefore electrophoresis proceeds very slightly in them.

Specific properties of aerosols are *thermophoresis* and *thermo-precipitation*, *i.e.*, the migration of dispersed aerosol particles away from a heated surface and the settling of the dispersed phase particles of aerosols on a cold surface. B. Deryagin explained these proper- ties by the fact that gas molecules travel away from a hotter surface at a higher speed and "push" the dispersed aerosol particles toward the colder parts of space.

Another specific property of aerosols is *photophoresis, i.e.*, the movement of dispersed aerosol particles toward a light source or away from one. Photophoresis is especially intensive for coloured dispersed particles of aerosols. Unlike thermophoresis and thermoprecipitation, photophoresis meanwhile has no substantiated theoretical explanation.

It is quite frequently necessary to prevent the formation of aerosols or destroy formed ones. This relates especially to mines, the production premises of flour mills, tyre plants, etc.

The simplest way of purifying air from dust is filtration. Here account must be taken of another feature of aerosols associated with the great diversity of particle sizes in their dispersed phase. When an aerosol passes through a filter, the large particles of the dispersed phase are detained at the entrance to its pores, while the very fine particles penetrate into these pores and owing to the high intensity of thermal motion collide with the pore walls and settle on them. Matters are different with particles of a moderate size. They are small enough to penetrate into the filter pores, but being "sluggish", they do not collide with the pore walls, do not settle on them, and pass through the pores. Consequently, a filter retains only definite particles in strict accordance with the size of its pores. With this in view, the complete separation of dust from gases requires that the latter be passed through a consecutive row of filters with filtering materials differing in structure.

Aerosols can also be destroyed by precipitating dispersed particles in an artificial force field produced by the centrifugal force in cyclones. The gas being purified, is whirled in the cyclone changes into a vortex flow. The centrifugal force throws the denser part of the dispersed phase toward the apparatus walls on which they settle. The purified gas emerges from the cyclone via its central pipe upward, while the dispersed phase particles merge on the apparatus walls into large aggregates and are discharged from the bottom of the cyclone. Generally several cyclones (a battery) are installed for the more complete separation of dust and dropping liquids from a gas.

The most effective way of separating dust and dropping liquids from gases is to destroy the aerosols in a strong electric field produced in an *electric filler* (Fig. 1.63). This process is based on the electrophoresis of disperse systems. But the particles of the dispersed phase of an aerosol have an electric charge that is not adequate for effective interaction with an electric field, *i.e.*, that is not adequate for electrophoresis. This is why a special corona electric discharge—a source of ions—is created in the apparatus." The dispersed aerosol particles pass through the zone of the corona discharge, adsorb ions in it, acquire an electric charge, and only after this change the direction of their motion in the electric field. The aerosol stratifies.

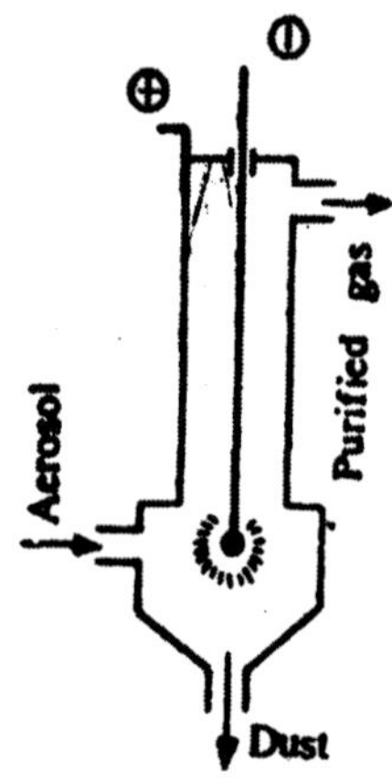

Fig. 1.63 : An electric filter.

Colloid Science

Of great importance for agriculture is the control of rain and hailstorms, based on the preventive destruction of hail and storm clouds. For this purpose, artificial crystallization centres are sprayed into them such as crystalline carbon dioxide, sand and other hygroscopic substances. These centres cause the aerosol particles to recrystallize and become consolidate.

MICELLIZATION

The diphilic nature of surfactant molecules, *i.e.,* the presence in them of a polar (hydrophilic) and non-polar (hydrophobic) parts has been a feature of their structure imparting special properties to these molecules. The diphilic nature was characterized very well by Hartley as a "split personality". It has been exactly the diphilic nature of surfactant molecules that underlies their tendency of gathering at phase interfaces, immersing their hydrophilic part in water, and isolating their hydrophobic part from it. This tendency ascertains their surface activity *i.e.,* their ability to be adsorbed at the water-air interface (Fig. 1.64a) or a water-oil one (Fig. 1.64b), to wet the surface of hydrophobic bodies (Fig. 1.64d), and form structures such as soap films (Fig. 1.64c) or lipid membranes (Fig, 1.64e).

With an increase in the asymmetry of the molecules (a growth in the length of the hydrophobic chain), their surface activity grows (*Traube's rule*), and accordingly, their special behaviour in a solution, different from that of simple salts, gets more pronounced. It manifests itself the most appreciably for long-chain surfactants with 10-20 carbon atoms in a chain that get characterized by an optimal balance of hydrophilic and hydrophobic properties. These substances, which are having many practical applications (for example, as flotation reagents, stabilizers, and detergents) have special properties in solutions that are of considerable interest.

At low concentrations these surfactants tend to form true solutions that disperse up to individual molecules (or ions). With a growth in the concentration, however, the duality of the properties of the molecules of such diphilic substances gives rise to their self-association in the solution. This results in the formation of what are called *micelles.*

The term micelle in this meaning was first introduced by McBain (1913). According to modern notions, micelles may be defined to be aggregates of long-chain diphilic surfactant molecules or ions formed spontaneously in their solutions at a definite concentration. The latter depends on the nature of the polar group and especially on the length of the molecule chain. Micelles can be characterized by the *aggregation number* (the number of molecules in a micelle) and the micellar mass (the sum of the *molecular masses* of the molecules forming a micelle).

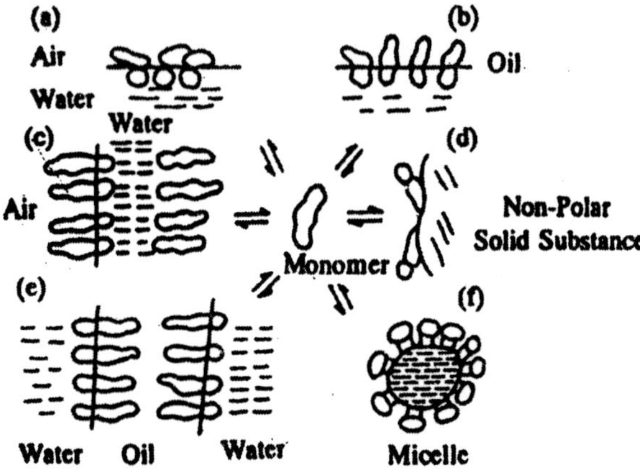

Fig. 1.64 : Illustration of the surface activity of diphilic substances and their ability of micellization :

a—air-water interface; b

—oil-water interface; c—soap films (formations of organized structures). d—adsorption on non-polar surfaces; e—formation of a bilayer (membrane model); f—micellization.

Micelles form by the cooperative binding of monomers to one another at concentrations exceeding a rather narrow region called the critical micellization concentration (CMC).

The latter is the concentration of a surfactant at which a large number of micelles form in its solution that have been in thermodynamic equilibrium with the molecules (ions), and a number of properties of the solution, sharply change,

This sharp transition to the CMC region for systems with flexible chains could be attributed to the cooperative nature of the self-association process. This makes aggregates containing many monomers considerably more stable than small particles. The CMC has been one of the most easily determined experimentally and useful quantitative characteristics of solutions of surfactants with flexible chains.

The methods of determining the CMC are based on the sharp change in the physiochemical properties of surfactant solutions (Fig. 1.65) in the region of the CMC.

For instance, the curves of the concentration dependence of the equivalent electrical conductance l = cx (Fig. 1.65) at low concentrations c are not different from the typical curves for strong electrolytes. But at a certain critical value (CMC) corresponding to a knee in the curve, the value of X sharply drops.

Similarly, on curves of the surface tensionσ versus log c (curve 2), a horizontal segment would be observed in the region of the CMC. The physical meaning of this plateau was initially not clear because by the Gibbs equation *i.e.*, $d\sigma = -\Gamma_i \, RT \, d \, ln \, c_i$, the value of the absorption Γ_i after growing on segment I and reaching the extreme value Γ_∞ on segment II seemed to acquire a zero value ($d\sigma/d \, ln \, c_i = 0$) on segment III.

The same sharp changes in the other properties take place at this characteristic concentration, especially in the detergency (curve 3), the osmotic pressure (curve 4), and in the refractive index. The scattering of light increases.

As micelles do not usually have a surface activity, the CMC can be regarded approximately as the equilibrium concentration at which the chemistry of surface phenomenon ends and colloid chemistry begins, *i.e.*, the chemistry of disperse systems.

Hydrophobic interactions have been regarded to be the thermodynamic driving force of micellization. The hydrocarbon part of a diphilic molecule gets forced out of the aqueous medium to avoid contact of the chain with water as much as possible. Because of this micelles form. Their internal part, the core consists of a liquid hydrocarbon (combined closely packed hydrocarbon chains), while their external part facing the aqueous solution are having polar groups.

Micellization has been a spontaneous process, *i.e.*, the change in the Gibbs potential $\Delta G = \Delta H - T\Delta S < 0$. But the main contribution to the value of ΔG has been made not by the change in the enthalpy, which has been insignificant in magnitude, but by the change in the entropy $T\Delta S$.

Indeed, the removal of the hydrocarbon "chains" of the diphilic molecules from the water into the micelles disorders the structure of the water, and the entropy of the system therefore grows ($\Delta S > 0$).

It is possible to explain the mechanism of micellization as follows. A growth in the concentration c is accompanied by an increase in the chemical potential of the surfactant μ_{sl} xpressing the tendency of the component to leave the solution. At low concentrations, the surfactant ions pass into the surface layer at the interface of the surfactant with the other phase, thereby lowering the free energy of the system.

Soon, however, the surface layer gets saturated. Now, with a further growth in the concentration, the system expels the hydrophobic chains from the water into the liquid "pseudophase"—a micelle, and separates it from the water by the hydrophilic shell of the polar groups. This peculiar phenomenon, often called self-adsorption, causes a gain in energy of 1.08kT (about 2600 J) per CH_2 group. This has been close to the work of adsorption at a water-air interface equal to about 3000 J.

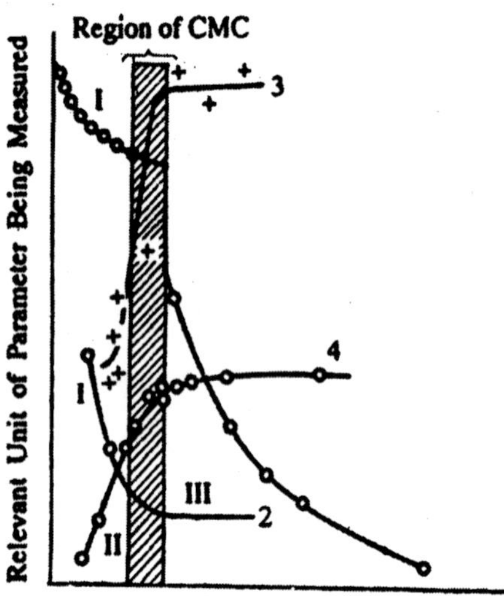

Fig. 1.65: Properties of solutions of a colloidal electrolyte (sodium dodecyl sulphate) 1. equivalent electrical conductance; 2. surface tension; 3. detergency; 4. osmotic pressure.

Hence, above the CMC, a true surfactant solution would transform into an ultramicroheterogeneous colloidal system (sol). With a growth in the concentration to above the CMC, the monomer concentration (c_m) remains almost constant ($\sigma \approx$const) with an increase in the number of micelles in the solution. The micellization process would be reversible, namely, dilution of a solution to c < CMC transforms the solution from a colloidal to a true one.

The generality of micellization and "orientation" phenomena like the formation of films on the surface of water and free films (a foam), molecular adsorption from solutions, and emulsification consists in that the tendency of a reduction in the thermodynamic potential gives rise to a define orientation that tends to lower the difference of the polarities. The formation of micelles decreases the value of G because of the combination of the oleophilic groups into a non-polar "drop" of oil. The latter gets covered, with a shell of hydrophilic groups like a protected emulsion or a globule of a polyelectolyte.

The longer a chain, the larger would be the gain in energy (at the expense of the withdrawal of the radicals from the water and their cohesion in micelle) and the lower would be the surfactant concentration needed for the formation of micelles. Hence, the CMC decreases with an increasing chain length (n_C). This qualitative conclusion will be treated further in a quantitative form.

When the energy balance is calculated for ionogenic surfactants called *colloidal electrolytes*, one must also take into account the expenditure of the electrical work of changing ($z\mathcal{F}\psi$) in the course of creating a micelle charge (the moving together of ions with like charges on its surface). It implies that for non-ionogenic surfactants with an identical chain length, the CMC must be lower because the expenditure of electrical work for them would be substantially reduced. This has indeed been confirmed experimentally.

Thus, at the CMC and above it, a thermodynamic equilibrium exists in a solution between the surfactant molecules (ions) and the micelles, which to a certain degree is similar to the formation of a new condensed phase. The aggregation numbers of the micelles (determined by the number of molecules in a micelle) have been not large enough (30-100 for most micelles) to regard the micelles as phase, but are adequate to consider a set of micelles as a "pseudophase".

This is why the modern thermodynamic interpretation of the properties of these systems successfully employs a homogeneous viewpoint (molecule $\rightleftharpoons$ associate equilibrium determined by the equilibrium constants, in terms of which important parameters of the system such as the CMC, the size of the micelles, and their distribution by size are expressed) and a heterogeneous one based on phase equilibrium determined by the equality of the values of $\bar{\mu}_i$ in the two coexisting phase.

The values of the CMC for most ionogenic surfactants have been found to range from 0.1 to 20 mmol/litre. They have been found to be depending on the chain length, the concentration of the counterions, additions of indifferent electrolytes, and other factors. The modern theory of micellization (the heterogeneous approach) helps us for establishing the laws of the change in the CMC depending on the indicated parameters. Suppose we consider the elements of this theory using the example of a singly charged ion.

In the state of equilibrium at the critical micellization concentration, the decrease in the chemical potential of an organic ion in its, transition from the dissolved state into a micelle should be equal to the electric work $K_g.\mathcal{F}\psi_1$ (per g-ion) of introducing a charge into the surface of a micelle :

$$\mu_i^{\alpha} - \mu_i^{m} = K_g \mathcal{F} | y_1 | \qquad ...(1)$$

where α and m refer to the pseudophases of the solution and the micelle, and K_g is a coupling factor.

If the state of the pure substance (micelle) is adopted as the standard one, then $a_i^m = 1$. Hence,

$$\mu_i^0 + RS \ln cc_{MC} = \mu_i^{0m} + K_g \mathcal{F} | \psi_1 | \qquad ...(2)$$

The values of $| \psi_1 |$ for micelles generally correspond to the condition $\mathcal{F} | \psi_1 | > 2RT$.

In this case, using the electric double layer theory, we get

$$\mathcal{F} | \psi_1 | = a - RT \ln c_i \qquad ...(3)$$

The difference of the standard chemical potentials is the work of adsorption, and it is

$$\mu_i^{0\alpha} - \mu_i^{0m} = c + bn_C \qquad ...(4)$$

where n_c refers to the chain length (expressed by the number of carbon atoms), and b = 2600 J mol of CH_2.

On introducing Eqs. (3) and (4) into (2) and combining the constants (at T = const), we obtain

$$\ln c_{CMC} = A' - B'n_C - K_g \ln c_i \qquad ...(5)$$

Let us consider two particular cases.

I. An indifferent electrolyte would be absent; $c_i = C_{CMC}$. Inserting this value into (5) and taking common logarithms, we obtain

$$\log C_{CMC} = A - Bn_C \quad ...(6)$$

This equation gives the dependence of the CMC on the chain length. It is consistent with the qualitative conclusion that the CMC lowers with a growth in nc.

II. In the presence of an indifferent electrolyte at n_C = const., we get

$$\log C_{CMC} = A'' - K_g \log c_i \quad ...(7)$$

The CMC gets decreased with an increase in the concentration of an indifferent electrolyte c_i because here $|\psi_1|$ diminishes, as well as the electrical work needed for the formation of a micelle.

A more detailed treatment of the question on the basis of statistical mechanics helps us to estimate the values of the constants A, B, and A". The theory has been found to be consistent with experimental values of the constant found from measurements of the CMC by various methods determining the points of inflection of the curves (the surface tension, electrical conductance, light scattering, etc.) are presented in Table 1.5.

Table 1.5 : Values of Constants in Eqs. (6) and (7) for Various Homologous Series of Surfactants.

Compound	*Temperature °C*	*Composition of medium*	*Constants*			
			A	B	D	E
Alkyl betaines	25	H_2O	2.85	0.47	2.28	0.20
Sodium Alkyl sulphates	25	H_2O	1.47	0.29	2.71	0.14
Sodium Alkyl sulphonates	25	H_2O	1.40	0.28	2.99	0.10
Alkyl trimethylammonium	25	H_2O	1.81	0.30	3.02	0.10
bromides	30	H_2O	2.02	0.32	2.94	0.10
		0.0125 M KBr	2.50	0.37	2.89	0.12
		2.025 M KBr	2.58	0.38	2.93	0.12
Quaternary pyridinium bromides	30	H_2O	1.57	0.28	2.95	0.11

The theory also gives rise to an expression for the micellar mass Mm depending on the chain length :

$$\log M_m = D + En_C \qquad ...(8)$$

The values of the constants D and E obtained experimentally are also included in Table 1 for various homologous series of surfactants. These data have been used for practical estimates of the CMC. Division of M_m by the molecular mass of the monomer would give the *aggregation number*.

STRUCTURE OF MICELLES

Micelles of Surfactants in Aqueous Solutions

Suppose we consider the structure of micelles using an ionic micelle as an example. According to modern notions, an ionic micelle (Fig. 1.66) has been a compact and, to a first approximation, spherical formation consisting of a liquid hydrocarbon core covered with a layer of polar (ionogenic) groups. The notions of spherical micelles were first of all introduced by Hartley.

The liquid state of hydrocarbon chains has been different, however, from the state of a bulk liquid phase typical, for example, of a drop of an emulsion. Because of the orientation of the polar groups, the entire micelle has been in the liquid-crystalline state similar to condensed films. Capillary pressure (arising because of the curvature of the micelle surface) that for fine spherical micelles reaches hundreds of atmospheres, also causes structural ordering of the hydrocarbon phase. Because of this, the diameter of a micelle core is less than the length of two completely extended chains.

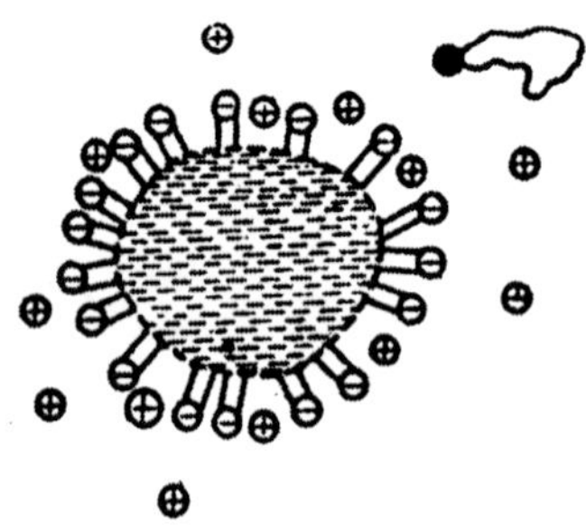

Fig. 1.66 : An ionic micelle in an aqueous solution : the circles with a minus sign are surfactant anions, and those with and a plus sign are counterions (cations).

The layer of polar (ionogenic) groups (together with the CH_2 group contacting with each of them) protrudes above the surface of the core by 0.2-0.5 nm is in the aqueous phase. Because of this, we speak of the "roughness" of a micelle. Roughness results in the fact that the counterions getting into the gaps between adjacent charged groups of a micelle are bound to them more firmly than the remaining ions of the diffuse layer.

Hence, the layer inner of the electric double layer of an ionic micelle gets formed by ionogenic groups of a surfactant, and the outer layer, by "bound" and free, diffusively arranged counterions. This binding, which completely explains the drop in the equivalent electrical conductance λ, can be regarded as incomplete dissociation or the formation of ion pairs, or as a Stern layer.

Because of this, the effective charge of an ion in a micelle would be K_g ze (or $K_g z.\mathcal{F}$ per g ion), where K_g is a coupling factor close in its meaning to the activity coefficient and usually equal to 0.2-0.6.

The presence of electric double layers on the surface of ionic micelles underlies their electrophoretic mobility and the electrical conductance of the solution. Binding of part of the counterions gives rise to a knee in the curve of λ versus c.

Spherical micelles form in a solution at concentrations close to the CMC. With a growth in the surfactant concentration, the micellar solution tends to pass through a number of equilibrium states characterized by a definite aggregation number, size and shape of a micelle.

For example, an increase in the surfactant concentration gives rise to the appearance of more complicated cylindrical and plate-shaped micelles (McBain micelles) and of a continuous gel-like structure of the system in the course of a consecutive transition through several pseudophases called mesomorphic phases and differing in their structure.

The application of the Gibbs-Curie principle to the analysis of polymorphic modifications gave rise to the relations determining the region of existence of micelles with various shapes.

Investigations using X-ray diffraction analysis and electron microscopy make it possible to advance the following notions on the structures of the mesomorphic phases.

With increasing the sufactant concentration, the number of spherical micelles in an isotrophic molecular-micellar solution grows (Fig. 1.67b). Next, the spherical micelles combine into rod-shaped aggregates like stacks of coins (Fig. 1.67c). The viscosity of the system sharply grows. These extended micelles further transform into a two-demensional heaxgonal continuous structure throughout the entire volume of the 'solution, forming a "middle" mesomorphic phase (Fig. 1.67d). With a further growth in the concentration, the system gets transformed into a lamellar mesomorphic phase. .The continuous structure in this phase gets formed by the parallel packing of extended flexible bimolecular layers with water interlayers that become thinner with an increase in the surfactant content. This structure is regarded to be similar to that of thin films.

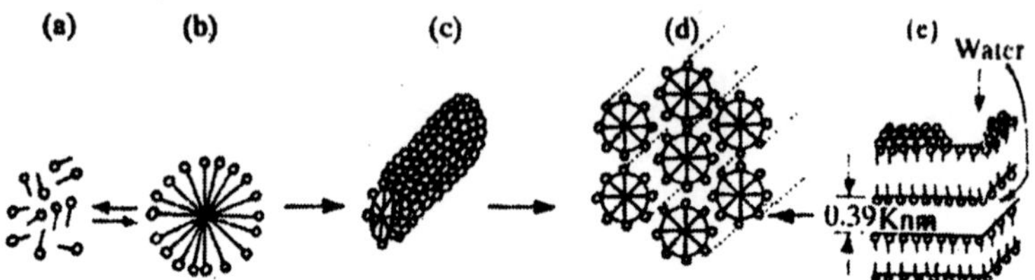

Fig. 1.67. Formation of structures in a surfactant solution; (a)—monomers; (b)—spherical micelle; (c)—cylindrical (rod-shaped) micelle (randomly oriented); (d)—hexagonal packing of cylindrical micelles; (e)—plateshaped micelle.

Actually, the concentration dependencies of the activation energy and the electrical conductance arc having absolutely the same shape for a structurized bulk phase of a colloidal electrolyte and free thin (black) films. In both cases, a phase transition would be attended by a sharp increase in E_a associated with vanishing of the "free" solution.

IONIC MICELLES

According to McBain colloidal electrolytes cannot be regarded as macromolecules as they are individual molecules of giant size in solution. They also differ from lyophobic colloids, because the latter are unstable. Large size anions, *e.g.*, $C_{17}H_{33}.COO^-$. RSO_3 etc. and cations like $RN(CH)_3^+$, aggregate to form ionic micelles of colloidal dimensions containing number of ions together containing appreciable water molecules. *So the ionic micelles are the aggregates formed in solution by colloidal electrolytes.*

The formation of micelles can also take place from neutral or non-ionic molecules, *e.g.*, polyethlene oxide.

Ionic micelles of sodium oleate : The formation of micelle by sodium oleate, $C_{17}H_{33}COO^-Na^+$ is astriking example of colloidal electrolytes. In this case there are two parts, one is tail, *i.e.*, hydrocarbon part $C_{17}H_{33}$ and the other is head COONa. The COONa is an ionisable lyophilic group which tries to go into water resulting into ions. The $C_{17}H_{33}$ part tends to go away from solution. But if the correcntration is increased the hydrocarbon part forms the aggregate. The micelle is therefore that of anions and may contain hundred or more oleate ions clamped together.

Critical micelle concentration (C.M.C) : In very dilute solutions, sodium and potassium oleate and other similar substances, remain as individual molecules ionising into positive and negative ions. According to Davies and Bury the concentration at which micelle concentration becomes appreciable, is termed as critical micelle concentration. At this concentration there is an abrupt change in the properties. It decreases with the increase of temperature. Every colloidal electrolyte has a definite value of C.M.C.

Types of Ionic micelles. Me Bain suggested the presence of more than one type of ionic micelles in a given solution of a colloidal electrolyte. Some of them are described below :

1. *Lamellar micelle :* It consists of double leaflets of soap molecules placed end to end and side by side. X-ray Study has revealed the existence of other kind of micelle in which molecules are laid end to end, side by side, as in lameller micelle. The difference is that lamellars repeat at a regular distance from each other and are separated by layer of water, depending upon concentration. *Hoffman's* investigation has shown that the molecules in the micelle are rotated at an angle of 55°. According to Stuff the modules are closely packed side by side in an irregular manner as in a liquid crystal (Fig. 1.68).

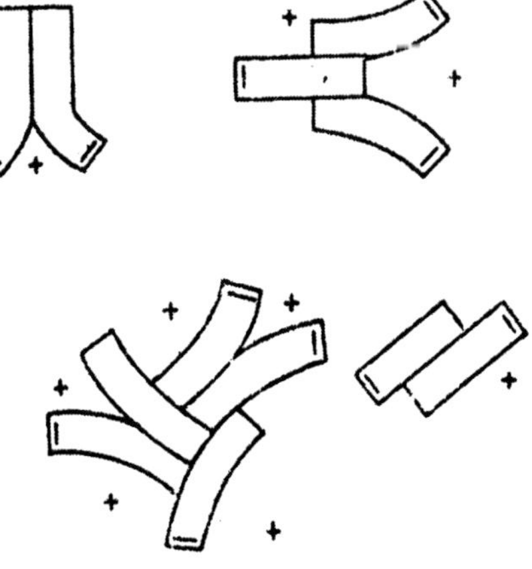

Fig. 1.68

2. *Spherical micelle* : According to *Hartley* ionic micelles can be spherical.

   Although this concept is employed but is open to criticism.

3. Ellipsoidal or cylindrical micelle : Klevens has suggested that the micelle may also be an elongated ellipsoidal or of cylindrical model.

MICELLIZATION IN NON-AQUEOUS MEDIA

In comparison with aqueous solutions, the lyophobic interactions of surfactants in non-aqueous media have been found to be much weaker. In non-polar organic solvents with a low permittivity e, the polar groups of diphilic molecules become lyophobic. This makes associates to form in which the core consists of polar groups, while the hydrocarbon chains of molecules are in the non-aqueous medium. Such aggregates are termed as inverted micelles.

Because of the low permittivity of non-polar solvents, ionogenic surfactants do not practically dissociate in them. The forces of dipole-dipole interaction between the ion pairs as well as possible hydrogen bonds have been responsible for the formation of a polar core. The presence of traces of water binding the polar groups have been found to facilitate micellization in non-aqueous media.

The theory of aggregation and models in non-aqueous media have been constructed by analogy with an aqueous solution. They have been based on the notion of monomer-micelle equilibrium and the value of the CMC. The analysis of recently accumulated experimental data has shown, however, that the association of surfactant molecules in non-polar media is having a number of features distinguishing it from association in an aqueous medium.

Among them are the presence of aggregates at a low surfactant concentration in the solution (10^{-6}–10^{-7} M) and the continuity of the aggregation process (stability of the aggregates with a small number of monomer units).

Consequently, doubt is cast on the existence of a single concentration at which micelles form, which deprives the definition of the CMC in such systems of any meaning. (Doubt has been also sometimes cast on the possibility of micellization in non-aqueous media due to the energy difficulties involved in combining polar groups into a core).

THE DETERMINATION OF CRITICAL CONCENTRATION OF MICELLE FORMATION

The methods are based on the fact that, when micelles are formed, all the properties of a surfactant solution change, and the change is the more sudden, the higher is the number of aggregation of molecules. It has been established nephelometrically that micelle formation causes a sharp increase in light scattering by a surfactant solution because it becomes heterogeneous.

McBain has shown that the concentration dependence of the osmotic coefficient for soap solutions is schematically expressed by a curve in Fig. 1.69. The curve, unlike a monotonic one that is typical of weak electrolytes which do not form micelles consists of three segments which correspond to (1) the precritical region, (2) a region in which the osmotic coefficient sharply changes with concentration, and (3) a post-critical region, where the curve is almost parallel to the concentration axis. The shape of the first segment of the curve changes at a point which corresponds to a diminution in the osmotic activity of a solution as a result of the formation of spherical micelles, *i.e.*, at a point which corresponds to CCMF.

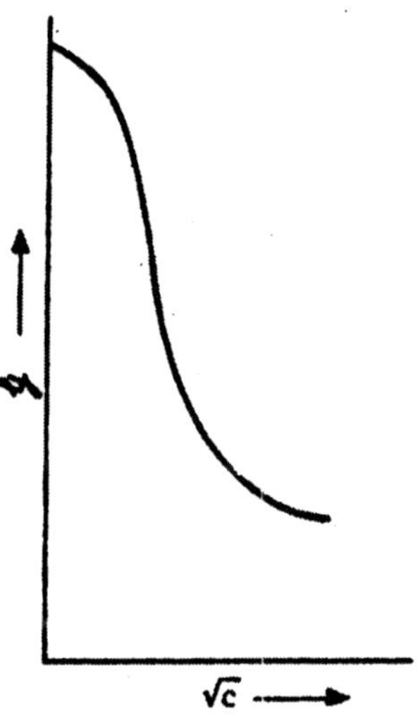

Fig. 1.69 : Dependence of the osmotic coefficient G on the concentration c of a soap solution.

The equivalent electric conductivity of a surfactant solution changes with concentration similarly to this curve. Fig. 1.70 shows typical concentration curves of equivalent electric conductivity for solutions of ordinary soaps. The initial curve segments, which correspond to the precritical region, are not plotted on the graph because of its small scale. However, we can see from the shape of the second and third segments that the equivalent electric conductivity of soap solutions containing 12 or more carbon atoms in a chain sharply drops in the region of low concentrations. Then, it reaches a minimum at a point, which corresponds to micelle formation, and later it somewhat increases again. 1036 Colloid Science as the molecular weight of surfactants increases, the minimum shifts towards lower concentrations and becomes more pronounced. According to Hartley, a drop in equivalent electric conductivity with an increase in the soap concentration of a solution is caused by the partial binding of counterions by ionized micelles. This binding occurs under the action of considerably strong electric forces which are due to ionic micelles formed in a system and bearing a high charge. It is not quite clear why equivalent electric conductivity increases after reaching the minimum.

Lastly, CCMF can be determined by a change in the surface tension of a colloidal surfactant solution when its concentration increases and as a result, the surface tension drops, reaching a minimum constant value at a point which corresponds to CCMF.

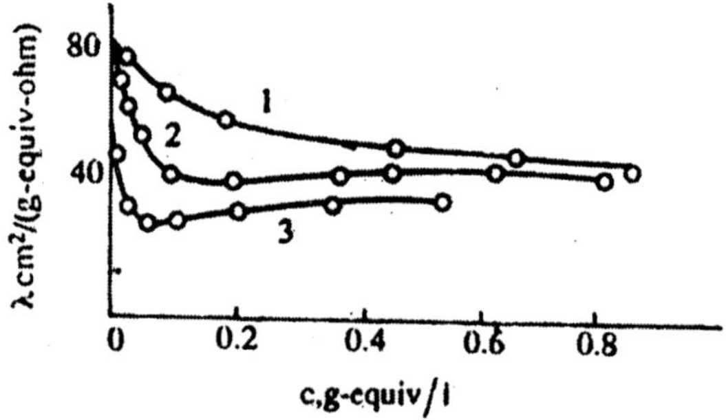

Fig. 1.70 : Dependence of equivalent electric conductivity λ of soap solutions on concentration c (at 18 °C): 1—potassium caprylate; 2—potassium laurate; 3—potassium oleate

Factors affecting the Critical Concentration of Micelle Formation

Several factors can affect CCMF in surfactant solutions. CCMF decreases as the molecular weight of a hydrocarbon chain of a surfactant grows because in this case, true solubility diminishes and the tendency of surfactant molecules to associate increases. The temperature effect on CCMF differs for ionogenic and non-ionogenic surfactants. For ionogenic surfactants. CCMF usually increases with temperature owing to the disaggregating action of the thermal motion of molecules. But this effect is not strong because it is weakened by hydrophobic interactions which cause an increase in the entropy of a system. Therefore, the effect of temperature on CCMF is the weaker, the more pronounced are the hydrophobic properties of soaps. For non-ionogenic surfactants, CCMF always decreases with an elevation of temperature because then hydrogen bonds between the oxygen atom of an ether and water molecules are broken, hydroxyethylene chains dehydrate, and their mutual repulsion, being a hindrance to aggregation, decreases.

Micelles can be formed not only in aqueous colloidal surfactant solutions but also in solutions of surfactants in hydrocarbons. In this case surfactant molecules in a micelle are oriented by their polar groups into a micelle, and by their hydrocarbon ends, towards a solvent. However, surfactants usually produce molecular solutions in ethanol because, by its polarity, ethanol is between water and hydrocarbons, and therefore it is a solvent for both polar and nonpolar parts of surfactant molecules.

SOLUBILIZATION IN SURFACTANT SOLUTIONS

Solubilization is an important property of surfactant solutions that is connected with their micellar structure.

When almost water-insoluble organic substances (aliphatic and aromatic hydrocarbons, oil-soluble dyes, etc..) are added to sufficiently concentrated surfactants solutions, these substances can dissolve colloidally, or *solubilize.* Almost clear, thermodynamically equilibrium solutions are formed as a result of such solubilization. A substance which dissolves in surfactant solutions is known as a *solubilizer.*

The solubilizing ability of various surfactants is very diverse. The amount of a colloidally dissolved organic substance in homologous series of surfactants increases as a hydrocarbon radical becomes longer, and also proportionally to the concentration of a surfactant solution in the region of spherical micelles; it grows sharply when plate-like micelles are formed. Solubilization depends also on the molecular structure of a solubilizer. For example, if hydrocarbons are used as solubilizers, colloidal solubility increases as their molecular weights decrease. Solubilization usually grows when polar groups are introduced into a solubilizer.

According to present views, solubilization is dissolution of organic substances in surfactant micelles. Solubilization mechanisms may be diverse. Non-polar hydrocarbons dissolve in a micelle nucleus (Fig. 1.71a), whereas polar organic substances (alcohols, amines) are oriented in micelles in a way that their hydrocarbon chains are directed into micelles, and polar groups, towards the aqueous phase (Fig. 1.71b). There is another mechanism of solubilizing non-ionogenic surfactants which contain polyhydroxyethylene groups. Molecules of a solubilizer (*e.g.*, phenol) do not penetrate deep into micelles, but are arranged at their periphery, between folded hydroxyethylene chains, apparently forming a hydrogen bond with the oxygen atom of the ether.

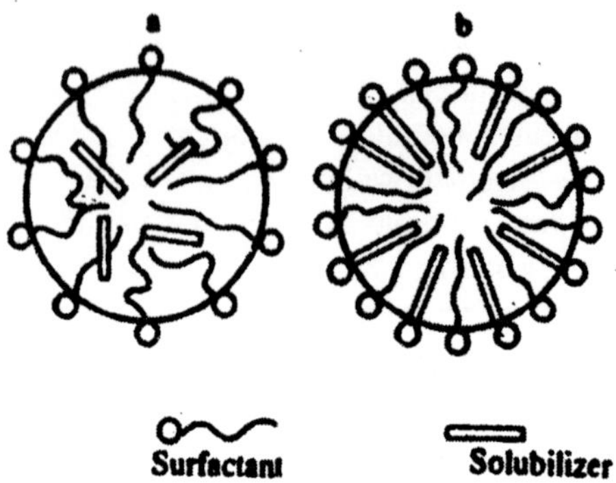

Fig. 1.71 : Schematic illustration of the solubilization of hydrocarbons (a) and polar organic substances (b) in surfactant micelles.

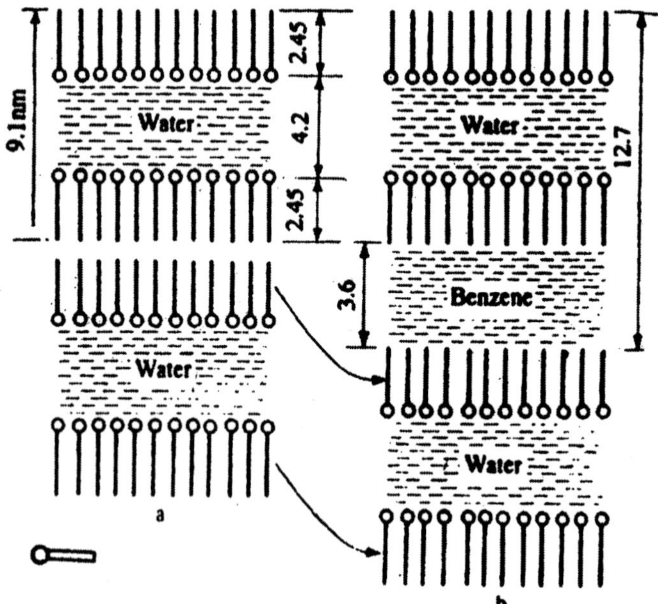

Fig. 172 : Solubilization of benzene in a sodium oleate micelle: a—micellar solution before solubilization; b—the same after solubilization

In solubilization, the micellar weight of surfactants grows due not only to the incorporation of solubilizer molecules, but also to an increase in the number of surfactant molecules in a micelle. Such a rearrangement of micelles occurs because the micelle nucleus becomes more hydrophobic upon hydrocarbon solubilization, and therefore the number of surfactant molecules constituting a micelle must also increase in order to maintain equilibrium.

In solubilization in plate-like micelles, the organic substance penetrates into a micelle, arranging itself between the hydrocarbon ends of soap molecules and thus drawing apart layers of molecular chains. Such a viewpoint is confirmed by X-ray diffraction analysis which has shown that the distance between soap molecules in a micelle increases upon solubilization. A change in the structure of plate-like micelles upon solubilization is represented in

The solubilization phenomenon is important in polymerizing unsaturated hydrocarbons in emulsions when latices are being synthesized. In this case the polymerization occurs mainly within or on the surface of soap micelles in which unsaturated hydrocarbon is solubilized, rather than in the drops of this hydrocarbon.

Reverse solubilization, i.e., the Colloidal dissolution of water in oils in the presence of the appropriate colloidally oil-soluble surfactants, is of great importance in the food industry, particularly in margarine production.

There is considerable interest in establishing the location within a micelle of the solubilized component. It can be seen that the environment changes from polar water to nonpolar hydrocarbon as we move radially toward the centre of a micelle. While the detailed structure of the various zones is disputed, there is no doubt that this gradient of polarity exists. Accordingly, any experimental property that is sensitive to molecular environment can be used to monitor the whereabouts of the solubilizate in the micelle. Spectroscopic measurements are ideally suited for determining the microenvironment of solubilizate molecules. The ultraviolet spectrum of solubilized benzene was used to explore the solvation of micelles. Here we take the hydration for granted and use similar methods to locate the solubilizate.

In addition to ultraviolet spectroscopy, resonance methods—such as electron spin resonance and nuclear magnetic resonance (NMR)—have been wieldy used to establish the site of solubilization. We shall briefly consider the use of proton NMR for such an investigation. Like the electron, the proton may have spin quantum numbers of $\pm 1/2$ resulting in two different nuclear quantum states. In a magnetic field these states differ in energy, the difference corresponding to the nuclear magnetic moment aligning with or against the magnetic field.

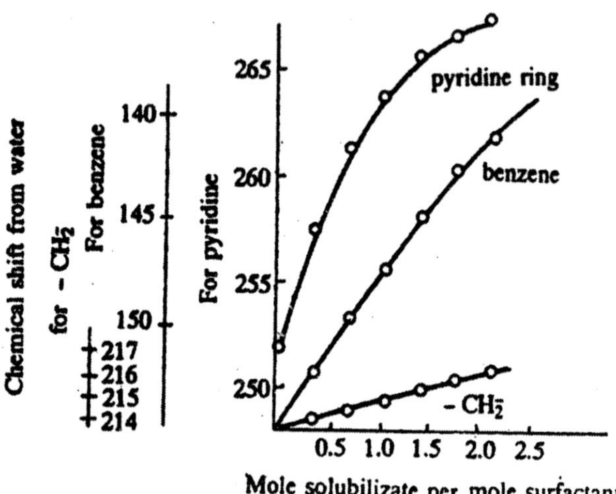

Fig. 1.73 : Chemical shifis of protons in benzene, pyridium rings and, methylene groups versus the ratio of moles of solubilized benzene to moles of hexadecyl pyridinium chloride.

If energy of the appropriate frequency is supplied, transitions from one state to the other are possible and the radiation is absorbed. The absorbed frequency measures the separation of the two states by the familiar formula $\Delta E = hv$, where h is Planck's constant. The energy separation of the affected states depends on the strength of the applied magnetic field as modified by the local environment. Thus even protons in different parts of a molecule absorb at slightly different frequencies, since the magnetic field each experiences is somewhat different from the applied field owing to the shielding effect of nearby electrons. Furthermore, neighbouring protons cause a peak in an absorption spectrum to be split into a multiplet with characteristic relative intensities. This aids the identification of peaks in an NMR spectrum. In addition to one part of a molecule influencing the NMR spectrum of another part, the medium in which the molecule is embedded also has an effect. Therefore the NMR spectrum of a solubilized molecule has the potential not only to reveal the location of a solubilized molecule in a micelle but also to give information about its orientation.

Some data measured for benzene solubilized in hexadecyl pyridinium chloride. The abscissa in the figure shows the extent of solubilization expressed as moles of solubilizate per mole of surfactant. The ordinate values show shifts in the resonance frequencies from water taken as an internal standard. Shifts of peaks arising from protons in the pyridinium ring, in benzene, and in methylene groups in the alkyl tail are shown versus the extent of solubilization.

The following points summarize these observations:

1. All peaks are shifted toward higher fields because of the diamagnetic effect of the solubilized benzene.
2. The slope of the shift versus extent of solubilization curve is much steeper for the protons on the benzene and pyridinium rings than for protons in methylenes in the tail. In fact, the first two increase in roughly parallel fashion.
3. Since the charged pyridinium ring must be at the surface, the benzene which parallels it in chemical shift must have a similar location.
4. The far less sensitive response of the chemical shift of the methylene protons to the extent of solubilization suggests that the core of the micelle is mainly composed of alkyl chains and is relatively unaffected by the benzene.

In another related study (using hexadecyltrimethyl ammonium bromide micelles), isopropyl benzene was solubilized and the chemical shifts of aromatic and alkyl protons were observed. The results suggest that the isopropyl benzene molecules are oriented such that the isopropyl groups are buried more deeply in the core of the micelle while the benzene ring is in the more hydrated palisade layer. This located the benzene in a relatively polar portion of the micelle.

DONNAN MEMBRANE EQUILIBRIUM

Ostwald noted some unusual facts when two solutions containing electrolytes are separated by membrane which is impermeable to one of the ions of the electrolytes. The impermeable ions are protein anions, congo red anions, etc. It was found that at equilibrium an unequal distribution occurs for the permeable (penetrable diffusable) ion on both sides of the membrane. The thermodynamic verification of the observed fact was formulated by *Donnan.*

At equilibrium the osmotic pressure of two solutions will be different and if two reference electrodes such as calomel electrodes are put into the two solutions, then a difference of potential will be set up. This type of equilibrium is known as *Donnan membrane equilibrium* or *Donnan effect* or *Gibbs-Donnan effect* and the potential so developed between two solutions as the Donnan membrane potential.

In order to explain the Donnan equilibrium we will describe the various components of systems.

1. *Internal solution :* It is a solution of an electrolyte containing non-penetrating ions of about colloidal dimensions and ion small enough to be penetrable, *e.g.*, congo red anions.
2. *External solution :* It is a solution of electrolyte containing both the penetrable ions *e.g.*, NaCl.
3. A semipermeable membrane. It separates the two solutions and is permeable to all except non-penetrable ions.

We shall now consider four different cases to which Donnan membrane equilibrium is applicable. In the first three cases, it is assumed that all ions are monovalent whereas in the last case the ions are polyvalent.

Case I : The electrolytes in the two solutions have an ion in common. Consider two electrolytes NaA and NaCl separated by a semipermeable membrane. A^- represents the non-penetrable anion. C_1 and C_2 are the respective concentrations of the ions in internal and external solutions initially.

Initially

Na^+	A^-	Na^+	Cl^-
c_1	c_1	c_2	c_2

Semipermeable Membrane

Since the internal solution does not contain Cl^-ions, therefore, some of the Cl^-ions will diffuse into it from the external solution. Suppose x moles/litre of Cl^-ions move from right to left solution. Since the two solutions must be electrically neutral, so x moles/litre of Na+ ion must also penetrate from right to left solution. Therefore, at equilibrium the conditions are

At equilibrium

Na^+	A^-	Cl^-	.	Na^+	Cl^-
$(c_1 + x)$	c_2	x	.	$(c_2 - x)$	$(c_2 - x)$

According to Donnan's theory of membrane equilibrium,

$$[Na^+]\,[Cl^-] = [Na^+]\,[Cl^-] \qquad ...(1)$$

Internal External

Thus, $x\,(c_1 + x) = (c_2 - x)\,(c_2 - x)$

$$x = \frac{c_2^2}{c_1 + 2c_2} \qquad ...(2)$$

From Eq. 1 (1) we also have

$$\frac{[Na^+]_{INT}}{[Na^+]_{EXT}} = \frac{[Cl^-]_{EXT}}{[Cl^-]_{INT}} = \gamma$$

or $$\gamma = \frac{c_2 - x}{x} \qquad ...(3)$$

Substituting the value of x from (2) in (3), we have

$$\gamma = \frac{c_2 - \dfrac{c_2^2}{c_1 + 2c_2}}{\dfrac{c_2^2}{c_1 + 2c_2}} = \frac{c_1 + c_2}{c_2} \qquad ...(4)$$

where γ represents the distribution ratio of NaCl solution between external and internal solutions. From this type of equilibrium, the following points should be noted :

(a) Presence of non-penetrable ion causes the unequal distribution of the penetrable ions between the two solutions.

(b) Total concentration of ions is greater in the internal solution than in the externals solution. In other words, the osmotic pressure of internal solution is greater than that of external solution.

(c) If c_1 be fixed and c_2 is increased, then y tends to decrease and finally reaches to unity.

Case II : The electrolytes on both the sides have no common ion. Consider a case when we have NaA on one side of the membrane and KCl on the other side. The state of equilibrium is represented by

Initially

$$\begin{array}{cc|cc} Na^+ & A^- & K^+ & Cl^- \\ c_1 & c_1 & c_2 & c_2 \end{array}$$

Suppose x moles/litre of K^+ ions move from external to internal solution. Let z moles/litre of Na^+ ions penetrate form internal to external solution and similarly y denotes the number of moles per litre of Cl^- ions that migrate from right solution.

The condition at equilibrium is then represented as

At equilibrium

$$\begin{array}{cccc|ccc} Na^+ & A^- & K^+ & Cl^- & Na^+ & K^+ & Cl^- \\ (c_1 - z) & c_1 & x & y & z & (c_2 - x) & (c_2 - y) \end{array}$$

For the electrical neutrality of the solution,

$$z = x - y \qquad ...(5)$$

But according to Donnan equilibrium,

$$\underset{\text{Internal}}{[Na^+ [Cl^-]} = \underset{\text{External}}{[Na^+] [Cl^-]} \qquad ..(5a)$$

$$(c_1 - z)\, y = z\,(c_2 - y) \qquad ...(6)$$

and

$$\underset{\text{Internal}}{[K^+] [Cl^-]} - \underset{\text{External}}{[K^+] [Cl^-]} \qquad ..(6a)$$

$$x.y = (c_2 - x)(c_2 - y) \qquad ...(7)$$

Substituting the value of z in (6), we have

$$(c_2 - x + y)\, y = (x - y)(c_2 - y)$$

$$x = \frac{y\,(c_1 + c_2)}{c_2} \qquad ...(8)$$

From eq. (7), we get

$$xy = (c_2 - x)\,(c_2 - y) \qquad ...(9)$$

$$x = (c_2 - y)$$

From eqs. (8) and (9), we have

$$c_2 - y = \frac{y(c_1 - c_2)}{c_2}$$

or
$$y = \frac{c_2^2}{c_1 + 2C_2} \qquad ...(10)$$

Putting the value of y from Eq. (10) in (9)

$$x = c_2 - \frac{c_2^2}{c_1 + 2C_2}$$

$$x = \frac{c_1c_2 + 2c_2^2 - c_2^2}{c_1 + 2C_2} = \frac{c_1c_2 + c_2^2}{c_1 + 2c_2} \qquad ...(11)$$

The value of z can be calculated by putting the values of x and y in the equation (5),

$$z = \frac{c_1c_2}{c_1 + 2c_2} \qquad ...(12)$$

From eqs. (5a) and (6a), we have

$$\frac{[Na^+]_{INT}}{[na^+]_{EXT}} = \frac{[K^+]_{INT}}{[K^+]_{EXT}} = \frac{[Cl^-]_{EXT}}{[Cl^-]_{INT}} = \gamma \text{ (distribution ratio)}$$

Putting the value of $[Cl^-]_{EXT}$ and $[Cl^-]_{INT}$ we get

$$\gamma = \frac{c_2 - y}{y}$$

Substituting the value of y from eq. (10), we have

$$\gamma = \frac{c_2 - \dfrac{c_2^2}{c_1 + 2C_2}}{c_2^2} = \frac{c_1 + c_2}{c_2}$$

It should be noted that the value of γ is same as in previous case. In this case also if c_1 is fixed and c_2 is increased then the f approaches to unity.

Thus, in this case, if the value of c_2 is very small as compared to that of c_1 then it means that there will be practically complete separation of the two ions of the electrolyte in the right solution.

Case III. One of the solutions is water alone : The Donnan membrane equilibrium is important in concentration with the phenomenon of membrane hydrolysis. In *membrane hydrolysis* we have solution of the salt NaA type separated by pure water.

The initial state of equilibrium is represented by

Initially

Na+	A-	:	H+	OH-
c_1	c_1	:		
Internal			External	

Suppose x mole/litre of Na^+ ions move from internal to external solution. In order to maintain electrical neutrality, an equivalent amount of OH^- ions must migrate from internal to external solution.

The condition of equilibrium is represented as :

At equilibrium

Na^+	H_2O^+	A^-	:	Na^+	OH^-
$(c_1 - x)$	x	c_1	:	x	x

The Solution on the left is acidic, whereas on the right it is alkaline.

Applying the Donnan's equilibrium, we have

$$\underset{\text{Internal}}{[Na^+]\,[OH^-]} = \underset{\text{External}}{[Na^+]\,[OH^-]} \qquad \text{...(14)}$$

We know that $K\omega = [H_3O^+]\,[OH^-]_{INT}$

or
$$[OH^-]_{INT} = \frac{K_\omega}{[H_3O^+]} = \frac{K_\omega}{x}$$

Kw is the ionic product of water. Substituting above in eq. (14).

$$(c_1 - x)\,.\,\frac{K_\omega}{x} = x^2 \text{ or } (c_1 - x)\,K\omega = x^3$$

The value of x is small as compared with that of c_1. Hence $(c_1 - x)$ may be regarded as nearly equal to c_1. Thus.

$$c_1\,K\omega = x^3$$

$$x = (c_1 K\omega)^{1/3} \qquad \text{...(16)}$$

This process of producing an excess of hydroxyl ions in the external solution is known as *'Membrane hydrolysis'*.

Case IV. When one of the electrolytes contains polyvalent ions :

In the above three cases, we have dealt with monovalent diffusion ions. In this case, the case of ion with unequal valencies will be taken up. Let us consider the initial state of equilibrium as :

Initially

$$\begin{array}{cc:cc} Na^+ & A^- & Ca^{++} & Cl^- \\ c_1 & c_1 & c_2 & c_2 \end{array}$$

At equilibrium, the condition will be

At Equilibrium

$$\begin{array}{cccc:ccc} Na^+ & A^- & Ca^{++} & Cl^- & Na^+ & Ca^{++} & Cl^- \\ (c_1 - z) & c_1 & x & y & z & c_2 - x & (2c_2 - y) \end{array}$$

In this case $z = 2x - y$ ...(17)

Applying the Donnan's equilibrium,

$$\frac{[Na^+]_{INT}}{[na^+]_{EXT}} = \frac{[Ca^{++}]_{INT}}{[Ca^{++}]_{EXT}} = \frac{[Cl^-]_{EXT}}{[Cl^-]_{INT}} = \gamma \quad ...(18)$$

From eqs. (15) and (16) the values of x, y, z may be calculated as in the case II. The equations will be of third degree.

Membrane Potential

When two solutions are separated by a membrane, a potential difference is set up due to the unequal distribution of ions.. This potential difference is known as membrane potential. This may be calculated for the first case :

$$Na^+ A^- : Na^+ Cl^-$$

At the equilibrium state, it becomes as follows :

$$Na^+ A^- Cl^- : Na^+ Cl^-$$

Internal External

Let E_1 be the positive potential in left solution and E_2 that of right solution. Suppose a very small quantity $F\delta n$ of positive electricity is transferred from right to left solution at a constant temperature.

Suppose there occurs virtual variation of equilibrium. Then the following terms will be involved :

(i) The decrease in free electrical energy may be given as follows :

$$F\delta n\ (E_1 - E_2)$$

(ii) The work obtained by the transfer of t+Sn moles of Na+ ion from right to left solution and $t_-\ \delta n$ moles of Cl^- ion from left to right may be given as follows:

$$t_+\ \delta n\, RT \log_e \frac{[Na^+]_{ext}}{[Na^+]_{int}} + t_-\ \delta n\ RT \log_e \frac{[Cl^-]_{int}}{[Cl^-]_{ext}}$$

where t_+ and t_- are the transport numbers of cation (Na^+) and anion (Cl^-) respectively.

When equilibrium is established, the decrease in free energy as given in stage (i) is equal to the amount of work done in stage (ii) *i.e.,*

$$F\delta n\ (E_1 - E_2) = t_+\delta n\ RT \log_e \frac{[Na^+]_{ext}}{[Na^+]_{int}} + t_-\delta n\ RT \log_e \frac{[Cl^-]_{int}}{[Cl^-]_{ext}}$$

where z is the concentration of gelatin ions. *i.e.,* of combined H^+. On applying the Donnan's equilibrium, we have

$$\text{But} \qquad t_1 + t_2 = 1 \text{ and } \frac{[Na^+]_{ext}}{[Na^+]_{int}} = \frac{[Cl^-]int}{[Cl^-]_{ext}} = \gamma\ (\text{say})$$

∴ The membrane potential Em is given as follows:

$$E_m = -(E_1 - E_2) = -\frac{RT}{F} \log_e \gamma = \frac{RT}{F} \log_e \left(\frac{1}{\gamma}\right)$$

The above is the relation for membrane potential. This relation is also valid for polyvalent ions.

If a general case is considered in which the diffusible chloride ion possesses the ion M^{z+} of valency z, then the membrane potential E_m is given by the following relation :

$$E_m = \frac{RT}{zF} \log_e \frac{[M^{z+}]_{ext}}{[M^{z+}]_{int}}$$

$$\frac{[M^{z+}]_{ext}}{[M^{z+}]_{int}} = \frac{[Cl^-]_{int}}{[Cl^-]_{ext}} = \frac{1}{\gamma}$$

$$E_m = \frac{RT}{zF} \log_e \left(\frac{1}{\gamma}\right)$$

This is the general expression for the membrane potential.

Applications : Donnan's membrane equilibrium finds number of applications in biological, technical and physico-chemical processes. A few of them are noted below :

1. *Swelling of gel* : It is in part due to Donnan membrane effect and consequent osmotic pressure. It can be minimised by taking the gel at the isoelectric point. For example it has been shown by Lloyd that gelatin in water undergoes a seven fold increase in volume but in HCl of pH 2.3 this increase is forty times.

Procter (1914) explained swelling of gels in the following manner.

According to him, there occurs an interaction between the gelatin and the acid to produce ionisable salts and the colloidal salts and the colloidal gelatin ions so produced are not able to diffuse through the membrane. As the gel is freely permeable to the $H+$ and Cl^- ions present, it is assumed that these ions must be distributed between the gel and the solution surrounding it in accordance with the Donnan's membrane equilibrium. He further assumed that these distributions result in an excess of osmotic pressure in the gel which would result the entry of water into the gel and due to which it swells.

The above facts were confirmed experimentally by Procter who also showed that they are in accordance with the Donnan

$$\begin{array}{ccccc} gH^+ & H^+ & Cl^- & : H^+ & Cl^- \\ z & y & y + z & : x & x \end{array}$$

Internal External

$$[H^+]\,[Cl^-] = [H^+]\,[Cl^-]$$

External Internal

$$x.x = y(y + z) \qquad \text{...(1A)}$$

or $y + z = x^2/y$

or $$z = (x^2 - y^2)/y \qquad \text{...(1B)}$$

Procter and Wilson argued that the swelling of the gel due to an acid takes place due to the high osmotic pressure of the gel which arises from the difference in concentration of diffusible ions. According to this there would occur maximum swelling of gel when the difference in concentration of diffusible ions is maximum. Difference in the concentration (C) of the diffusible ions may be given as follows,

$$C = y + (x + z) - 2x \qquad \text{...(2)}$$

On substituting the value of z from Eq. (1B) in (2), we get

$$C = 2y + \frac{x^2 - y^2}{y} - 2x$$

or

$$C = \frac{(x - y)^2}{y} \qquad \text{...(3)}$$

From Eq. (3), the value of C may be calculated. If the values of C are plotted against the pH of the external solution, it is seen that the value of C will be maximum at pH 2.44 and 2.65.

From the theory of Procter and Wilson, it is also evident that the degree of swelling of pure gel depends on the final pH of the solution in which it is kept but it should be independent of the solution. However if the gelatin is having salt impurities, the degree of swelling will no: only depend upon the final pH of the solution but also on the final volume of acid added. This was later confirmed by Northrop and Kunitz (1927-29).

2. *Tanning of Leather* : It was explained by Procter and Wilson on the basis of Donnan's theory of membrane potential. According to this theory, the process of tanning is a combination of tannin with hide fibre. The main constituent of fibre is collagen" which is a colloidal substance. The main aim of this theory is to give a satisfactory explanation of the influence of acids, bases and salts upon the tanning process.

When the tannin is dissolved in water, it forms a colloidal solution. in which the particles are negatively charged. The surface layer surrounding the particles must contain a definite concentration of positive ions held by electrochemical attraction to the negatively charged particles of tannin.

Suppose the concentration of tannin particles is $[T^-]$ and that of positive ions held by electrochemical attraction to tannins is $[M^+]$. Suppose some amount of electrolyte MN is added to the solution, then in the bulk of the solution $[M^+] = (N^-)$, In the surface layer there will be certain concentration of M^+ and N^- ions. At equilibrium, an unequal distribution of the electrolyte takes place. At this point of equilibrium, no work will be done during a small reversible change which is made at constant temperature and pressure. Suppose δn molecules are transferred from the bulk solution to the surface layer. Then, the work done in this transfer will be as follows :

$$\delta n \ RT \log_e \frac{[M^+]_1}{[M^+]_2} + \delta n \ RT \ \log_e \frac{[N^-]_1}{[N^-]_2} = 0$$

or $$\log_e \frac{[M^+]_1}{[M^+]_2} = -\log_e \frac{[N^-]_1}{[N^-]_2}$$

or $$\frac{[M^+]_1}{[M^+]_2} = \frac{[N^-]_2}{[N^-]_1}$$

or $$[M^+]_1 [N^-]_1 = [M^+]_2 [N^-]_2$$

where $[M^+]_2$ denotes the concentration of ions in the bulk solution and $[M^-]_1$ the concentration of ions in the surface layer.

If x denotes the concentration of positive or negative ions in the bulk of the solution, y denotes that of –ve diffusible ions in the surface layer and z denotes the concentration of positively charged ions held by electrochemical attraction by tannin, then using the Eq. (14), we get

$$x^2 = y\ (y + z) \qquad ...(4)$$

The unequal distribution of the ions in the surface layer and in the bulk gives rise to a difference of potential (E) which may be given as follows :

$$E_m = \frac{RT}{F} \log_e \frac{[N^-]_2}{[N^-]_1} = RT \log_e \frac{x}{y} \qquad ...(5)$$

Eq. (4) may be written as follows :

$$y^2 + yz - x^2 = 0$$

or $$y = \frac{-z \pm \sqrt{(z^2 + 4x^2)}}{2}$$

or $$\frac{x}{y} = \frac{2x}{-z \pm \sqrt{(z^2 + 4x^2)}} \quad ...(6)$$

On substituting Eq. (6) in Eq. (5). we get

$$E_m \frac{RT}{F} \log_e \frac{2x}{-z \pm \sqrt{(z^2 + 4x^2)}} \quad ...(7)$$

If x is increased so much that z may be neglected in comparison to x, then Eq. (7) will become as follows:

$$E_m \frac{RT}{F} \log_e \frac{2x}{\sqrt{4x^2}}$$

The value of E_m will diminish with the increase in the concentration of the electrolyte in solution. As soon as the difference in potential becomes sufficiently small by the addition of the electrolyte, a condition is established which would favour the coalescence of the colloidal particles and under these conditions the tannin is precipitated.

3. From the Donnan's theory we know that larger is the initial concentration of the non-penetrating anion greater will be the inequality in the concentrations of the penetrating anion and of the penetrating electrolyte as a whole. Such situations are extremely common biological cells. *The red blood cells* have walls through which ions of NaCl can penetrate but haemoglobin ions, present within these cells cannot penetrate. The concentration of NaCl in red blood cells is only 70% of the NaCl concentration in the blood plasma which constantly surrounds the red blood cells.

4. An unequal distribution of particles on the two sides leads to the development of Osmotic pressure. If the molecular weight of a protein is measured by the osmotic pressure, it will come out wrong. However, the Donnan membrane effect disappears, at the isoelectric point of a protein or in the presence of high salt concentrations.

5. This theory is helpful in the calculation of cataphoretic velocity.

6. Donnan's theory is applicable to the dyeing of animal fibres by acid dyes.

7. It has also been used in the determination of molecular weight by ultracentrifuge.

8. Donnan equilibrium provides a method for the measurement of adsorption of ions by the particles in inorganic colloidal system.
9. The Donnan's theory also offers an evidence for the origin of the fluid present in the cavities of eyes.
10. In biological field, Donnan's theory relates to the distribution of diffusible ions between the red blood corpuscles and the serum of the blood.

APPLICATIONS OF COLLOID SCIENCE

This branch of chemistry is becoming increasingly important day by day. Colloids play an important role in our daily life. Human body itself is colloid. Protoplasm, the building material of plant cells and animal tissues, and the blood that flows through our veins are colloidal in nature. Fats pass through intestines in the form of emulsion.

Recently various synthetic high polymers reported are nothing but colloids. Examples are polystyrene, silicones, PVC, synthetic rubber, etc.

The problem of heterogeneous catalyst has been fully explained by colloid chemistry. The important applications of colloid science are noted below :

1. *Medicines :* Colloidal medicines being very finely subdivided can be easily adsorbable by the body and they are thus more effective. For example Argvrol—an eye lotion is a colloidal solution of silver protected by gelatin. Colloidal gold and calcium are used as tonics. Colloidal sulphur is an effective germ killer. Codliver oil used as tonic is an emulsion. Milk of magnesia, an emulsion, it used for stomach troubles.
2. *In Industry :* (a) *Purification of water :* Negatively charged colloidal clay particles present in river water are removed by coagulation by addition of alum. The positive Al^{3+} ions bring about the precipitation.

   (b) *Tanning of leather :* Tanning is a process which imparts hardness to leather. Skin and hides are positively charged colloidal sols of proteins. These are coagulated by addition of negatively charged sol of tannin present in tree bark. By this process, the leather gets hardened and becomes fit for further use.

Salts of chromium have also been used in place of tannin. The process is then known as *chrome tanning*. The leather obtained by chrome tanning is more soft and can take up polish easily.

(c) *In laundary* : The cleansing action of soap is explained by formation of an emulsion with the dirt and grease of the dirty clothes (Fig. 1.74). The emulsion is easily washed away by water.

(d) *Rubber industry* : Latex obtained from the rubber trees is an emulsion of negative rubber particles dispersed in water. On heating with sulphur, vulcanised rubber used for manufacturing cycle and motor tyres, tubes, etc. is formed. Rubber particles can also be deposited electrically on articles placed at anode in an electrolytic cell containing latex.

(e) *Removal of carbon from smoke : Cottrell's precipitator* : Carbon particles in air are colloidal in nature and carry negative charge. Air from chimney of an industrial plant before letting out into atmosphere is passed through Cottrell's precipitator maintained at high potential difference (about 50,000 volts). The removal of carbon particles from air involves the principle of electrophoresis. Carbon particles get precipitated by losing their charge and, thus, the air which finally comes out is free from carbon particles.

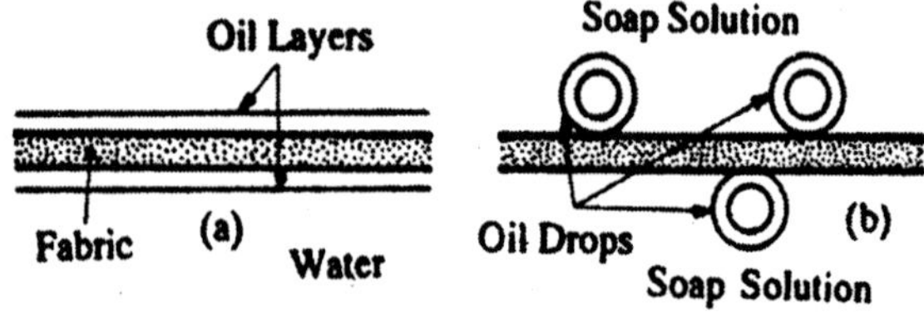

Fig. 1.74 : Cleansing action of soap.

(f) *Sewage disposal* : Sewage is a colloidal system where dirt, mud, etc. are dispersed in water. Sewage in big cities is passed through big tanks fitted with electrodes. Dirt particles which are colloidal in nature lose their charge and get settle down. The dirt deposited at the electrodes is removed and used as manure.

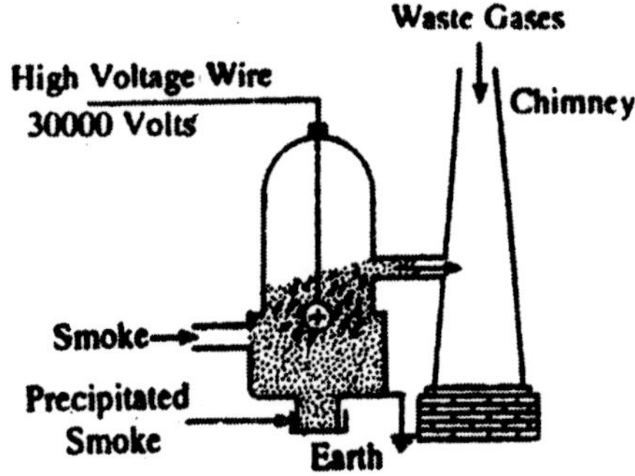

Fig. 1.75

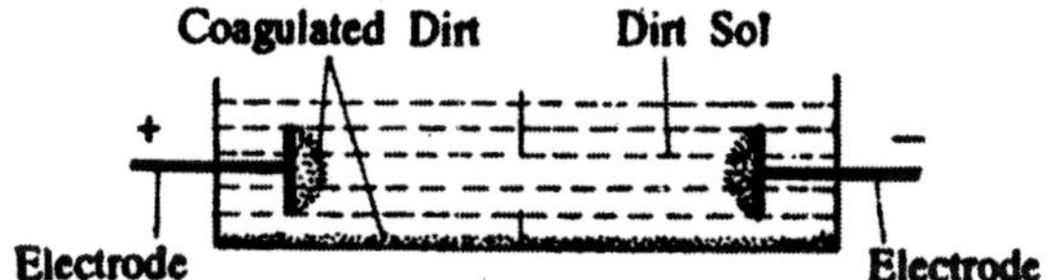

Fig. 1.76 : Sewage disposal.

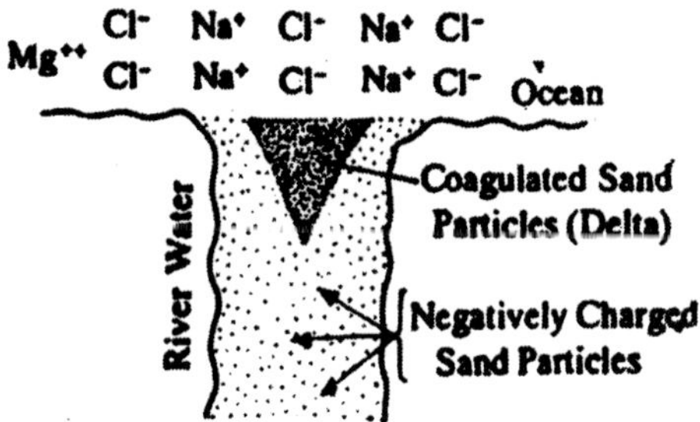

Fig. 1.77 : Formation of Delta (Illustration)

(g) *Ceramic industry* : The earth and clays are typical gels containing aluminium silicate and iron hydroxide mixed with organic colloids. The properties like plasticity, hydration, peptisation, etc. in ceramic clays are due to the presence of colloidal substance.

(h) *Dyeing industry* : The cotton, silk or wool fibres almost possess the get structure. Dyes and mordants employed for dyeing are all colloids. For example, $Al(OH)_3$ $SnCl_2$ tannin, etc. are well known examples of mordants.

(i) *Manufacture of Acheson's graphite* : Acheson's graphite is a lubricant and is colloidal in nature. In its manufacture, the graphite is grounded in presence of tannin (protective colloid) and the colloidal sol of graphite dispersed in water is formed. The lubricating solution of graphite depends, upon its degree of dispersion.

(j) *Photographic plates and papers* : Sensitive emulsions are deposited on photographic plates; films are the fine suspensions of silver bromide in gelatin. AgBr dispersed in gelatin gives rise to a photographic emulsion.

3. *In Nature* : (a) Blue colour of the sky ; It is due to the scattering of blue light by colloidal dust particles in air.

(b) *Tails of comets* : Comets fly with very high velocity and it has been found that solid particles arc left behind. These particles scatter light and are visible as the tyndall cone which forms the tail of the comet seen in the sky.

(c) *Formation of delta* : Formation of delta at the meeting place of river with sea is due to the coagulation of negatively charged clay particles by the positively charged Na, K or Mg ions present in sea water (Fig. 1.77).

(d) *Rain* : Rain is caused by the aggregation of colloidal water particles dispersed in air. The coagulation of the colloidal water in air may be caused by the cold air or when the oppositely charged clouds meet.

It is possible to cause artificial rain by throwing in the air electrified sand or some colloidal solution having particles, with charge opposite to that on the clouds.

(e) *Blood* : Blood is a colloidal system, having albuminoid substance as the dispersed phase carrying a negative charge. Bleeding can be stopped by the application of alum or ferric chloride. By the addition of Al^{+++}, or Fe^{+++}, the coagulation of the blood takes place and bleeding stops.

(f) *Articles of daily use* : Milk which is a complete food is an emulsion of fat in water (o/w type) stabilised by casein.

Butter is an emulsion of water dispersed in fat (w/o type). Ice cream is ice dispersed in cream.

4. Analytical Applications

(a) Qualitative and quantitative Analysis : (i) In volumetric analysis, hydrophilic colloids alter end point, *e.g.*, in a titration of HCl and NaOH, the amount of deviation in the end point increases with increasing amounts of colloids. In the determination of silver by Mohr's method (precipitation titration), the phenomenon of adsorption is involved.

(i) *In gravimetric analysis* crystalline precipitates are needed and the procedure thus adopted gives rise to the formation of sufficiently bigger particles because smaller particles may pass through the filter paper. In order to prevent coprecipitation and post precipitation, it is necessary to precipitate fairly dilute solutions in hot conditions.

(ii) *In qualitative analysis* organic matters should be destroyed, otherwise they will form protective colloids and prevent the precipitation of various precipitates. The solubility of $Zn(OH)_2$ in an excess of NaOH, in the separation of Zn from Mn is almost a case of peptisation of the sol. Similar is the case in the dissolution of $Cr(OH)_3$ by NaOH solution.

(b) *Identification of traces of noble metals* : Noble metals such as platinum metal and gold, silver etc., when present in the colloidal form produce very bright and intense colours. For example purple of cassius indicates the presence of colloidal gold. According to *Donnan* the dilute solution of noble metals are reduced in the common borax beads to sols. Ag salts and Pt salts give blue yellow, violet beads respectively. This method is a very sensitive test and more sensitive than the spectre-analytical methods.

(c) *Distinction of natural from artificial honey* : Natural honey is also colloidal in nature. The distinction between the two is made by *Ley's test*. It consists in treating few drops of the honey with an ammoniacal $AgNO_3$ solution. In the case of pure honey, a metallic silver produced assumes a reddish yellow colour due to the traces of albumin or ethereal oil, which act as protecive colloids.

A dark yellow or greenish precipitate is formed with artificial honey.

5. Miscellaneous Applications

(a) *Fertility of soils :* Good fertile soils are due to the presence of plants assimilable materials in the colloidal form in the soil.

(b) *In warfare :* Smoke screens in war are obtained by dispersing colloidal TiO_2 particles in air. Animal charcoal is used in gas masks for the adsorption of poisonous gases.

(c) *Metallurgy of alloys :* In the metallurgy of alloys there are wide applications of colloidal processes. The properties of alloys are modified by the presence of colloidal substances.

(d) *Concentration of ores :* During froth flotation process employed for the concentration of sulphide ores, emulsion is formed.

In addition to the above there are numerous other applications of colloid science. *In fact there is hardly any field in our life, which is untouched by the colloids.*

2

ELECTROLYTIC CONDUCTANCE AND ELECTROLYTIC TRANSFERENCE

INTRODUCTION TO CONDUCTANCE

The metallic conductors as well as electrolytes obey Ohm's law, which states that "*The strength of current (I) flowing through a conductor is directly proportional to the potential difference (E) applied across the conductor and inversely proportional to the resistance (R) of the conductor.*" Thus, the mathematical form of Ohm's law becomes as

$$I = E/R.$$

In the relation to electrolytes, the term conductance (C) is used more frequently than resistance. Conductance implies the ease with which electric current can flow through a conductor. It is therefore, defined as the reciprocal of resistance,

Conductancc – 1/R.

Units: It is expressed in units of reciprocal or inverse ohms, $ohms^{-1}$ or mhos.

Resistance

The resistance R of a conductor is

(i) directly proportional to its length (l cm) and

(ii) inversely proportional to its area of cross of cross section.

Hence $$R \propto \frac{l}{a} \text{ or } R = \rho \frac{l}{a} \qquad ...(1)$$

where ρ is a constant depending on the nature of the material of the conductor, and is called the "*specific resistance*" or "*resistivity.*"

If l = 1 cm and a = l sq. cm, then ρ = R ohms.

Hence, specific resistance may be defined as:

"*The resistance of a uniform column of the material of the conductor having a length of 1 cm and a cross section of 1 sq. cm.*"

Units: As we know that

$$\rho = R\frac{a}{l} = \frac{\text{Ohm (cm)}^2}{\text{cm}} = \text{ohm. cm} \qquad ...(2)$$

Specific Conductance

The specific conductance of a conductor is the reciprocal of specific resistance, and is generally denoted by K (Greek, Small Kappa). Equation (2) can be written as

$$R = \frac{1}{K}\frac{1}{a} \text{ ohms} \qquad \left[\therefore \rho = \frac{1}{K}\right] ...(3)$$

$$K = \frac{1}{R} \times \frac{1}{a} \text{ ohms}^{-1} \text{ com}^{-1}$$

If l = 1, and a = 1 sq. cm. equation (3) becomes as

$$K = \frac{1}{R} = \rho$$

In the case of solution, the definition will be the same as of a conductor. In simple words, it is defined as:

"*The conductivity offered by a solution of length 1 cm area of unit cross section.*"

Units: We know that K = 1/ρ

Hence the unit of K will be

$$\frac{1}{\text{ohm. cm.}} = \text{ohm}^{-1} \text{ cm}^{-1} = \text{mho/cm.}$$

Equivalent Conductance

It may be defined as,

"*The conductance of a solution containing 1 gm. equivalent of an electrolyte when placed between two sufficiently large electrodes which are 1 cm apart.*" It is denoted by λ_r, where V is the volume in c.c. containing one gram equivalent of electrolyte dissolved in it and is measured in reciprocal ohm, or mho.

Molecular Conductance

It is defined as,

"*The conductance of a solution containing 1 gm mole of the electrolyte when placed between two sufficiently large electrodes placed one cm apart.*"

It is denoted by symbol μ, and is measured in mhos,

Relation Between Specific Conductance and Equivalent Conductance

Consider a rectangular metallic vessel with opposite sides exactly 1 cm apart [Fig. (2.1)]. If now 1 c.c. of solution is placed in this vessel, the area of the opposite faces of the cube converted by the solution will be 1 sq. cm. The measured conductance of the solution will be its specific conductance. If 1 c.c. of the solution is placed in the above vessel containing 1 gm. equivalent of the electrolyte, then by definition, the measured conductance will also be equal to the equivalent conductance,

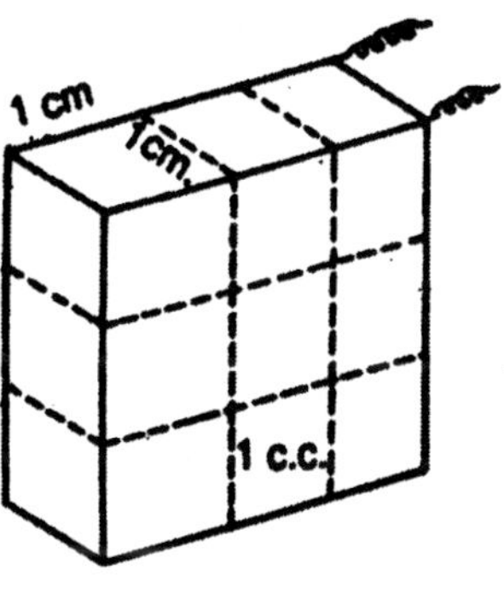

Fig. 2.1

$$\lambda = K$$

Now if 9 c.c. of pure solvent are added so that the total volume now occupied by the solution is 10 c.c.; the (because 10 c.c. of the diluted solution still contains 1 gm. equivalent of the electrolyte). But by definition of the specific conductance, the measured conductance of the diluted solution will be ten times the specific conductance,

$$\therefore \qquad \lambda = K \times 10$$

Similarly if the solution is diluted 100 times its volume, the measured conductance will still be equivalent conductance, but will be 100 times the specific conductance.

Thus, $$\lambda_v = K \times 100$$

In general, if the solution contains 1 gm equivalent of the electrolyte dissolved in V c.c. of the solution, then

$$\lambda_r = K \times V$$

∵ Equivalent conductance = specific conductance × volume of solution in c.c. containing 1 gm. equivalent of the electrolyte.

Relation between Molecular conductance and Specific Conductance: As discussed above, a similar type of the relation also exists between molecular conductance and specific conductance *i.e.*,

$$\mu_r = K \times V$$

or Molecular conductance = [Specific conductance] × [Volume of solution in c.c. containing 1 gm. mole of electrolyte]

Effect of Dilution

(i) *Conductance:* The conductance of s solution is due to the presence of ions in solution. Since on dilution the degree of dissociation of electrolyte in increased and more ions are produced in solution, it is thus expected that the conductance of a solution increase on dilution.

(ii) *Specific Conductance:* Specific conductance depends upon the number of ions present per c.c. of the solution. Since on dilution the degree of dissociation increases but the number of ions per c.c. decreases, it is therefore expected that the specific conductance of a solution should decrease on dilution.

(iii) *Equivalent Conductance:* The value of equivalent conductance increases on dilution. This increase is due to the fact that equivalent conductance is the product of specific conductance and the volume V of the solution containing 1 gm equivalent of the electrolyte. Since the decreasing value of specific conductance is more than compensated by the increasing value of V, the value of λ_y will increase with dilution.

(iv) *Molecular Conductance: As* discussed in equivalent conductance, the molar conductance of an electrolyte increases with dilution.

(v) The variation of equivalent conductance, with dilution is shown in Fig. (2.2) where the equivalent conductance of different dilutions is plotted against $\sqrt{(c)}$, where c is the concentration of the solution in moles per litre.

From Fig. (2.2), it is evident that the equivalent conductance tends to become maximum as the concentration decreases or dilution increases.

However the maximum conductivity in case of strong electrolytes like KCl is reached at relatively higher concentrations (or lower dilutions) that the case of weak electrolytes like CH_3COOH.

Equivalent Conductance at Infinite Dilution

The maximum value of the equivalent conductance of an electrolyte obtained at high dilutions is known as the equivalent conductance at infinite dilution. It is denoted by λ_∞. By infinite dilution we mean that the solution is so dilute that any further dilution does not produce any charge in the equivalent conductance of the solution.

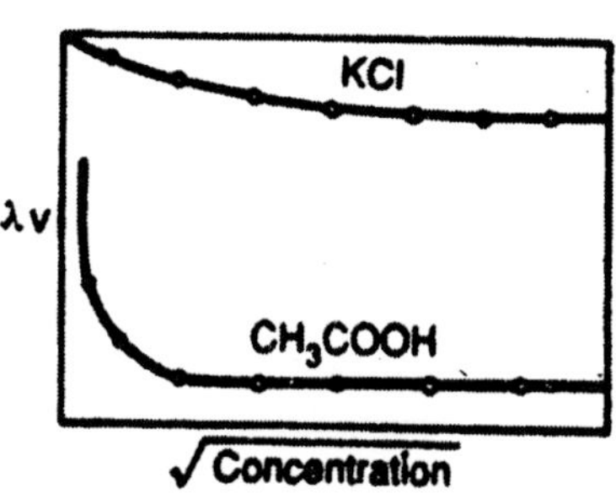

Fig. 2.2

Measurement of Specific Conductance

We know $$R = \rho \times \frac{l}{a}$$

$$\frac{1}{\text{obs. conductance}} = \frac{1}{\text{sp. conductance}} \times \frac{l}{a}$$

$$\text{Sp. conductance} = \text{Obs. conductance} \times \frac{l}{a}$$

$$= \text{Obs. conductance} \times \text{cell constant.}$$

The ratio *l/a* is known as *cell constant* and is generally denoted by *x*. It depends upon the area of cross section of electrodes and the distance between them. Thus, the determination involves the measurement of conductance of the solution and of the cell constant.

Measurement of Conductance: The conductance of a solution can be expressed by

$$\frac{1}{R} = x(c_1\lambda_1 + c_2\lambda_2 + c_3\lambda_3 \ldots + c_l\lambda_l)$$

Here c_i terms are the concentrations of various ions in solution, λ_i terms are numerical constants characteristic of ions and x is a proportionality factor that takes account of the geometry of the cell.

Because of interionic interactions and solvent-ion interactions, the values of λ_i depend slightly on concentration, and extrapolation to infinite dilution is required to obtain values that are physical constants of ions. Thus with proper calibration, a single measurement can give the concentration of any electrolyte.

The conductance of a solution can be measured by finding out the resistance by means of a Whetstone bridge arrangement. However when D.C. current is used in the solutions, the following difficulties may arise:

(i) During the conduction of D.C. current, there starts electrolysis which results in a change in the concentration of the solution.

(ii) Products of electrolysis get accumulated at the electrodes and set up a back E.M.F. Above mentioned difficulties are overcome by using A.C. current in the following manner.

(i) *The polarisation produced due to A.C. current in one direction is neutralised by the effect of current in other direction.*

(ii) *No electrolysis takes place.*

When an A.C. current is used, the following modifications are made:

(i) *An A.C. current of frequency fo 1000-4000 cycles per second is generally used, an oscillator has to be used.*

(ii) *An alternating current cannot be employed to detect the null-point and is therefore replaced by a Head phone.*

(ii) *In order to minimise the polarisation still further, the platinum electrodes in the cell are coated with finely divided platinum. This gives rise to a larger surface to the electrodes and consequently the polarisation is retarded.*

Method : The apparatus used for the purpose is a modified form of the Wheatstone bridge and is shown in Fig. 2.3.

(i) The solution whose conductance is to be measured is taken in the conductivity cell. The conductivity cell consists of electrodes which are platinum coated with finely divided platinum. These are welded to platinum wires which are sealed through two glass tubes. The tubes are firmly fixed in an ebonite cover so that the distance between the electrodes is not altered during the experiment.

The copper wires of the circuit dip in mercury placed in glass tubes to complete the circuit. Some other types of conductivity cells are shown in Fig. 2.4.

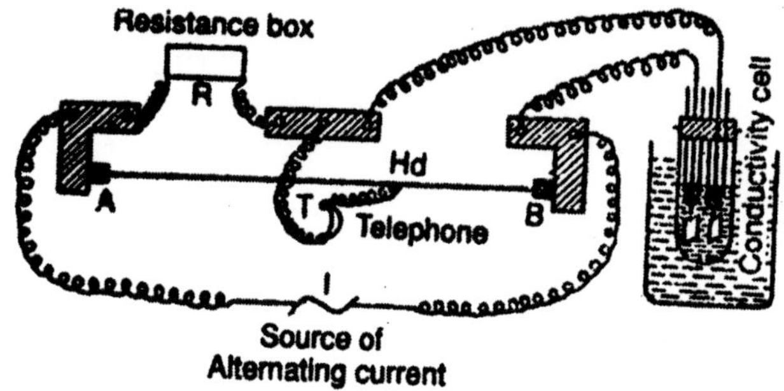

Fig. 2.3 : Measurement of Eq. conductance.

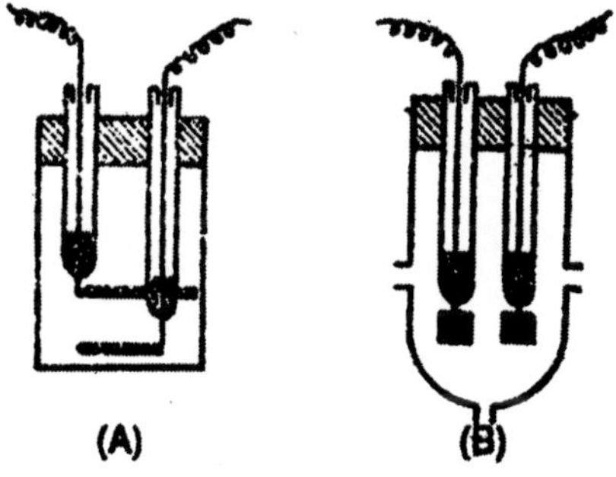

Fig. 2.4

(ii) The cell is connected to a resistance box R on one side and to a long thin wire AB stretched along a scale on the other side.

(iii) Some known resistance 'R' is taken out of the resistance box. An alternating current of about 1000 - 4000 cycles per second is passed through the solution. The sliding contact d is moved on the wire AB so that minimum or no sound is heard in the head phone. Then,

$$\frac{\text{Resistance of the solution}}{\text{Resistance R}} = \frac{\text{Length Bd}}{\text{Length Ad}}$$

Thus, knowing R, Ad and Bd the resistance of the solution can be calculated.

$$\therefore \text{Conductivity of the solution} = \frac{1}{\text{Resistance of the solution}}$$

Determination of Cell Constant: The cell constant is determined by substituting the value of sp. conductance of N/50 KCl solution at 25°C as accurately determined by Kohlrausch (0.002765) and the value of conductance is observed with the given cell in the relationship:

$$\text{Cell constant} = \frac{0.002765}{\text{obs. conductance}}.$$

Measurement of Equivalent Conductance: Knowing specific conductance as above, the equivalent conductance of the solution can be measured by measuring the specific conductance K with volume V in c.c. containing 1 gm. equivalent of the electrolyte

$$\lambda_r = K \times V.$$

APPLICATION OF CONDUCTIVITY MEASUREMENTS

(a) Basicity of Organic Acids : From an examination of the conductances of the sodium salts of a number of organic acids, Ostwald (1887) discovered the empirical relation

$$\lambda_{1024} - \lambda_{32} = 10.8b \qquad \text{...(1)}$$

where λ_{1024} and λ_{32} are equivalent conductances of the salt at 25°C at dilutions of 1024 and 32 litres per equivalent, respectively and *b* is the basicity of the acid.

Procedure : To prepare N/32 solution of the sodium salt of an acid, a N/16 NaOH solution is accurately prepared. A known volume of it, say 50 ml is neutralised by adding dropwise a concentrated solution of the acid, using an indicator paper. When the neutralisation is complete, the volume is made upto 100 ml. This gives N/32 solution of the salt and the N/1024 solution is prepared by a careful dilution of the solution.

Measure the resistance of N/32 and N/1024 solution of the taking all the usual precautions. Using cell constant, calculate the equivalent conductances at both the dilutions. Then, calculate the basicity of the acid using the formula given above.

Limitations

(i) This rule applies to some organic acids but does not succeed with inorganic acids,

(ii) This method fails when applied to very weak acids whose salts are considerably hydrolysed in solution.

(b) Degree of Hydrolysis : When the salt of a weak acid or base is dissolved in water, hydrolysis occurs, so that the conductance of the solution is now partly due to the ions of the salt and partly due to the ions of the salt and partly due the ions (H^+ or OH^-) of the acid or base formed by hydrolysis. An excess of the weak acid or base in the presence of its salts can be regarded as completely unionised. Therefore, addition of weak acid or base to the salt solution suppresses the hydrolysis but the ionisation will be unaffected.

Aniline hydrochloride ($C6H_5NH_3^+Cl^-$), the salt of a weak base and a strong acid, is hydrolysed in water giving (1 – α) equivalent of unhydrolysed salt and α equivalents of free acid and free base, where α is the degree of hydrolysis.

$$\underset{1-\alpha}{C_6H_5NH_3^+} \rightleftharpoons \underset{\alpha}{C_6H_5NH_2} + \underset{\alpha}{H^+}$$

The weak undissociated base aniline does not contribute to the total equivalent conductance λ of the solution, which may, therefore, be written as

$$\lambda = (1-\alpha)\lambda_c + \alpha\lambda_{HCl}$$

or
$$\alpha = \frac{\lambda - \lambda_c}{\lambda_{HCl} = \lambda_c} \qquad ...(2)$$

which λ_c and λ_{HCl} are the equivalent conductances of the unhydrolyzed salt and free acid at the same concentration. λ_c as has been pointed out, can be determined by measuring the equivalent conductance in the presence of excess aniline.

The value of λ_{HCl} at concentrations M/32, M/64, M/128 are known to be respectively 393, 399 and 401. Thus α being determined from eq. (2) the hydrolysis constant K_h is easily obtained from the following relation:

$$K_h = \frac{\alpha^2 C}{1-\alpha}$$

(c) Solubility of a Sparingly Soluble Salt : This is based on the simple fact that the saturated solution of a sparingly soluble salt is so dilute that the electrolyte present there in may be regarded as completely ionised. The relationship between concentration C (in gm. equivalents per !itre), the equivalent conductance λ, and the specific conductance K, is given by the equation

$$\lambda = K \times \frac{1000}{C} \qquad ...(3)$$

The solution being saturated, C represents the solubility (in gram equivalents per litre). Since the solution (though saturated with respect to the almost insoluble salt) is so dilute, it may be regarded as being practically ionised at infinite dilution. Hence, λ may be put equal to λ_∞. This as we know, is equal to $\lambda_{(cation)} + \lambda_{(anion)}$, the figures for which are available from published tables. Hence the solubility of the sparingly soluble salt, C, is

$$C = \frac{100\ K}{\lambda_{(cation)} + \lambda_{(anion)}} \text{ gm equivalent/1000 ml} \qquad ...(4)$$

So, the only experimental work required is to determine the conductance of a saturated solution.

(d) Ionic Product of Water, K_w : This can be calculated from the specific conductance of pure water, as determined experimentally, and its equivalent conductance, as calculated from the λ_0 values of its two ions. The equivalent conductance at infinite dilution λ_∞ is defined by the equation

$$\lambda_\infty = 100 \frac{K_\infty}{C} \text{ mhos cm}^2 \qquad ...(5)$$

where K_∞ is the specific conductance at infinite dilution and C the concentration in gm. equivalent per litre. The most accurate experimental value of specific conductance of purest water so far available is 5.54 $\times 10^{-8}$ mho cm^{-1} at 25°C. This may be put equal to K as pure water may be regarded as a solution of ionised water at infinite dilution.

The values of equivalent conductances of the two ions of water at infinite dilution are:

$$\lambda_\infty(H^+) = 349.8 \text{ mhos cm}^2$$

$$\text{and } \lambda_\infty(OH^-) = 198.6 \text{ mhos cm}^2$$

Hence the equivalent conductance at infinite dilution, λ_∞, of water may be equal to 349.8 + 198.6 *i.e.* 548.4 mhos cm^2. Substituting the values of K_∞ and λ_∞ in equation (5), we get value of C for each of the ions of water:

$$[H^+] = \frac{1000 \times 5.54 \times 10^{-4}}{548.4} = 1.01 \times 10^{-7}$$

$$[OH^-] = \frac{1000 \times 5.54 \times 10^{-4}}{548.4} = 1.01 \times 10^{-7}$$

Hence the ionic product of water.

$$K_w = [H^+][OH^-]$$
$$= 1.01 \times 10^{-7} \times 1.01 \times 10^{-7} = 1.02 \times 10^{-14}$$

(e) Degree of Dissociation of Weak Electrolyte : The degree of dissociation of ionisation α, of a weak electrolyte is given by

$$\alpha = \frac{\lambda_v}{\lambda_\infty} = \frac{\text{Equivalent conductance at the given dilution}}{\text{Equivalent conductance at infinite dilution}}$$

λ_r for a weak electrolyte is determined experimentally, while the λ_∞ is calculated with the help of Kohlrausch's law.

(f) Transport Number : See transport number.

CONDUCTANCE IN NON-AQUEOUS SOLVENTS

Various substances (generally in liquid form) other than water which are used for dissolving the other compounds are known as *non-aqueous solvents*. In earlier studies of measurement of conductance, water was generally utilised due to:

(a) its wide liquid range

(b) ease of handling

(c) its plentiful occurrence in nature

(d) its unusual solvating characteristics and

(e) its high dielectric constant

The work on the measurements of conductances or substances in non-aqueous solvents first appeared in 1892.

Later on a large number of papers appeared in 1897 when Cady started his studies on liquid ammonia as a solvent. The conductance of non-aqueous solutions depend upon the following two factors:

(a) Effect of Dielectric Constant : The dielectric constant of a solvent can affect the dissociating power. This effect was realised by Thomson and Nernst and is known as Thomson-Nernst rule. According to this rule,

"The force exerted between the charged particles say e_1 and e_2 is inversely proportional to the dielectric constant (D) of the medium and to the square of the distance (d) between the particular."

$$F = \frac{1}{D}\frac{e_1 e_2}{d^2}$$

From the above relation it follows that E is the amount of energy required to separate two ions of charges e_1 and e_1 which are d cm apart. If the dielectric of the solvent is large the energy required to separate two ions will be smaller and therefore, the ionisation can take place easily.

Thus, formic acid, hydrogen cyanide, nitromethane, ethylene glycol, methyl alcohol, etc., with a high value of dielectric constant should behave as good ionising solvents whereas the solvents with a low dielectric constant such as hydrocarbons and halogenated hydrocarbons should be poorest ionising solvents.

In actual practice, we observe that the above expectations are in agreement with experimental observations which may be seen from the following Table 2.1.

Table 2.1

Solvent	*Dielectric Constant*	*Degree of Ionisation in 2000 litres dilution* $(C_2H_5)_3N$ *Picrate*	$(C_3H_5)_1NH$ *Picrate*	$(C_2H_5).NH_2.HCl$
H_2O	81.1	0.99	0.97	0.97
CH_3OH	35.4	0.94	...	0.92
C_2H_5OH	25.4	0.89	...	0.88
CH_3CN	36.4	0.95	0.90	0.31
$(CH_3)_2CO$	20.7	0.88	0.96	0.10
C_2H_5N	12.4	0.74	0.53	0.04
$CHCl_3$	10.9	0.47	0.005	0.00

In media of low dielectric constant, one observes, that for dilute solutions, the value of λ_r diminishes (in the usual manner) with increasing concentration of the solute. However, a minimum value of λ_r is attained and again there occurs an increase in λ_r as concentration increases. The dielectric constant of a solvent depends upon its dipole moment. Fajan has proposed that the ionisation is large in those solvents which contain a large cation and a small anion.

(*b*) Effect of Viscosity: In 1906, Walden reported the quarternary ammonium salts like $(C_2H_5)_6NI$ can be dissolved in a number of solvents other than water. He also studied the conductance of solutions of those compounds and gave a relation connecting equivalent conductivity of the solution, viscosity and dielectric constant of the medium.

$$\lambda_\infty \eta = \text{constant} \qquad ...(1)$$

where λ_∞ is the equivalent conductance of a solution in non-aqueous solvents and η is the viscosity of this solvent. Equation (1) is known as Walden's rule which may be stated in words as:

"*The product of limiting equivalent conductance of a given solute and viscosity of the pure solvent is constant.*"

The constant in equation (1) is independent of temperature and nature of the solvents and the value of $(C_2H_5)_4NI$ in more than 30 solvents at 0°C is 0.70, *i.e.*,

$$\lambda_\infty \eta = 0.70 \qquad ...(2)$$

The above rule of Walden does not apply to water and glycol as solvents. For water, $\lambda_\infty \eta = 1$ and the value of constant was observed to vary with the nature of the solute. From equation (1) it follows that one can calculate the equivalent conductance of a solute at infinite dilution from the viscosity of the solvent and the value of constant which can be determined.

Limitations of Walden's Rule

(1) In general, Walden's rule is not universally valid. It is not applicable to highly associated solvents and also not to the solvents of high viscosity.

(2) The value of constant in Walden's rule equation (1) is found to vary with variation of temperature.

Another Walden's Relation

Walden also showed that there is a relation among $(\lambda_\infty - \lambda_r)$, viscosity and dielectric constant. His mathematical relation is

$$\lambda_\infty - \lambda_r = \frac{K_e}{D\eta} \qquad ...(3)$$

where K_e is constant which depends only on the nature of the solute and its concentration but not on the nature of the solvent. He reported that the value of K_c is approximately given by

$$K_e \sim 51.5\, C^{0.45} \quad ...(4)$$

where C is the concentration of the solute in gm. equivalent per litre.

Walden also reported that equation (1) may be modified for strong electrolytes in various solvents such as

$$= \frac{41.5}{D} C^{1/2} \quad ...(5)$$

Equation (5) can be used successfully to interpret the conductivity of dilute solution of strong electrolytes in non-aqeuous solvents of relatively high dielectric constants.

For example, CH_3OH(D = 30.3): C_2H_5OH (D = 24.1); $C_6H_5NO_2$ (D = 35.5); NH_3 liquid at – 40°C (D = 23).

Onsagar's Equation

This may be written in the form

$$\lambda_r = \lambda_\infty - (A + B\lambda_\infty \sqrt{C}) \quad ...(6)$$

where A and B are constants dependent only on the nature of the solvent and the temperature. When the above equation was applied to non-aqueous solutions, it was observed to be in full agreement with theoretical requirements for the chlorides, and thiocynates of the alkali metals in methyl alcohol solution. Other electrolytes, such as nitrates, tetralkyl ammonium salts and salts of higher valency, however, exhibit appreciable deviations. These discrepancies become more pronounced for the solvents of law dielectric constants. In order to explain these deviations, equation (6) was modified to:

$$\lambda_r = \alpha[\lambda_\infty - (A + B\lambda_\infty)\sqrt{(\alpha C)}\,] \quad ...(7)$$

Equation (7) may be written as

$$\lambda_r = \alpha \lambda_r' \quad ...(8)$$

where

$$= \lambda_\infty - (A + B\lambda_\infty \sqrt{(\alpha C)}\,) \quad ...(9)$$

From equation (9), it follows that is smaller than λ_r for solvents of low dielectric constants (non-aqueous solvents). From equation (8), it follows that for medium of low dielectric constant the degree of dissociation α is numerically equal to $\lambda_r/$ instead of λ_r/λ_∞ as proposed by Arrhenius.

CONDUCTOMETRIC TITRATIONS

Introduction: These can be applied to all titrations in which there occurs a sharp change in the conductivity at the end point. *The determination of the end point of a titration by means of conductivity measurement is termed as conductometric titration.*

In a conductometric titration the titrant is added from the burette and the conductivity readings corresponding to various increments of titrant are plotted against the volume of titrant. Two curves will be obtained which will intersect each other at a point, called "*end point*" or "*equivalence point*".

When the reaction is not quantitative, there is some curvature in the curve near the end point which may be due to hydrolysis, dissociation of the product or appreciable solubility in case of precipitation reactions.

It should be borne in mind that the resistance, and thus the conductance of an electrolytic solution is profoundly affected by changes in temperature. The resistance decreases by about 1% to 2% for each degree increase in temperature and for this reason it is desirable to carry out a conductometric titration at an approximately constant temperature. If absolute measurements are to be made a constant temperature bath is, of course, necessary.

The concentration of the titrant must be 10 times as the solution being titrated. This is done to keep the volume charge small. If it cannot be done, a correction to the reading must be applied, *i.e.*,

Actual conductivity = $\left(\dfrac{v + V}{V}\right)$ × observed conductivity where v is the volume of titrant or reagent added and V is the original volume.

Advantages of Conductometric Titrations

Among the many advantages that conductometric titrations possess over the ordinary volumetric methods, we may mention the following:

(i) They are particularly useful in the case of coloured or turbid liquids, where ordinary indicators cannot function.

(ii) They can be employed for the analysis of dilute solutions and also for very weak acids.

(iii) No special precautions are necessary as the titration approaches the equivalence point, since the latter is found graphically.

(iv) It is not necessary to measure the actual conductance, as we may use any quantity that is proportional to it, *e.g.*, the reading on a ?Wheatstone bridge or resistance box. The same can be directly plotted against the volume of the alkali solution used up.

Apparatus for Conductometric Titrations

(a) *Measuring Circuits:* In order to prevent concentration changes due to reaction at the electrodes, the resistance of solution is generally measured with alternating currents and frequencies from 60 to 10000 c/sec have been used for ordinary conductance measurements, although, work at high frequencies finds may applications. Most of the circuits used are of the Wheatstone bridge type.

Stock has published an excellent critical review of the apparatus used for conductometric titrations, specially for microtitrations.

(b) *Commercial Apparatus :* Fairly simple commercially available equipment is accurate enough for most conductometric titrations. Its main advantages are that it is convenient and can be set up and used without any knowledge of electricity or electronics.

(c) *Electrodes and Cells :* A number of titration cells have been recommended of which those of Britton and Kano are the most practical. The simplest titration cell consists of a dipping electrode, a beaker of the appropriate size and a mechanical stirring device. For precipitation titrations, the electrodes are mounted vertically to prevent excessive amounts of precipitate from collecting n the electrode surfaces.

Because a large effective surface area minimises polarisation, most conductivity electrodes are plated with a layer of platinum black. The platinisation of electrodes also tends to decreases the cell capacitance and allows the establishment of a sharp balance point. On the other hand, platinised electrodes have a greater tendency to adsorb substances from solution. Difficulties are encountered with dilute solution or solution containing surface active agents; bright platinum of lightly platinised electrodes are used in such cases.

(d) *Special Methods :* Although conductance is generally measured by means of a.c. apparatus, it has been shown by a number of workers that d.c. can be used provided that the cell has two unpolarisable electrodes, such as Ag/AgCl electrode. Eastmen has showed that direct current method is capable of yielding results that agree to 10% with those obtained with the usual a.c. method.

Differential Conductance Titration

Such titration have been investigated by Daval and Duval. They used two cells connected in two branches of a Wheatstone bridge, both cells contained identical sample solution and both were titrated with the same reagent, but one cell always had 0.05 to 0.1 ml more titranti in it than the other.

The titration curves for each cell were plotted on graph and their point of intersection gave one end point.

Types of Conductometric Titrations

1. Acid-base Titrations

(a) *Conductometric titrations of strong acids and strong bases: A* typical strong acid/strong base titration may be exemplified by:

$$H^+ + Cl^- + Na^+ + OH^- \rightarrow Na^+ + Cl^- + H_2O$$

The net result being the replacement of hydrogen ions by sodium ions. As we know that equivalent ionic conductance of hydrogenise is much greater than that of the sodium ion, so that the conductance of the solution at the beginning of titration is large but decreases up to the equivalence point; this is shown in Fig. 2.5. At the equivalence point, the conductance is equal to that of a pure solution of sodium chloride and is at a minimum. Beyond the equivalence point, the conductance increases in direct proportion to the excess of sodium hydroxide added.

The slope of the first branch of the titration curve, and thus sharpness with which the end point may be established, depends on differences between the equivalent ionic conductance of the hydrogen ion and the cation of the base used, in this case the sodium ion. As the equivalent ionic conductance of the hydrogen ion is much greater than that of any other cation, strong acids can almost always be successfully titrated to a conductometric end point.

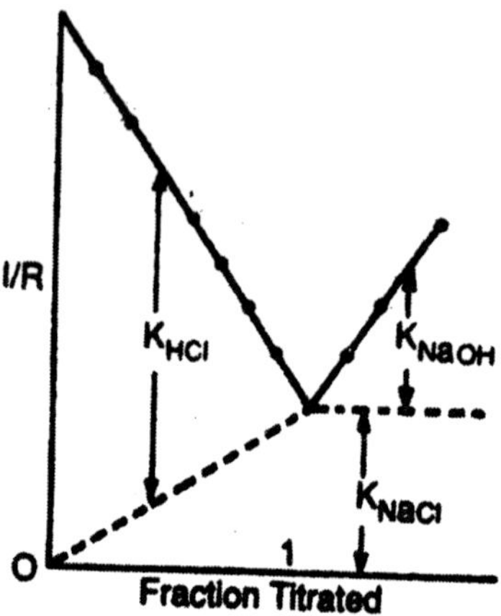

Fig. 2.5

On *special important advantage of* conductometric titration is that the *relative change in conductance is almost independent of the concentration of strong acid titrated.* Thus dilute solution can be titrated with almost the same accuracy as concentrated solution. The titration curve is displaced along the 1/R axis, but the sharpness is not generally affected. In this respect, conductometric acid-base titrations differ from potentiometric titrations.

Although the titration involving strong acids with strong bases was among the first conductometric titrations attempted, the conductometric method is not much used for such determinations, because a large number of indicator methods are available, even coloured solutions are now mostly titrated potentiometrically with a glass electrode.

(b) *Strong acid with a weak base:* Let us consider the titration of HCl (strong acid) with NH_4OH (weak base). When ammonium hydroxide is added to HCl, the conductivity decreases first due to the replacement of the fast moving H^+ ions by slow moving NH_4^+ ions.

$$H^+Cl^- + NH_4OH \rightarrow NH_4^+ + Cl- + H_2O.$$

After the end point is passed, further addition of NH_4OH does not change the conductance because NH_4OH, a weakly ionised electrolyte, has a very small conductivity compared with that of the acid or salt (Fig. 2.6 and 2.7).

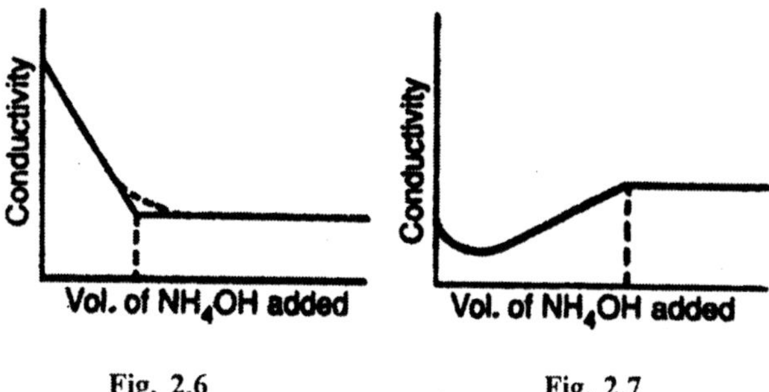

Fig. 2.6 **Fig. 2.7**

(c) *Weak acid with a strong base:* Let us consider the titration of the weak acid like CH_3COOH with strong base NaOH. When a small amount of NaOH is added to CH_3COOH, the conductivity decreases initially and then increases with the further addition of NaOH.

$$CH_3COO^- + H^+ + Na^+ + OH^- \rightarrow CH3COO^- + Na^+ + H_2O$$

When the neutralisation of acid is complete, further addition of alkali introduces excess of OH^- ions. The conductance of the solution, therefore, begins to increase more sharply.

(d) *Conductometric titration of weak acid and weak bases:* When weak acids are titrated with strong bases, curves such as 1 and 2 in (Fig. 2.8) are obtained. Before the equivalence point the conductivity is governed by the extent of ionisation (and by the concentration) the conductivity first decreases as the anion formed suppresses the ionisation, passes through a minimum, and then increases upon the equivalence point because of the conversion of non-conductivity weak acid into its conducting salt.

Extensive information about the effect of the dissociation constant on the titration curves of both weak acids and bases has been compiled by Britton. Bruni and Eastman have carried out mathematical analysis of the titration curves of weak electrolytes.

Near the equivalence point, the graphs of such titrations are often curved because the incompleteness of reaction allows extra hydroxide ions (or hydrogen ions if a weak base is being titrated) to be present, thus increasing the conductance; the more incomplete the reaction, the more is curvature.

Conductometric titrations utilise measurements far enough removed from the end point for a fair amount of curvature near the end point to be tolerated. In fact many titration reaction that are too incomplete for potentiometric end point detection can be used conductometrically. For instance, boric acid can be titrated conductometrically with sodium hydroxide whereas the pH values change at end point is not large enough for potentiometric or indicator methods to be used. Weak acids with dissociation constants as small as 10^{-10} can readily be determined, especially in fairly concentrated solution. (Fig. 2.9).

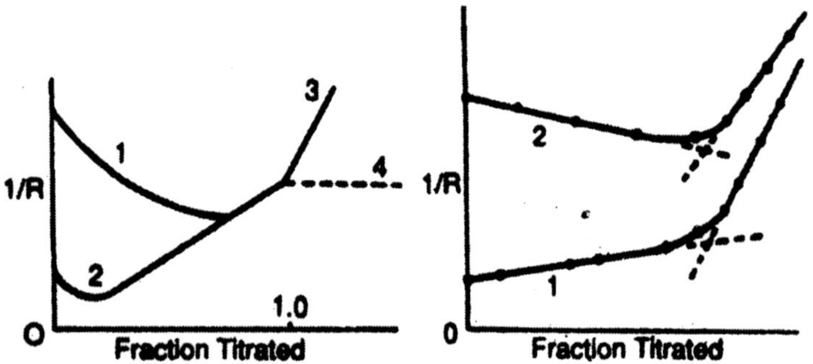

Fig. 2.8 : Conductometric titrations of weak acids and weak bases.

Fig. 2.9 : Replacement conductometric titrations.

End-point Calculation Method: After the conductometric titration curve has been plotted in usual way, estimate the end point by drawing two straight lines (the usual graphical method). Arbitrarily choose values for f_1, f_2, and f_3 such that f_1 and f_2 are before the end point and f_3 is after the end point and calculate f_4 by using the appropriate formula.

For a weak acid/weak base titration:

$$\frac{f_2 + f_1 - 2f_1f_2}{(1 - f_2)(1 - f_1)} = \frac{f_4 + f_3 - 2}{(f_2 - 1)(f_4 - 1)}$$

For a strong acid/weak base titration,

$$\frac{1 - f_2f_1}{(1 - f_2)(1 - f_1)} = \frac{f_4 + f_3 - 2}{(f_2 - 1)(f_4 - 1)}$$

Draw straight lines through f_1 and f_2 and through f_3 and f_4. The intersection of these lines is the new end point. Repeat the process using the new estimated end point. The new end points found in this way will converge rapidly so long as the first estimated end point was not very greatly in error.

2. Replacement Titration

When a strong acid reacts with the sodium or potassium salts of week acids, the weakest acid will be replaced first. Such a replacement reaction is exemplified by

$$HCl + CH_3COONa \rightarrow NaCl + C_3H_3COOH$$

The investigation of the replacement of weak acid from its salt was first investigated by Dutoit.

Analogous reaction for the replacement of the weak bases from their salts by strong bases can, of course, be cited.

Kolthoff has studied this type of reaction extensively, and curves for two of Kolthoff's system are shown in (Fig. 2.9). Curve I shows the increase in conductivity during the replacement of acetic acid by hydrochloric acid to form sodium chloride; the increase is due to the fact that the mobility of chloride ion is greater than that of acetate ion. After the equivalence point, the conductively increases more rapidly, owing to the additions of excess of hydrochloric acid.

Kolthoff has calculated that accurate titrations with 0.01 M solutions are possible if the dissociation constant of the liberated acid is less than 5×10^{-5}; Lapshin has, however, titrated the salt of weak acids and bases whose dissociation constants are greater than 10^{-4}.

The main application of replacement titrations is in the determination of alkaloids. As alkaloids are weak bases that are often isolable in water, indicator methods do not work well.

3. Precipitation Titrations

Precipitation titrations can be effectively followed by conductometric methods, though not so well as acid-base titrations which are characterised by sharp breaks because both the hydrogen ion and hydroxyl ion have very high equivalent ionic conductances. In precipitation titrations, one pair of ions is substituted for another and as one experimenter has a choice of reagents, good results can usually be obtained. If a cation is to be precipitated, titrant whose cation has the smallest possible mobility is selected and, if an anion is to be precipitated, a titrant whose anion has as small a mobility as possible, in this way the maximum possible change in conductance during a titration is assured.

Let us consider the titration between $AgNO_3$ and KCl. The reaction involved may be represented as

$$K^+ + Cl^- + Ag^+ NO_3^- \rightarrow K^+ + NO_3^- + AgCl.$$

In the early stages of the titration , addition of silver nitrate, the conductance does not change very much because the Cl^- ions are replaced by NO_3^- ions; both have almost same ionic conductances. After the end point is passed, the excess of the added salt causes a sharp increase in conductance. The end point of the titration can thus be determined (Fig. 2.10).

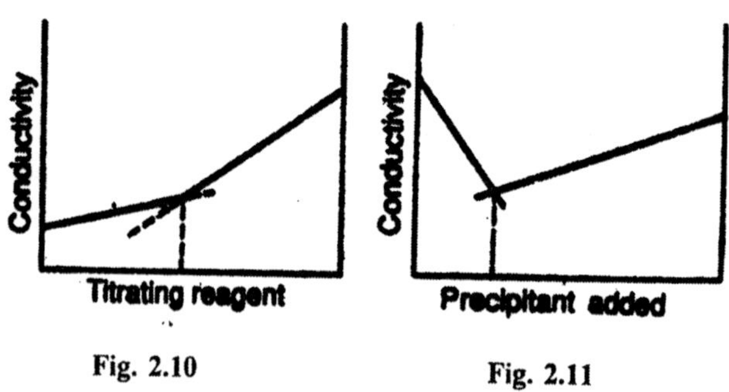

Fig. 2.10 Fig. 2.11

If both the products of the reaction are sparingly soluble, the curve obtained takes up the form as shown in Fig. 2.11., An example of this is the reaction between $MgSO_4$ and $Ba(OH)_2$. The reaction is as follows:

$$Mg^{2+}SO_4^{2-} + Ba_2^+ + 2OH^- ® Mg(OH)_2 + BaSO_4$$

In this type of titrations, conductivity decreases in the beginning but increases after the end point has passed.

Precipitate titrations cannot be performed very accurately. The main reasons for this are as follows:

(i) Due to slow separation of the precipitate.

(ii) Due to the removal of titrated solute by adsorption on precipitate.

Some efforts have been made to eliminate the errors that are due to adsorption, slowness of precipitation and electrode fouling, and are common to a number of conductometric precipitation titrations carrying out the titrations at elevated temperature, *e.g.*, in boiling solution, this apparently gives accurate results but there is insufficient evidence to judge these methods critically.

4. Complexometric Titrations

Conductivity has long been used in the study of complex formation. Werner, for example, supported many of the postulates of his coordination theory with conductivity measurements. Under controlled conditions conductivity measurements can be used to show the type of electrolyte, *e.g.* univalent, a complex salt gives in solution.

Conductometric titrations have also been applied to the study of complex salts; for instance, Job titrated roseo-cobaltic sulphate $[Co(NH_3)_5H_2O]_2(SO_4)_3$ with barium hydroxide, finding two breaks in the titrations curve.

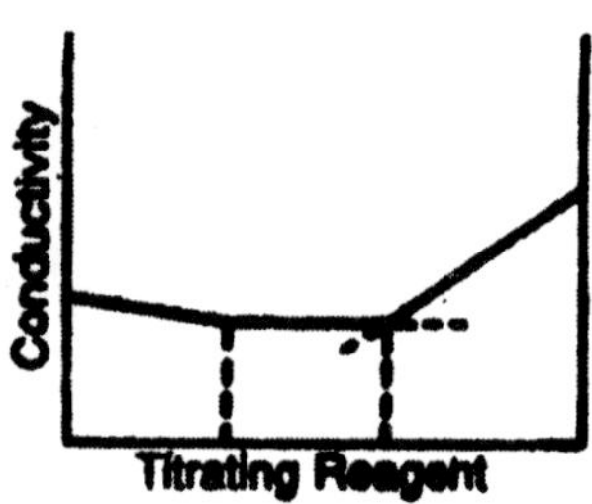

Fig. 2.12

Let us illustrate the complexometric titration by titrating KCl with $Hg(ClO_4)_2$. This is shown in Fig. 2.12. This curve shows two breaks one due to the formation of $HgCl_4^{2-}$ and the other one at the end of the reaction due to the formation of K_2HgCl_4.

$$Hg(ClO_4)_2 + 4KCl \rightarrow HgCl_4^{2-} + 2K^+ + 2KClO_4$$

$$HgCl_4^{2-} + 2K^+ \rightarrow K2HgCl_4$$

THEORY OF ARRHENEOUS

Arrhenius (1887) successfully put forward a theory of which Clausius's theory was a fore runner. The postulate of this theory known as theory of electrolytic dissociation are as follows:

(a) When dissolve in water, every electrolytic yields two types of charged particles called ions. Those carrying a positive charge are known as cations, while those carrying negative charge are known as anions.

 However, according to the modern view, a solid electrolyte is a combination of charged ions, bound by electrostatic lines of forces. When dissolved in solution, these lines of forces are cut and the ions separate.

$$\underset{}{AB} \rightarrow \underset{\text{caution}}{A^+} + \underset{\text{anion}}{B^-} \qquad \text{(old view)}$$

$$A + B^- \rightarrow A^+ + B^- \qquad \text{(Modern view)}$$

(b) Ions present in solution recombine to form unionised electrolyte and there is a state of equilibrium to which the law of mass action can be applied.

Therefore, $A^+B^- \rightleftharpoons A^+ + B^-$

According to the law of mass action,

$$K = \frac{[A^+][B^-]}{[A^+B^-]}$$

where K is known as ionisation or dissociation constant. The fraction of the total number of molecules present as ions in solution is known as degree of dissociation.

(c) Each ion that is formed as a result of electrolytic dissociation produce the same effect on colligative properties (like osmotic pressure, lowering of vapour pressure, elevation of boiling point etc.) as an undissociation molecule. In simpler words, each ions behaves osmotically as a molecule. For example, an electrolyte yielding there ions per molecule will exert thrice the normal effect of the electrolyte, had the electrolyte not dissociated.

(d) On the passage of electricity, cautions and anions move towards the cathode and anode, respectively. This movement of ions towards the electrodes constitutes the electric current through the solution. This explains the process of electrolysis.

(e) The properties of the electrolytes in solution are the properties of the ions. All cobalt salt solution are pink, because each solution contains cobalt ions which are pink in colour. Similarly, all chromium salt solution are green, as chromium ions are green.

Facts explained by Arrhenius theory-Arrhenius theory could successfully explain all the phenomena concerning electrolytes. It explains the anomaly of the electrolytes towards the van't Hoff theory of dilute solutions. Besides these, the degree of dissociation of electrolytes calculated on the basis of this theory conform well with the values obtained by the other methods. All these facts led to the universal acceptance of this theory.

(a) *Explanation of electrolysis :* On assuming the presence of ions in an electrolytic solution, the conduction of electricity through them can be explained. When a difference of potential is applied across two metal electrodes dipping in the solution of an electrolyte, the cations and anions move gain of electrons, as shown below:

$$A^+B^- \rightleftharpoons A^+ + B^-$$

$$B^- \rightarrow B + e$$

$$A^+ + e^- \rightarrow A.$$

The electron liberated at the anode travels from the anode the cathode outside the cell, through the connecting wire. This electron is now available for the conversion of A^+ ion at the cathode. The circuit is completed and the electricity flows through the solution. These ions after losing their charges are liberated as atoms as molecules. This explains the chemical reaction caused due to the passage of electricity. This process is known as electrolysis. Solutions of non-electrolytes do not produce ions, hence they are unable to conduct electricity.

(b) Variation of Equivalent Conductivity with Dilution : As the degree of dissociation of an electrolyte increases on dilution, therefore, we can easily explain the increase in equivalent conductivity on dilution, as it depends upon the number of ions. At a certain dilution known as infinite dilution, equivalent conductivity acquires a maximum value.

This can be explained on the basis that at a certain dilution, the dissociation of an electrolytes is complete. In other word, the number of ions produced is maximum and does not increase further on dilution. Hence, a constant maximum value of equivalent conductivity is reached at infinite dilution.

(c) Heat of Neutralisation : We know that the heat of neutralisation of strong acids and strong bases is about 13.7 k cals. This fact can also be explained on the basis of electrolytic dissociation. As strong acids and bases are almost completely ionised, the process of neutralisation can be represented as:

$$\underbrace{H^+ + A^-}_{\text{from acid}} + \underbrace{B^+ + OH^-}_{\text{from base}} \rightarrow \underbrace{A^- + B^+}_{\text{from electrolyte}} + H_2O + 13.7 \text{ k cals}$$

or $H^+ + OH^- \rightarrow H_2O + 13.7$ k cals

Thus, the heat of neutralisation is simply the heat of formation of water from its ions and since this reaction is involved in all neutralisation processes, the heat evolved is always the same.

In the case of a weak acid or weak bases or both the heat neutralisation is less than 13.7 k cals. This is because the weak substance is dissociated for complete neutralisation and the dissociation process requires energy and is therefore, endothermic. The energy is taken from the system itself. Thus, the net heat of neutralisation envolved is less than 13.7 k cals.

(d) *Common ion Properties :* We know that solutions of electrolytes with a common ion show similar physical and chemical properties. Thus, we know that Cu^{+2}, MnO_4^-, CO^{+2} ions are blue, purple, pink, respectively. Thus, any solution of the salt containing these ions will possess the same colour and will also have the same absorption spectrum. Similarly, all cyanides are poisonous, as CN^- ions are poisonous in nature.

(e) *Ionic Reaction :* Evidence for the existence of ions in aqueous solutions can be had by a number of reactions in inorganic chemistry. In qualitative analysis, the radicals are tested as long as they are in the form of ions. Thus, Ag^+ ion can be tested if it is present in the solution, by adding HCl solution.

$$AgNO_3 \rightarrow Ag^+ + NO_3^-$$

$$HCl \rightarrow H^+ + Cl^-$$

$$Ag^+ + Cl^- \rightarrow AgCl \text{ (white ppt.)}$$

If silver ions do no exist in the solution, we will not get a white precipitate on adding HCl solution. Thus, when one of the ions combines with a molecule to form a complex ion, the solution does not give the test of that ion. If silver exists as $NaAg(CN)_2$, then no test no test for Ag^+ ions is obtained as the compound gives only Na^+ and $Ag(CN)_2^-$ ions on dissociation.

Similarly, an acid which gives all the tests for hydrogen ion when dissolved in water, does not give the same tests when dissolved in organic solvents. This is because the acid does not ionise in organic solvents to give the hydrogen ions.

(f) *Colligative Properties of Electrolytic Solutions in Water* : On the basis of Arrhenius theory, one can explain the anomalous behaviour of electrolytes in aqueous solution. van't Hoff showed that colligative properties like osmotic pressure, lowering of vapour pressure etc. for electrolytes were higher than those for non-electrolytes. It was seen that the elevation in boiling point due to the presence of one gram mole of an electrolyte dissolved in 1000 gm of water for a non electrolytes (like cane sugar, urea etc.) was 0.52°, while that for electrolytes like NaCl, HNO_3, KCl was found of approach a value of 1.34° (twice that of 0.52°) at large dilutions. Similarly, for electrolytes like Na_2SO_4, $BaCl_2$ etc. the value was about 1.56° (thrice that of 0.52°) at large dilutions. In order to explain this anomaly, van't Hoff introduced a new factor known as van't Hoff factor (i), which is defined by,

$$i = \frac{\text{Observed elevation of boiling point}}{\text{Theoretical elevation of boiling point}}$$

In general, we can say that

$$i = \frac{\text{Observed colligative property of an electrolyte}}{\text{Theoretical colligative property of an electrolyte}}$$

Factors influencing ionisation—

(a) *Nature of Solute* : Strong electrolytes are completely ionised in solution and such electrolytes are obtained by the interaction between strong acids and weak bases are not so strongly ionised.

(b) *Nature of Solvent* : The solvent cuts the lines of forces binding the two ions and hence breaks them. Greater the capacity of the solvent, greater will be the ionisation of the electrolyte.

This tendency of solvent is measured in terms of dielectric constant, which is defined as, 'the capacity to weaken the forces of attraction, between the electrical charges immersed in that solvent.'

(c) *Temperature* : The degree of ionisation of an electrolyte in solution is proportional to the temperature. At higher temperatures, the increased molecular velocities cut the forces of attraction between the ions to a greater extent and thus cause increased ionisation.

(d) *Concentration* : The degree of ionisation of an electrolytes is inversely proportional to the concentration. This is due to the fact that at higher dilution, the solvent molecules make more molecules of the solute to break in to ions.

On substituting the value of λ_v, λ_∞ and V, the value of K_a can be calculated. At 20°C the value of K_a for acetic acid is found to be 1.8×10^{-5}.

KOHLRAUSCH'S LAW

The values of equivalent conductances at infinite dilution for certain pairs of electrolytes at 18°C are given in Table 2.2

Table 2.2

Pairs of electrolytes with the same anion	λ_0 *Ohm*$^{-1}$	*Difference Ohm*$^{-1}$	*Pairs of electrolytes with the same cation*	λ_0 *Ohm*$^{-1}$	*Difference Ohm*$^{-1}$
KF	111.2		KCl	130.0	
		21.1			18.8
NaF	90.1		KF	111.2	
KCl	130.0		NaCl	108.9	
		21.1			18.8
NaCl	108.9		NaF	90.1	

From the above table it follows that

(i) When potassium is replaced by sodium in any of the electrolytes, there is always a difference of 21.1 ohm^{-1} in equivalent conductances at infinite dilution irrespective of the nature of the anion with which they are associated.

(ii) When chlorine is replaced by fluorine, there is always a difference of 18.8 ohm^{-1} in equivalent conductances at infinite dilution irrespective of the nature of the cation with which they are associated. The above observations led Kohlrausch (1875) to formulate a law known after his name as Kolhrausch's law. This law can be stated as follows:

"*The equivalent conductance at infinite dilution for any electrolyte is the sum of equivalent conductances for the cation and anion at infinite dilution.*"

Mathematically, Kohlrausch's law can be put as

$$\lambda_\infty = \lambda_\infty^+ + \lambda_\infty^-$$

where λ_∞^+ and λ_∞^- are the equivalent conductances of the cation and anion respectively at infinite dilution.

Applications of Kohrausch's Law

(i) *Calculation of Equivalent Conductance at Infinite Dilution for Weak Electrolytes:* Equivalent conductance at infinite dilution can be calculated by a graphical method. But the graphical method is not applicable to weak electrolytes as they are very slightly ionized. But Kohlrausch's law can be used to determine the value of λ_∞ for weak electrolytes. For example, equivalent conductance of acetic acid at infinite dilution can be calculated from the equivalent conductances at infinite dilution of hydrochloric acid, sodium acetate and sodium chloride as follows:

$$\lambda_\infty HCl = \lambda_\infty H^+ + \lambda_\infty Cl^- \quad \text{(say x)} \quad ...(1)$$

$$\lambda_\infty CH_3COONa = \lambda_\infty Na+ + \lambda_\infty CH_3COO^- \quad \text{(say y)} \quad ...(2)$$

$$\lambda_\infty NaCl = \lambda_\infty Na^+ + \lambda_\infty Cl^- \quad \text{(say z)} \quad ...(3)$$

Add equations (1) and (2) and subtract (3), we get

$$\lambda_\infty HCl + \lambda_\infty CH_3COONa — \lambda_\infty NaCl$$

$$= \lambda_\infty H^+ + \lambda_\infty Cl^- + \lambda_\infty Na^+ + \lambda_\infty CH_3COO^- — \lambda_\infty Na^+ - \lambda_\infty Cl^-$$

$$x + y - z = \lambda_\infty H^+ + \lambda_\infty CH_3COO^-$$

$$= \lambda_\infty CH_3COOH$$

or $\lambda_\infty CH_3COOH = x + y - z.$

Since the values of x, y and z are experimentally determined, we can easily calculate the value of λ_∞ for acetic acid.

(ii) *Calculation of Degree of Dissociation of Weak Electrolytes :* The value of λ_0 helps to calculate the degree of dissociation and the dissociation constants of weak acids and weak bases.

For example, the degree of dissociation α_c of a weak electrolyte at the concentration c equivalents per litre may be given by the following relation:

$$\alpha_c = \frac{\lambda_c}{\lambda_0}$$

where λ_c is the equivalent conductance of electrolyte a$^+$ concentration c and λ_0 is the equivalent conductance of the same electrolyte at infinite dilution. Hence, measurement of λ_c permits evaluation of α_c if λ_0 known.

(iii) *Determination of Solubility of Springly Soluble Salts* : For ordinary purposes ionic salts like AgCl, $BaSO_4$, $CaCO_3$, Ag_2CrO_4, $PbSO_4$, PbS, $Fe(OH)_3$, etc. are regarded as insoluble, but they do have a *very small* but *definite* solubility in water. The solubility of such sparingly soluble salts is obtained by determining the specific conductivity (K) of a saturated salt solution. Since only a very small amount of salt is present in solution. The equivalent conductivity at such high dilution can practically be taken as λ_∞, *i.e.*,

For very sparingly soluble salts; $\lambda_r = \lambda_\infty = K.V$

where V is the volume in cm3 containing 1 g equivalent of salt.

The value of λ_∞ can be calculated with the help of Kohlrausch's law *i.e.*,

$$\lambda_\infty = \lambda_+ + \lambda_-$$

Thus, knowing the values of K and λ_∞, the value of V can be calculated:

$$V = \frac{\lambda_r}{K} \simeq \frac{\lambda_\infty}{K} = \frac{\lambda_+ + \lambda_-}{K}$$

But V cm^3 of saturated solution contains = 1 g eqt. of salt

$\therefore$ 1000 cm^3 (or litre) = solution contains = 1000/V g eqt. of salt

Hence, solubility of salt,

$$S = \frac{1000}{V} \text{ g equivalent/litre}$$

$$= \frac{1000\,E}{V} \text{ g/litre} = \frac{1000 \times E \times K}{(\lambda_+ + \lambda_-)} \text{ g/litre}$$

where E is the equivalent weight of salt.

DISSOCIATION CONSTANTS OF PHOSPHORIC ACID

Phosphoric acid (H_3PO_4) is a tribasic acid and it ionises in three stages and consequently it has *three* dissociation constants, one each for each stage. Thus

$$H_3PO_4 \overset{K_1}{\rightleftharpoons} H^+ + H_2PO_4^- \qquad \text{...(i)}$$

$$H_2PO_4^- \underset{}{\overset{K_2}{\rightleftharpoons}} H^+ + HPO_4^{2-} \quad ...(ii)$$

$$HPO_4^{2-} \underset{}{\overset{K_3}{\rightleftharpoons}} H^+ + PO_4^{3-} \quad ...(iii)$$

$$\because \quad K_1 = \frac{[H^+][H_2PO_4^-]}{[HPO_4^{3-}]};$$

$$K_2 = \frac{[H^+][HPO_4^{2-}]}{[H_2PO_4^-]}$$

and $$K_3 = \frac{[H^+][PO_4^{3-}]}{[HPO_4^{3-}]}.$$

The dissociation of acids (i), (ii) and (iii) into H^+ ions becomes more and more difficult due to increasing negative charge on the acid molecule. Hence $K_1 > K_2 > K_3$. The experimental values are:

$$K_1 = 1.1 \times 10^{-2}; K_2 = 2 \times 10^{-7},$$

and $$K_3 = 3.6 \times 10^{-12}.$$

EVIDENCES IN FAVOUR OF THE THEORY

(i) *Validity of Ohm's Law (E = RC) in Electrolytic Solutions and Hence Explanation of Electrolysis:* The electrolysis starts even when a current of very small e.m.f. is applied. This can happen *only if ions are already present before electrolysis.* This fact is also supported by the fact that electrolytes obey Ohm's law which go to show that *whole of the current passed is utilized only in overcoming the resistance of the ions to move* and no part of it is utilized in breaking the molecules into ions. Thus, on passing electric current during electrolysis, the ions already present in solution move towards opposite electrodes and on reaching the respective electrodes they are discharged and are either deposited there or evolved as a gas or undergo secondary changes.

(ii) Heat of Neutralization: Whenever one gram equivalent of a strong acid (like HCl, H_2SO_4, HNO_3, etc.) in solution is neutralized with 1 gram equivalent of a strong base (like NaOH, KOH, etc.) in solution, the same quantity of heat, viz. 13,700 calories is produced. This is explained on the basis of ionic theory that strong acid and strong bases are nearly completely ionized in solution and the heat of neutralization is nothing but the heat

of formation of water molecules from H^+ ions (of acid) and OH^- ions (of base). Hence, this value is always the same irrespective of the nature of the strong acid and base used.

$$(H^+ + Cl^-) + (Na^+ + OH^-) = (Na^+ + Cl^-) + H_2O + 13{,}700 \text{ cals}$$

or $$H^+ + OH^- = H_2O + 13{,}700 \text{ cals.}$$

However, the heat of neutralization of a weak acid or weak base is less. For example, when one 1 gram equivalent of acetic acid is neutralized by 1 gram equivalent of NaOH, 13,400 cals. of heat are evolved. This difference (13,700 – 13,400 = 300 cals) is nothing but the *heat of dissociation of weak acid, i.e.,* the heat required to dissociate the weak acid, acetic acid, which is not fully dissociated.

(iii) *Presence of Ions in the Solid Salts:* X-ray diffraction studies have shown that ions are also present in solid salts. Thus, a crystal of sodium chloride does not contain NaCl units, but it consists of each Cl^- ion surrounded by six Na^+ ions and each Na^+ ion surrounded by six Cl^- ions. *The solid crystal is not a good conductor, because the crystal lattice restricts the movement of ions.* However, when such a crystal is dissolved in a suitable solvent, the attractive forces between the ions weaken and the crystals fall apart into free Na^+ and Cl^-, which can move freely under the influence of electricity and hence the salt solution becomes good conductor.

(iv) *Colour of Solution :* Certain compounds possess colour in solution. *The colour of solution is due to the presence of coloured ions.* Thus, Cu^{2+} ions are blue and any solution of its salt like $Cu(NO_3)_2$ or $CuSO_4$ will always show blue colour on *sufficient dilution.* Similarly, CO^{2+} ions and Fe^{2+} ions are pink and yellow coloured respectively and their presence impart these colours to solution.

(v) *Ionic Reactions:* The reactions of electrovalent compounds are nothing but the reactions of ions in solution. Thus, the formation of white precipitate of AgCl, when NaCl solution is reacted with $AgNO_3$ solution is nothing but the reaction of Ag^+ ions and Cl^- ions.

$$(Na^+ + Cl^-) + (Ag^+ + NO_3^-) = (Na^+ + NO_3^-) + AgCl\downarrow$$

or $$Cl^- + Ag^+ \rightarrow AgCl\downarrow$$

A solution of $FeCl_3$ gives a brown precipitate of ferric hydroxide; while a solution of $K_3Fe(CN)_6$ does not give a brown precipitate on addition of NH_4OH, because the latter does not yield Fe^{3+} ions which may combine with OH^- ions to form precipitate of $Fe(OH)_3$.

(vi) *Strength of Electrolytes:* On the basis of extent of ionization the *compounds such as* HCl, NaOH, NaCl, *etc., which are almost fully ionized in water, are called strong electrolytes; while those (like* CH_3COOH, NH_4OH), *which have low degree of ionization, are called weak electrolytes* However, these terms are relative as the degree of ionization varies in different solvents. Thus, HCl is a strong electrolyte in aqueous solution, but it is very weak electrolyte in non-aqueous solvents like CCl_4. On the other hand, NH_4OH is a weak electrolyte in aqueous solution, but will be strong electrolyte in acetic acid medium.

(vii) *Colligative Properties:* Colligative properties like osmotic pressure, depression in freezing point, elevation in boiling point and lowering of vapour pressure are a function of the number of particles (molecule as well as ions) of the solute in solution. Thus, a decimolar solution of urea has been found to extent the same osmotic pressure as a decimolar solution of sugar. On the other hand, a decimolar solution f NaCl under the same conditions gives approximately the double value of osmotic pressure This clearly shows that the *number of particles present in HCl solution is approximately double the number of particles present in solution of urea or sugar*. This is explained by the fact that urea and sugar, being non-electrolyte, do not dissociate; while HCl, being an electrolyte, undergoes ionization to give approximately double number of particles (HO^+, Cl^- ions).

Similarly, the abnormal values of depression in freezing point, elevation in boiling point in case of electrolytes can only be explained on the basis of theory of ionization.

(viii) *Heat of Precipitation:* The heat of precipitation of AgCl formed by the addition of 1 g equivalent of any soluble chloride to 1 g equivalent of Ag salt solution is always the same. This is explained on the basis of ionic theory according to which the heat of precipitation of AgCl is nothing but the heat of combination of Ag^+ ions and Cl^- ions to give AgCl precipitate

$$Ag^+ + Cl^- \rightarrow AgCl + x \text{ cals.}$$

(ix) *Increase of Conductivity on Dilution:* The conductivity of a solution is due to the presence of ions formed by dissociation. As we know, with the increase of dilution, the degree of ionization of electrolyte and the total number of ions in solution increases and heat conductivity also increases.

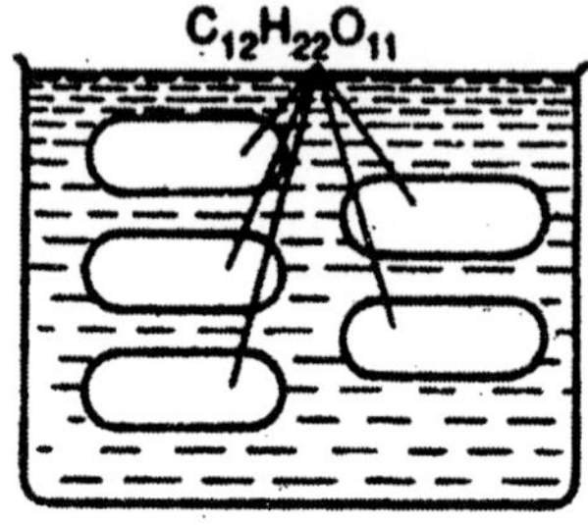

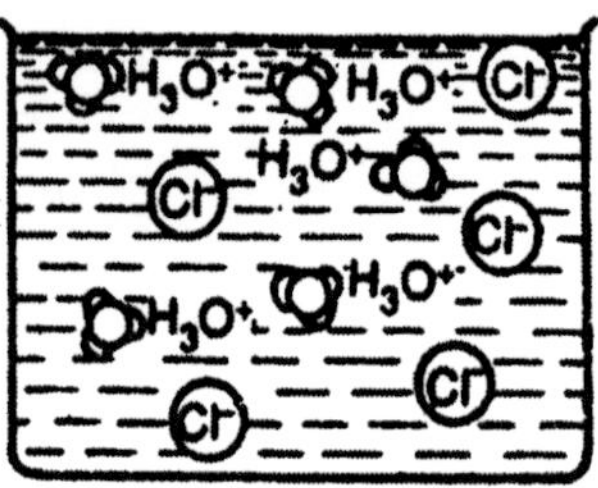

(a) Sugar Molecules produce only 5 particles in solution

(b) Five HCl Molecules produce 10 particles when dissolved

Fig. 2.13 : Colligative effect on electrolytes and non-electrolytes.

(x) *Ostwald's Dilution Law:* The Ostwald's dilution law is based on the theory organization applied to weak electrolytes. The validity of Ostwald's dilution formula in weak electrolytes is most reliable proof of the dissociation fo molecules into ions.

(xi) When a tube containing NaI is whirlled very rapidly, the farthest end is found be + very charged. This is explained on the basis of ionic theory that NaI is present more as Na^+ and I^- ions and the lighter Na^+ ions move more rapidly to the ends as compared heavy I^- ions.

LIMITATIONS OF ARRHENIUS THEORY

Although the Arrhenius theory was quite successful in explaining the behaviour of wear electrolytes, yet it could not explain the various observed facts about the strong electrolytes some important limitations of this theory are as follows:

(i) Ostwald's dilution law which is based on Arrhenius theory fails completely where applied to strong electrolytes.

(ii) As strong electrolytes (KCl, NaOH, etc.) conduct electricity in molten states also, it means that their ionisation must have taken place even in the absence of water.

(iii) Arrhenius theory simply assumes the existence of ions in solution. It does not explain how and why ions are formed in the solution.

(iv) Arrhenius theory simply assumes the independent existence of ions in solution but does not take into account that the mobility of ions can be changed by electrostatic forces between oppositely charged ions lying side by side, particularly in solutions of moderate concentrations.

(v) It was observed that while there was good agreement between the values of α obtained by conductivity measurements and also from the study of colligative properties such as osmotic pressure, etc., in the case of univalent electrolytes like HCl, KCl, etc., but there were marked discrepancies in the value of bivalent electrolytes like $MgSO_4$, $CuSO_4$, etc.

(vi) The heat of neutralization of HCl and NaOH was found to be independent of concentration but their degree of dissociation as measured by conductivity measurements or study of colligative properties was found to decrease with increase in concentration. Thus, the heat of neutralisation should also decrease with increase in concentration.

(vii) Arrhenius theory could not explain the variation of transport number with concentration.

(viii) The values of equilibrium constant of strong electrolytes were found to vary with the concentration of the electrolytes.

STRONG ELECTROLYTES THEORY

In order to explain the above mentioned anomalies, the various theories were put forward from time to time but none of these could explain the behaviour of strong electrolytes. However, the accepted theory is that of Debye and Huckel (1923). This theory was later developed by Onsagar on the basis of interionic concept.

Debye-Huckel Theory

The failure of Oswald's dilution law in case of strong electrolytes has been explained satisfactorily both qualitatively and quantitatively by Debye-Huckel (1923). *This theory is based upon the following assumptions:*

(i) *All strong electrolytes are completely ionised in all dilutions i.e., their degree of ionisation is 100%.*

(ii) *The ratio l_v/l_∞ for strong electrolytes does not represent the degree of ionisation but it is only a conductivity ratio.*

(iii) *The increase in conductivity (l_v) of a strong electrolyte solution on dilution, is due to increase in the mobility of ions.*

(iv) *The two ions present in the solution obey the Coulomb's law, i.e.,*

$$F \propto \frac{Q_1 Q_3}{r^2}$$

where Q_1 and Q_2 are the charges on the ions and r is the distance between them.

(v) *Each ion is surrounded by an ionic atmosphere of oppositely charged ions. To this ionic atmosphere, solvent molecules are then attached. Thus, a positive ion is surrounded by negative ions and vice-versa. This ionic atmosphere may be assumed to be formed in the following manner.*

Let us consider a positive ion situated at A. Suppose there is small volume element dv at the end of a radius vector from the point 'A'. It is assumed that the magnitude of the distance 'r' is of the order of less than about 100 times as the diameter of the ion.

Due to the thermal movement of ions, sometimes there occurs an excess of positive ions and sometimes an excess of negative ions in the volume dv. If average time is taken, it will be found that there will be a negative charge density in the volume dv if a positive charge is assumed to be present at A. In this way every ion may be assumed to be present at A. In this way every ion may be assumed to have an ionic atmosphere of opposite sign. The net charge of the ionic atmosphere is equal in magnitude and opposite in sign to that of central ion.

Debye and Huckel took the interionic attraction into consideration and calculated the ratio of activity to the concentration of on ion in the dilution solution. This was done as follows:

Electrical potential at any point is defined as "*the work done in bringing a unit charge from infinity to the particular point.*"

If ψ is the potential at any given point in the vicinity of a positive ion, then the work done in bringing a positive ion of valency Z_+ and carrying a charge ε to that point is given by $Z_+\varepsilon\psi$.

Similarly, the work done in bringing a negative ion is Z, εψ.

At an appreciable distance from the given ion, the value of ψ may be taken as zero. At this zero potential, if n_+^o and n_-^o are the concentrations of positive and negative ions per unit volume then by the Maxwell-Boltzmann's distribution law for particles in a field of varying potential, we can write

$$n_+ = n_+^0 e^{-(Z_+ \varepsilon\psi/KT)} \qquad ...(1)$$

$$n_- = n_-^0 e^{-(Z_- \varepsilon\psi/KT)} \qquad ...(2)$$

where n_+ and n_- are the concentrations of positive and negative ions at the given point under condition, K is the Boltzmann's constant.

It can be understood very easily that there are on an average more negative ions that positive ions in the vicinity of any positive ion and vice-versa. Thus it can be said that every ion is surrounded by an oppositely charged ionic atmosphere, *i.e.*, ions of opposite signs predominate in the ionic atmosphere.

The density of electricity ρ_e at any point may be taken equal to the excess of positive or negative electricity per unit volume at that point, *i.e.*,

$$\rho_e = (n_+Z_+\varepsilon) - (n_-Z_-\varepsilon)$$

$$\rho_e = \left[n_+^0 e^{-(Z_+\varepsilon\psi/KT)} Z_+\varepsilon\right] - \left[n_-^0 e^{-(-Z_-\varepsilon\psi/KT)} Z_-\varepsilon\right] \qquad ...(3)$$

[From Eqs. (1) and (2)]

For a uni-univalent electrolyte $Z_+ = Z_- = 1$ and $n_+^0 + n_-^0 = n$, where n represents the number per c.c. of ions of either kind in the bulk of the solution. Thus, equation (3) becomes as:

$$\rho_e = n\varepsilon\left(e^{-\varepsilon\psi/KT} - e^{\varepsilon\psi/KT}\right) \qquad ...(4)$$

If an assumption is made that εψ/KT is small in comparison with unity, it is thus possible to write the experimental series and neglecting higher powers of εψ/KT of equation (4). Therefore, equation (4) becomes as

$$\rho_e = -\left(\frac{\varepsilon^2\psi}{KT}\right)2n \qquad ...(5)$$

In the general case when Z_+ and Z_- are not necessarily unity, and if the solution may contain several kinds of ions, equation (5) may take the following form

$$\rho_e = -\frac{\varepsilon^2\psi}{KT}\Sigma n_1 Z_1^2 \quad ...(6)$$

where n_1 and Z_1 represent the number per unit volume and valence of the ions of the *i*th kind. The summation is taken over for all the types of ions present in the solution and equation (6) is applicable irrespective of the number of different kinds of ions.

In order to solve Eq. (6) for ψ it essential to introduce Poisson's equation, a relationship between ρ_e and ψ, and this equation in rectangular coordinates is

$$\frac{\partial^2\psi}{\partial x^2}+\frac{\partial^2\psi}{\partial y^2}+\frac{\partial^2\psi}{\partial z^2} = -\frac{4\pi\rho_e}{D} \quad ...(7)$$

Converting equation (7) into polar coordinates and putting $\partial\psi/\partial\theta$ and $\partial\psi/\partial\phi$ as zero, equation (7) becomes as

$$\frac{1}{r^2}\frac{\partial}{\partial r}\left(r^2\frac{\partial\psi}{\partial r}\right) = -\frac{4\pi\rho_e}{D} \quad ...(8)$$

where D is the dielectric constant of the medium and $\partial\psi/\partial\theta$ and $\partial\psi/\partial\phi$ are zero because the distribution of potential about any point in the electrolyte must be spherically symmetrical, and consequently independent of the angles θ and ϕ.

Substituting the value of ρ_e from equation (6) in (8), we get

$$\frac{1}{r^2}\frac{\partial}{\partial r}\left(r^2\frac{\partial\psi}{\partial r}\right) = \frac{4\pi\varepsilon^2}{DKT}\psi\Sigma n_1 Z_1^2 \quad ...(9)$$

or

$$\frac{1}{r^2}\frac{1}{\partial r}r^2\left(\frac{\partial\psi}{\partial r}\right) = k^2\psi \quad ...(10)$$

where

$$k = \left(\frac{4\pi\varepsilon^2\Sigma n_1 Z_1^2}{DKT}\right)^{1/2} \quad ...(11)$$

The general solution of equation (10) is given by

$$\psi = \left(\frac{A_e^{-kr}}{r}+\frac{B_e^{-kr}}{r}\right) \quad ...(12)$$

where A and B are integration constants.

Calculation of A and B

As r increases, ψ decreases. At an infinite distance from the given ion, *i.e.*, r, ψ must approach zero *i.e.*, constants A and B must be zero. Hence, equation (12) consequently becomes as

$$\psi = \frac{Ae^{-kr}}{r} \qquad ...(13)$$

The value of A can be calculated by the fact that when k = 0, the concentration becomes zero and, therefore, the potential is simply that of a single ion in the absence of any other charges, *i.e.*,

$$\psi = \frac{Z_1\varepsilon}{Dr} \qquad ...(14)$$

Also k = 0; therefore equation (13) becomes as

$$\psi = \frac{A}{r} \qquad ...(15)$$

Eliminating ψ between Eqs. (14) and (15), we obtain

$$\frac{Z_1\varepsilon}{Dr} = \frac{A}{r}$$

or

$$A = \frac{Z_1\varepsilon}{D} \qquad ...(16)$$

Substituting the value of A in equation (13), we get

$$A = \frac{Z_1\varepsilon\, e^{-kr}}{Dr} \qquad ...(17)$$

Expanding equation (17), we obtain

$$\psi = \frac{Z_i\varepsilon}{Dr}\left(1 - kr + \frac{k^2r^2}{2!} - ...\right) \qquad ...(18)$$

As k is a function of the concentration, it means that the higher powers of kr may be neglected for very dilute solution. Hence, equation (14) becomes as:

$$\psi = \frac{Z_i\varepsilon}{Dr}(1 - kr)$$

$$= \frac{Z_i\varepsilon}{Dr} - \frac{Z_i\varepsilon k}{D} \qquad ...(19)$$

From equation (19), it follows that

(i) The term $Z_i\varepsilon/Dr$ represents the potential at a distance r due to a single ion of charge Z_i ε in medium of dielectric constant D, and (ii) the term $-Z_i\varepsilon k/D$ is then potential due to the other ions, *i.e.*, those forming the ionic atmosphere of the given ion. It is this extra potential which is related to the extra free energy of the ionic solution.

Physical Significance of k

The value of 1/k is regarded as the equivalent radius of the ionic atmosphere, *i.e.*, it possesses the dimensions of length and is generally termed as *Debye Length.* It is of the order of 10^{-8} cm of ordinary solutions. From equation (11), it follows, that the actual value of k depends upon the (i) the concentration of the solution, and (ii) the valances of the ions. For one molar aqueous solution of univalent electrolyte at 25°C, 1/k = 3.1 Å.

Energy of a charged body: It is given as:

$$\text{Energy of a charged body} = \frac{1}{2} \times \text{charge} \times \text{potential.}$$

Therefore, for an ion of charge (= $Z_i\varepsilon$), the energy possessed by virtue of its ionic atmosphere is given by

$$E_I = \frac{1}{2}(Z_i\varepsilon)\left(-\frac{Z_i\varepsilon k}{D}\right)$$

$$= \frac{1}{2}\frac{Z_i^2\varepsilon^2 k}{D} \qquad \text{...(20)}$$

where $(-Z_i\varepsilon k/D)$ is the potential due to the ionic atmosphere of the given ion [see equation (12)]. The corresponding energy for 1 gm. ion is obtained on multiplying by the Avogardo number N, so that

$$E_t = -\frac{NZ_i^2\varepsilon^2 k}{2D} \qquad \text{...(21)}$$

According to the definition of chemical potential, the chemical potential of a particular ion in an ideal solution is given by

$$\mu_i = \mu_i^0 + RT\log_e x_i \qquad \text{...(22)}$$

where x_i is its mole fraction in the given solution. For a non-ideal solution, one can write

$$\mu_i = \mu_i^0 + \log_e a_i$$

$$= \mu_i^0 + RT\log_e(x_i f_i) \qquad [\because \; a_i = x_i f_i]$$

$$= \mu_i^0 + RT\log_e x_i + RT\log_e f_i \qquad \text{...(23)}$$

where a_i is the activity and f_i is the activity coefficient.

The difference between Eqs. (23) and (22) is RT $\log_e f_i$ which is equal to the difference in the free energy change accompanying the addition or removal of 1 gm ion of the ionic species from a large volume of real and dilute solution. Thus, the difference of free energy, RT $\log_e f_i$, is regarded as equivalent to the electrical energy of the ion due to ionic atmosphere [Eq. (20)]. Hence equation (20) becomes as:

$$RT \log_e f_i = \frac{NZ_i^2\varepsilon^2}{2D} \quad ...(24)$$

$$-\log_e f_i = +\frac{NZ_i^2\varepsilon^2 k}{DRT} \quad ...(25)$$

Debye Huckel Limiting Law: From equation (11), we have

$$k = \left(\frac{4\pi\varepsilon^2 \Sigma n_i Z_i^2}{DRT}\right)^{1/2} \quad ...(26)$$

If in the above equation n_i is replaced by c_i N/1000 and R/N written for K, equation (26) becomes as:

$$k = \left(\frac{4\pi N^2\varepsilon^2}{1000\ DRT}\Sigma c_i Z_i^2\right)^{1/2}$$

$$= \left(\frac{8\pi N^2\varepsilon^2}{1000\ DRT}\mu\right)^{1/2} \quad ...(27)$$

where $\mu = \frac{1}{2}\Sigma c_i Z_i^2$

Substituting equation (27) in (25), we get

$$-\log_e f_i = +\frac{N^2\varepsilon^2}{R^{3/2}}\left(\frac{2\pi}{1000}\right)^{1/2}\frac{Z_i^2}{(DT)^{3/2}}\sqrt{(\mu)} \quad ...(28)$$

$$\text{or } -2.301 \log f_i = \frac{N^2\varepsilon^2}{R^{3/2}}\left(\frac{2\pi}{1000}\right)^{1/2}\frac{Z_i^2}{(DT)^{3/2}}\sqrt{(\mu)} \quad ...(29)$$

$$\text{or } \log f_i = -\frac{N^2\varepsilon^2}{2.303\ R^{3/2}}\left(\frac{2\pi}{1000}\right)^{1/2}\frac{Z_i^2}{(DT)^{3/2}}\sqrt{(\mu)} \quad ...(30)$$

The values of universal constants N, ε, π and R as well as the numerical quantity substituted in eq. (30) to yield the following equation

$$\log f_1 = -1.823 \times 18^6 \frac{Z_i^2}{(DT)^{3/2}} \sqrt{(\mu)} \qquad ...(31)$$

For a given solvent and temperature, D and T have definite values which may be inserted; equation (31) then takes the general form

$$\log f_1 = -AZ_i^2 \sqrt{(\mu)} \qquad ...(32)$$

where A is constant for the solvent at the specified temperature.

Discussion: Equation (32) is known as Debye-Huckel limiting law. Some important points about this law are described below:

(i) This law expresses the variation of the activity coefficient of an ion with the ionic strength of the medium.

(ii) This law is called limiting law because it can only work if infinite dilution is approached.

(iii) From equation (32), it follows that activity coefficient of an ion should decrease with increasing ionic strength of the solution.

Limitation: Equation (32) cannot be tested in this form because there does not appear to be any experimental method of evaluating the activity coefficient of a single ionic species.

Mean Activity of an Electrolyte

Let us consider a binary electrolyte which dissociates into v ions, v^+ and v_-. Now the mean activity coefficient $f_{\neq}$ is related to individual ion activities by the following relation:

$$f_{\neq} = v\sqrt{(f_+ v + f_- r^-)}$$

or

$$\text{lof } f_{\neq} = \frac{v^+ \log f_+ + v^- \log f_-}{v^+ + v^-}$$

If Z_+ and Z_- are the valencies of cations and anions respectively then $v^+z^+ = v_-z_-$, then

$$\log f_{\neq} = \frac{Z_- \log f_- + Z_+ \log f_-}{Z_+ + Z_-}$$

Substituting the above value in eq. (32), we get

$$-\log f_{\neq} = AZ_+Z_- \sqrt{(\mu)} \qquad ...(33)$$

Equations (32) and (33) define a law which is known as Debye-Huckel Limiting law which is applicable to ideal solutions. According to this law,

"The departure from ideal behaviour in a given solvent is equal to the ionic strength of the medium and the valencies of ions and is not dependent upon their chemical nature."

DEBYE-FALKENHAGEN EFFECT, CONDUCTANCE UNDER HIGH A.C. FREQUENCIES

Debye and Falkenhagen examined the conductance behaviour of a solution of a strong electrolyte by applying alternating currents of different frequencies. They predicted that *if the frequency of alternating current is high* so that *the time of oscillation is small* in comparison with the relaxation time of the ionic atmosphere, the asymmetric effect will be virtually absent. In other words, the ionic atmosphere around the central ion will *remain symmetrical.* The retarding effect due to asymmetry may, therefore, be entirely absent and the conductance may be higher. The conductance of a solution, therefore, should vary with the frequency of the alternating current used. The higher the frequency, the higher the conductance, evidently. This effect, also known as dispersion of conductance, has been verified experimentally. The conductance remains independent of the frequency of alternating current upto 10 cycles per second. But with further increase in frequency, the conductance starts increasing towards a certain limiting value indicating complete absence of asymmetric effect.

WIEN EFFECT, CONDUCTANCE UNDER HIGH POTENTIAL GRADIENTS

Speed of an ion in an electric field varies with the applied potential gradient. Thus, under a potential gradient of about 20,000 volt per cm, an ion may have a speed of about 100 cm per sec. The ion, therefore, should *pass several times through the thickness of the ionic atmosphere during the time of relaxation.* The moving ion, therefore, will be almost free from the effect of the oppositely charged ionic atmosphere. The ion will be moving so fast that there will be no time for the ionic atmosphere to be built up. The asymmetry and electrophoretic effects, under these circumstances, may be negligibly small or even absent. Thus, the conductance of a strong electrolyte in aqeuous solution increases to a certain limiting value with increase in potential gradient applied. This observation had been verified experimentally by Wien much before the development of the theory of strong electrolytes and is known as the Wien effect.

TESTS OF DEBYE-HUCKEL THEORY

The Debye-Huckel theory may be used in the following ways:

(i) If $-\log f_{\neq}$ (33) is plotted against $\sqrt{(\mu)}$ for a dilute solution, straight lines should pass through origin as shown in Fig. (2.14) with the slope of the curve AZ_+Z_-. This has been epxerimentally verified.

(ii) In 1921 Lewis and Randall gave an empirical rule, according to which the mean activity in dilute solution is same in solutions of same ionic strengths. It has been observed that the equation (33) is in agreement with the above empirical rule of Lewis and Randall.

(iii) Different workers determined activity coefficients for univalent electrolytes by various methods and found that activity coefficients are the function of square root of the concentration. For such substances, ionic strength is equal to the concentration. These results are obeying equation (33).

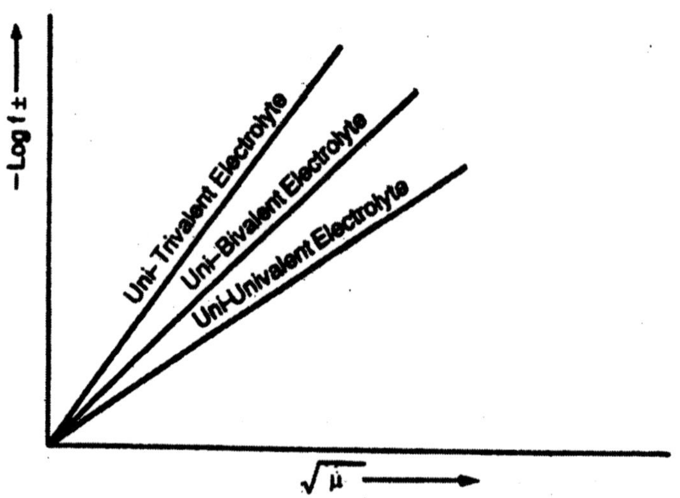

Fig. 2.14

Debye-Huckel—Onsagar's Equation

In the case of strong electrolytes the value of λ_v is much less than λ_∞. It was assumed to be due to the following reasons.

(i) Relaxation Effect: Due to inter-ionic forces each ion has a tendency to be surrounded by ions of opposite charge called the *ionic-atmosphere.* Let us suppose a negative ion be surrounded by the ionic atmosphere of positive ions. When an E.M.F. is applied, the negative ions move towards anode where by the ionic atmosphere of positive ions is left behind to disperse while a new ionic atmosphere is under formation in front of it. But the new ionic atmosphere is not formed at the same rate at which the old disperses and the latter takes more time called "*the relaxation time*".

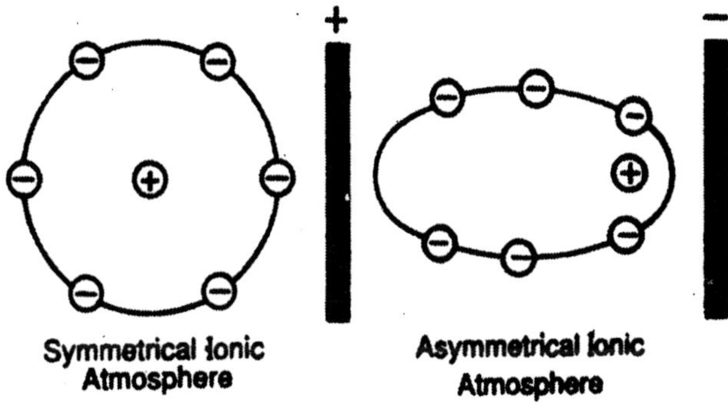

Fig. 2.15

It shows that in the rear of the moving ion there will be always an excess of ions of opposite sign. The ion, therefore, will always be dragged back. The effect thus decreases the mobility of the ion and is known as relaxation effect or asymmetric factor. Onsagar (1927) showed that the value of relaxation force may be given by

$$\text{Relaxation force} = \frac{\varepsilon^3 Z_i K}{6DkT} wV$$

where ε, Z, k and K have their usual significance while that of w and V are given below.

(ii) *Electrophoretic Effect:* The solvent molecules attached to ionic atmosphere move in direction opposite to that of central ion. Thus, they cause friction due to which the mobility of the central ion is retarded. This effect is called electrophoretic effect. On the basis of Stoke's law, Debye-Huckel calculated the following expression for the electrophoretic force on an ion of *i*th kind.

$$\text{Electrophoretic force} = \frac{KVK_i}{6\pi\eta} \cdot eZ_l$$

where η = viscosity coefficient of the medium.

K_i = coefficient of frictional resistance of the solvent opposing the motion of *i*th kind, and

On the basis of the above arguments Debye-Huckel and Onsagar derived the following expression, w is defined by

$$w = Z_+Z_- \frac{2q}{1+q^{1/2}} \qquad ...(34)$$

and the value of q is given by

$$q = \frac{Z_+Z_-(\lambda_+ + \lambda_-)}{(Z_+ + Z_-)(Z_+\lambda_- + Z_-\lambda_+)} \qquad ...(35)$$

It is now possible to equate to forces acting on an ion of the *i*th kind when it is moving through a solution with a steady velocity u_i; the deriving force due to applied electric field is $\varepsilon z_1 V$ and this is opposed by the frictional force of the solvent, K_iU_i, together with the electrophoretic and relaxation forces; hence

$$\varepsilon Z_i V = K_iU_i + \frac{\varepsilon z_1 K}{6\pi\eta} K_i V + \frac{\pi^3 Z_i K}{6DKT} wV \qquad ...(36)$$

Dividing throughout by K_iV and rearranging, we get

$$\frac{U_i}{V} = \frac{\varepsilon Z_i}{K_i} - \frac{\varepsilon Z_i K}{6\pi\eta} - \frac{\varepsilon Z_i K}{6DKT} \cdot \frac{w}{K_4} \qquad ...(37)$$

If the potential gradient is taken as one volt per cm. *i.e.*, $V = \frac{1}{300}$, then

$$U_i = \frac{\varepsilon Z_i}{300\ K_i} - \frac{\varepsilon K}{300}\left(\frac{Z_i}{6\pi\eta} + \frac{\varepsilon^2 Z_i}{6DkT}\frac{w}{K_i}\right) \qquad ...(38)$$

At infinite dilution K is zero and so equation becomes

$$U_i^o = \frac{\varepsilon Z_i}{300\ K_i}$$

But $\lambda_i = FU_i^o$

$$\therefore \quad \frac{\varepsilon Z_i}{300\ K_i} = \frac{\lambda_i^o}{F} \qquad ...(39)$$

Again $$U = \frac{\lambda_i}{\alpha F} \qquad ...(40)$$

Substituting equations (39) and (40) in (38), we get

$$\frac{\lambda_i}{\alpha F} = \frac{\lambda_i^o}{F} - \frac{\varepsilon K}{300}\left(\frac{Z_i}{6\pi\eta} + \frac{\varepsilon}{6DkT}\frac{\varepsilon Z_i}{K_i}w\right) \qquad ...(41)$$

When the electrolyte is completely ionised, it means that $\alpha = 1$. Therefore, the above equation becomes as

$$\lambda_i = \lambda_i^o - \frac{\varepsilon K}{300}\left(\frac{Z_i}{6\pi\eta}F + \frac{300\varepsilon}{6DkT}\lambda_i^o w\right) \qquad ...(42)$$

We also know $\frac{\varepsilon Z_i}{K_i} = \frac{300\lambda_i^o}{F}$. Therefore, the above expression reduces to

$$\lambda_i^o = \lambda_i^o - \left[\frac{29.15 Z_i}{(DT)^{1/2}\eta} + \frac{9.90\times10^5}{(DT)^{3/2}}\lambda_i^o w\right]\times\sqrt{(c_+Z_+^2 + c_-Z_-^2)} \qquad ...(43)$$

But $c = c_i z_i$, therefore

$$\lambda_i = \lambda_i^o - \left[\frac{29.15 Z_1}{(DT)^{1/2}\eta} + \frac{9.90\times10^5}{(DT)^{3/2}\eta}\cdot\lambda_i^o w\right]\times\sqrt{[c(Z_+ + Z_-)]} \qquad ...(44)$$

We know, equivalent conductance of the electrolyte is equal to the sum of conductances of the constituent ions and so it follows from equation (44) that

$$\Lambda = \Lambda_0 - \left[\frac{29.15\,(Z_+ + Z_-)}{(DT)^{1/2}\eta} + \frac{9.90\times10^5}{(DT)^{3/2}}\Lambda_o w\right]\sqrt{[c(Z_+ + Z_-)]} \qquad ...(45)$$

In uni-univalent electrolyte, $Z_+ = Z_- = 1$ and $w = 2 - \sqrt{2}$, equation (45) becomes as

$$\Lambda = \Lambda_0 = \frac{82.4}{(DT)^{1/2}\eta} + \frac{8.20\times10^5}{(DT)^{3/2}}\Lambda_0\Big]\sqrt{c} \qquad ...(46)$$

For a uni-univalent electrolyte, equation (46) becomes as

$$\Lambda = \Lambda_0 - [A + B_{\Lambda 0}]\sqrt{c} \qquad ...(47)$$

where A and B are constants dependent upon the nature of solvent and the temperature, thus

$$A = \frac{82.4}{(DT)^{1/2}\eta} \qquad ...(48)$$

$$B = \frac{8.20 \times 10^5}{(DT)^{3/2}} \quad ...(49)$$

Testing of Onsagar's Equation (49) : When equivalent conductance is plotted against the square root of concentration, a straight line should be obtained with slope $A = B\ \lambda_0$ as shown in the Fig. 2.16. These values closely agree with experimental data. Onsagar equation is obeyed at the concentration of about 2×10^{-3} equivalent per litre. This equation is also applicable to non-aqueous solvents.

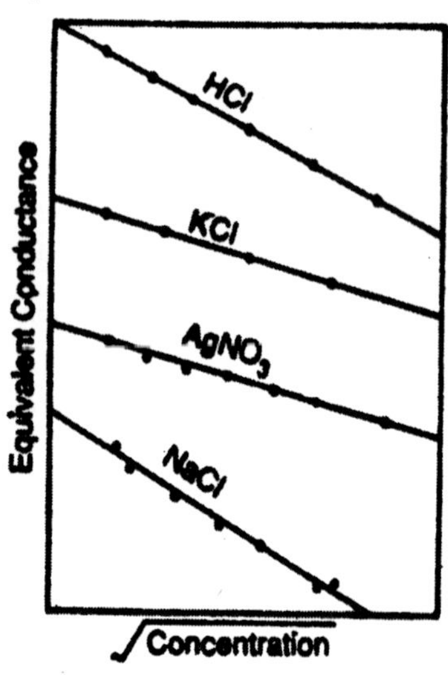

Fig. 2.16

Applications of Debye-Huckel Equation: The various forms of the equations resulting from the Debye-Huckel theory find practical application in the determination of activity coefficients and make possible the determination of thermodynamic data. Two important cases will be considered here.

1. *Determination fo thermodynamic equilibrium constants:* Let us consider as an example the dissociation of a 1 : 1 weak electrolyte AB

$$AB \rightleftharpoons A^+ + B^-$$

The thermodynamic dissociation constant k_T is given by

$$k_T = \frac{[A^+][B^-]}{[AB]} \frac{\gamma_A^+ \gamma_B^-}{\gamma_{AB}}$$

$$k_T = k \frac{\gamma \pm^2}{\gamma AB}$$

$$k_T \sim k\gamma\pm^2 \quad ...(1)$$

where k is the concentration or conditional dissociation constant. For a weak electrolyte in dilute solution γ_{AB} for the undissociated, and therefore non-ionic species is very nearly unity. Taking logarithms of Equation (1) and substituting for $\gamma\pm$ form the limiting law expression, we obtain

$$\log k = \log k_T + 2A\sqrt{\mu} \qquad ...(2)$$

k_T may therefore be determined from measured values of over a range of ionic strength values and extrapolating the k versus $\sqrt{\mu}$ plot to $\sqrt{\mu}$ = 0. This is a general technique for the determination of all types of thermodynamic equilibrium constants, *e.g.*, solubility, stability and acid dissociation constants.

2. *Effect of ionic strength on ion reaction rates in solution:* In the treatment of ionic reactions by Bronsted and Bjerrum an equilibrium is considered to exit between reactant ions and a critical complex, the latter bearing close resemblance to the activated complex of the theory of absolute reaction rates. Thus for the reaction scheme,

$$A_z^A + B_z^B \rightleftharpoons X_z^{(A} + {}_z^{B)} \pm \rightarrow \text{Products} \qquad ...(3)$$

We may write, for the per-equilibrium

$$k = \frac{[X\pm]}{[A][B]} \frac{\gamma \pm}{\gamma_A \gamma_B} \qquad ...(4)$$

Omitting charges for clarity so that the rate, v with which A and B react may be expressed by

$$v = k[A][B] = k_0[A][B] \frac{\gamma_A \gamma_B}{\gamma \pm} \qquad ...(5)$$

where
$$k = k_0 \frac{\gamma_A \gamma_B}{\gamma \pm} \qquad ...(6)$$

k_0 being the specific rate in infinitely dilution

where
$$\frac{\gamma_A \gamma_B}{\gamma \pm} = 1 \qquad ...(7)$$

In logarithmic form equation (7) becomes

$$\log k = \log k_0 + \log \gamma_A + \log \gamma_B - \log \gamma_\pm \qquad ...(8)$$

in which activity coefficients may be expressed by the following equation,

$$\log k = \log k_0 - \frac{A\sqrt{\mu}}{1 + Ba\mu}[Z_A^2 + Z_B^2 - (Z_A + Z_B)^2] \qquad ...(9)$$

$$= \log k_0 + \frac{2AZ_A Z_B \sqrt{\mu}}{1 + Ba\sqrt{\mu}} \qquad ...(10)$$

$$\log k \sim \log k_0 + 2AZ_A Z_B \sqrt{\mu} \qquad ...(11)$$

in very dilute solution,

or $$\log k \sim \log k_0 + 1.02\, Z_A Z_B \sqrt{\mu} \qquad ...(12)$$

for water as solvent at 298 k.

These last equations take account of the salt effect observed for reactions between ions, the slopes of graphs of log k/k_0 versus $\sqrt{\mu}$ being very close to those predicted by Equation (3) at low concentration. At higher concentrations, deviations from linearity occur and these are particularly noticeable for reactions having $Z_A Z_B = 0$, *e.g.*, for a reaction between an ion and a neutral molecule. According to Eq. (12), such reactions should show no variation of rate with ionic strength and this is indeed the case until about $\mu = 0.1$. Above this point, increasing ionic strength does cause the rate to vary.

STRENGTH OF AN ACID

According to the *ionic theory of ionization:*

(i) An acid is a substance which furnishes H^+ ions by itself or when H^+ dissolved in water.

(ii) The strength of an acid depends upon the concentration of free H^+ ions.

(iii) With the dilution, the degree of ionisation of an acid increases, thus increasing H^+ ions also.

(iv) An infinite dilution all acids are *completely ionised* and hence *they behave nearly equally strong at infinite dilution.*

(v) As degree of ionisation of an acid depends on its dissociation constant, so, the *dissociation constant of acid is the measure of its strength.*

(vi) Mineral acids like HCl, NHO_3, H_2SO_4, etc., which are largely ionised at all dilutions, are called *strong acids*, which oxalic, acetic, formic, benzoic acids, etc., which are *feebly ionised*, are called *weak acids.*

Methods of Comparing Relative Strengths of Two Acids

1. Thomson's Thermal Method

This method is based on the fact that when a mixture of two acids is allowed to react with a base, the quantity of heat evolved will be the measure of their relative strengths. Suppose we want to compare the strength of HCl and H_2SO_4.

Let the heat of neutralization of 1 g equivalent of HCl by NaOH

$$= x \text{ cal}$$

and heat of neutralization of 1 g equivalent of H_2SO_4 by NaOH

$$= y \text{ cal}$$

Now treat mixture containing 1 g equivalent of each HCl and H_2SO_4, with 1 g equivalent of NaOH and let the heat produced

$$= z \text{ cal}$$

The two acids will react with NaOH in the ratio of their relative strengths, the stronger acid taking more part in neutralizing it. Suppose n g equivalent of HCl and (1 – n) g equivalent of H_2SO_4 have neutralized 1 g equivalent of NaOH; then

Heat evolved due to neutralization of n g equivalent of HCl by NaOH+ heat evolved due to neutralization of (1 – n) g equivalent of H_2SO_4 by NaOH

$$= z \text{ cal}$$

or $$nx + (1 - n)y = z$$

or $$n = \frac{z - y}{x - y}$$

Knowing n, the relative strengths of HCl/H_2SO_4 = n/(1 – n) can be calculated.

2. Conductivity Method

The strength of an acid at any dilution depends upon its degree of dissociation. So, the relative strengths of two acids at same dilution is equal to the ratio of their degrees of ionisation. For a weak acid, the degree of ionisation is given by conductivity ratio. Thus

For acid I: $$\alpha_1 = \frac{\lambda_v}{\lambda_\infty}$$

For acid II: $$\alpha_2 = \frac{\lambda'_v}{\lambda'_\infty}$$

But at infinite dilution

$$\lambda_\infty = \lambda'_\infty$$

$$\therefore \quad \frac{\alpha_1}{\alpha_2} = \frac{\lambda_v}{\lambda'_v} = \frac{\text{Strength of acid I}}{\text{Strength of acid II}}$$

Thus, *the relative strengths of two weak acids is equal to the ratio of their equivalent conductivities at the same dilution.*

3. Dissociation Constants Method

(For weak acids only).

For acids HA_1 and HA_2 at some dilution (V), we have

$$K_1 = \frac{\alpha_1^2}{(1-\alpha_1)V}; \quad K_2 = \frac{\alpha_2^2}{(1-\alpha_2)V}$$

For weak acids $(1 - \alpha_1)$ and $(1 - \alpha_2)$ may be taken as unity.

$$\therefore \qquad \frac{\alpha_1^2}{V} = K_1; \quad \frac{\alpha_2^2}{V} = K_2$$

or

$$V = \frac{K_1}{\alpha_1^2} = \frac{K_2}{\alpha_2^2}$$

or

$$\frac{\alpha_1}{\alpha_2} = \sqrt{\left(\frac{K_1}{K_2}\right)} = \frac{\text{Strength of acid I}}{\text{Strength of acid II}}$$

Thus, *relative strength of two weak acids is equal to the square root of the ratio of their dissociation constants.*

4. Ostwald's Volume Method

In this method change in volume is noted instead of noting the heat evolved. Let x cm^3 be the change in volume when 1 equivalent of each of NaOH and HA_1 react and y cm^3 be the change in volume when 1 equivalent each of NaOH and HA_2 react. Let z cm^3 be the change when 1 equivalent of NaOH and a solution containing 1 equivalent of each HA_1 and HA_2 are mixed. Calculating in the same manner,

$$n = \frac{z-y}{x-y}$$

$$\therefore \qquad \frac{\text{Strength of } HA_1}{\text{Strength of } HA_2} = \frac{n}{(1-n)}$$

5. The Catalytic Method

It is well known that H^+ ions act as catalyst in many chemical reactions. The hydrolysis of esters and the inversion of cane sugar are catalysed by H^+ ions. To compare the strength of two acids HA_1 and HA_2, therefore, the velocity constants k_1 and k_2 of above first order reactions are determined in presence of these acids and the strengths are compared as

$$\frac{\text{Strength of HA}_1}{\text{Strength of HA}_2} = \frac{k_1}{k_2}$$

The strength of two bases can also be compared by the above mentioned methods. The conductivity method and comparison of dissociation constant method can be used as such. In catalytic process, we shall choose the base catalysed reactions such as condensation of acetone. In the Thomson's thermal and Ostwald's volume methods the two bases are neutralised by strong acids, say nitric acid.

DISSOCIATION CONSTANTS OF POLYBASIC ACIDS

Polybasic acids contain two or more ionisable hydrogen. They always dissociate in steps. The degrees of dissociation at different steps are different. Consider for example, the dissociation of phosphoric acid. It is a tribasic acid and ionises in three steps.

(i) $H_3PO_4 \rightleftharpoons H^+ + H_2PO_1^-$; $K_{a_1} = 7.5 \times 10^{-3}$ M

(ii) $H_2PO_4^- \rightleftharpoons H^+ + HPO_4^{2-}$; $K_{a_2} = 6.2 \times 10^{-7}$ M

(iii) $HPO_4^{2-} \rightleftharpoons H^+ + PO_4^{-3}$; $K_{a_3} = 4.8 \times 10^{-3}$ M.

The values of dissociation constants, K_{a_1}, K_{a_2} and K_{a_3} for the above three successive steps are taken from data book. These dissociation constants can be calculated in terms of the concentration, $[H^+]$,$[H_2PO_4^{1-}]$ and $[HPO_4^{2-}]$ and $[PO_4^{3-}]$ as follows:

$$K_{a_1} = \frac{[H^+][H_2PO_4^-]}{[H_3PO_4]}$$

$$K_{a_2} = \frac{[H^+][HPO_4^{2-}]}{[H_2PO_4^-]}$$

$$K_{a_3} = \frac{[H^+][HPO_4^{3-}]}{[HPO_4^{2-}]}$$

The ionisation constant in these three stages successively indicates that

$$K_{a_1} > K_{a_2} > K_{a_3}$$

The reason for the decrease in the dissociation constant values is that in the first dissociation the positively charged proton comes from a neutral molecule while in the second it is detached from a negatively charged molecule and in the third dissociation, it is detached from a doubly negatively charged molecule. The presence of negative charges makes the process difficult for the loss of proton.

COMMON ION EFFECT

Consider a simple weak electrolyte AB which ionises in solution as

$$AB \rightleftharpoons A^+ + B^- \qquad ...(1)$$

On applying the law of mass action, we get

$$K = \frac{[A^+][B^-]}{[AB]} \qquad ...(2)$$

Now, suppose to the above solution a strong electrolyte CB or AD *i.e.*, an electrolyte having A^+ or B^- as common ion with the electrolyte AB is added. These will ionise as

$$CB \rightleftharpoons C^+ + B^-$$

$$AD \rightleftharpoons A^+ + D^-$$

Suppose we have added AD to the equilibrium (1). Thus, the concentration of A in solution is increased. In order to keep K constant [equation (2)] at the given temperature, some of the A^+ added must combine with B^- to form unionised AB *i.e.*, the above equilibrium (i) is shifted backwards. In other words, the degree of ionisation of AB is suppressed. *This decrease in ionisation of a weak electrolyte by the addition of a strong electrolyte having an ion common with the weak electrolyte is known as common ion effect.*

For example, addition of ammonium chloride suppresses the ionisation of ammonium hydroxide and hydrochloric acid suppresses the ionisation of hydrogen sulphide.

Applications of Common Ion Effect

(i) *In Analytical Chemistry*: In II group of qualitative analysis, the ionisation of weak acid H_2S

$$H_2S \rightleftharpoons 2H^+ + S^{2-}$$

is suppressed by the presence of strong acid HCl having a common ion H^+

$$HCl \overset{\longrightarrow}{\rightleftharpoons} H^+ + Cl^-$$

Due to the suppressed degree of dissociation of H_2S, the concentration of S^{2-} ions produced is low which causes the precipitation of sulphide of II group cations only.

Similarly, in III group, the ionization of weak base NH_4OH,

$$NH_4OH \rightleftharpoons NH_4^+ + OH^-$$

is suppressed by the prior addition of NH_4Cl which furnishes a high concentration of common ion NH_4^+.

$$\longrightarrow$$

$$NH_4Cl \rightleftharpoons NH_4^+ + Cl^-$$

Due to this low dissociation of NH_4OH, the OH^- concentration is quite low. This low OH^- ion concentration is sufficient only to precipitate hydroxides of Al^{3+}, Cl^{3+} and Fe^{3+}.

(ii) *Purification of Common Salt:* NaCl, as obtained from natural sources, is usually contaminated with small amounts of deliquescent impurities like $CaCl_2$ and $MgCl_2$. The purification of chloride solution is effected by passing HCl gas into a saturated solution of the impure sodium chloride solution is effected by passing HCl gas into a saturated solution of the impure sodium chloride.

The addition of HCl (containing Cl^-) causes the precipitation of only NaCl, but impurities remain in solution since the solution is not saturated so far as calcium and magnesium chlorides are concerned.

(iii) *Salting out of Soap:* In the soap manufacture, a solution of common salt is added to solution of soap (sod. stearate) to obtain solid soap. Addition of common salt (NaCl) increases the value of $[Na^+] \times [stearate^-]$ in solution until the solubility product of soap is exceeded, whereby soap gets precipitated.

(iv) *In Gravimetric Analysis:* When excess of precipitant (HCl or H_2SO_4) is added to solution containing Ag^+ or Ba^{2+} ions, complete precipitation of AgCl or $BaSO_4$ occurs due to the presence of excess of common ion Cl^- or SO_4^{2-}.

SOLUBILITY PRODUCT

Consider the *saturated* solution of a sparingly soluble binary electrolyte AB. There exist two equilibria simultaneously between undissolved solid, dissolved (but unionised) molecules and free ions, *i.e.*,

$$\underset{\text{(Undissolved solid)}}{AB} \overset{K_1}{\rightleftharpoons} \underset{\text{(Dissolved, but unionsed)}}{AB} \overset{K_2}{\rightleftharpoons} \underset{\text{(Ions in solution)}}{A^+ + B^-}$$

Applying the law of mass action, it follows that

$$\frac{[A^+][B^-]}{[(AB)_d]} = K_1; \qquad \frac{[(AB)_d]}{[(AB)_d]} = K_2$$

∴ On multiplying, we get

$$\frac{[A^+][B^-][AB)_d]}{[(AB)_d][AB)_s]} = \frac{[A^+][B^-]}{[(AB)_s]} = K_1 \times K_3$$

But the concentration of solid AB is constant, so that

$$[A^+][B^-] = \text{a constant} = K_s$$

where the constant K_s is known as the solubility product of AB.

In general, for an electrolyte, A_xB_y, which ionises as

$$A_xB_y \rightleftharpoons xA^{y+} + yB^{x-}$$

the solubility product is given by

$$K_s = [A^{y+}]^x [B^{x-}]^y$$

∴ where $[A^{y+}]$ and $[B^{x-}]$ are the ionic concentration in a *saturated solution.* Thus, solubility product of an electrolyte may be defined as the *maximum product of the concentrations of its constituent ions (expressed in g-ion per litre) in its solution, when each ionic concentration term being raised to the number of times the ion occurs in the equation representing the solution of 1 molecule of the electrolyte.* For example for

$$Ag_2CrO_4 \rightleftharpoons 2Ag^+ + CrO_4^{2-}$$

the solubility product, $K_s = [Ag^+]^2[CrO_4^{2-}]$.

Relation Between Solubility and Solubility Product

Consider the general form of electrolyte A_xB_y which dissociates as

$$\underset{S}{A_xB_y} \rightleftharpoons \underset{xS}{x[X^{y+}]} + \underset{yS}{y[X^{x-}]}$$

Let S gram mole/litre be the solubility of the electrolyte, then

$$[A^{y+}] = x \,.\, S \quad \text{and} \quad [B^{x-}] = y \,.\, S$$

$$\therefore \quad K_s = [A^{y+}]^x[B^{x-}]^y$$

$$= [x \,.\, S]^x[y \,.\, S]^y = x^x \,.\, y^y \,.\, S^{x+y}$$

For example in case of $BaCl_2 \rightleftharpoons Ba^{2+} + 2Cl^-$; x = 1 and y = 2

$$Ks \rightleftharpoons 1^1 \,.\, 2^2\, S^{1+2} = 4S^3.$$

Analytical Applications of Solubility Product

From the relation:

$$K_s = [A^+]^x[B^-]^y$$

it is clear that:

(a) a solution in which $[A^+]^x[B^-]^y$ is *less than* K_s, the solution is unsaturated and more A_xB_y can be dissolved in it,

(b) a solution in which $[A^+]^x[B^-]^y$ is *equal* to K_s, *is just saturated*, and

(c) if anything is done to the solution which tends to make $[K^+]^x[B^-]^y$ *greater than* K_s, then solid A_xB_y will be precipitated.

From the above discussion it is evident that if the product of concentration of ions in a saturated solution of an electrolyte becomes greater than the solubility product of the electrolyte, it will precipitate out. In qualitative analysis, separation and identification of metal ions such as Cu^{2+}, Zn^{2+}, etc., is possible only due to difference in solubility products of their sulphides.

The above concepts of solubility product have become extremely useful in getting a clear picture of different processes in analytical chemistry. A few cases are illustrated below:

1. Qualitative Analysis of Cationic Mixtures

The identification of individual species of cations present in an unknown mixture of electrolytes involves a preliminary segregation of the cations into selected classifiable groups.

A simple system devised for this purpose is based on the selective precipitation of sparingly soluble compounds of the cations from aqueous solution of the electrolyte mixture.

In its most simplified form, for the more common cations only, these compounds fall into five groups. Within each group the anion is the same for the different cations and the solubility products are small and do not differ too greatly from one another:

Group I: The sparingly soluble chlorides derived from Pb^{2+}, Hg_4^{2+} and Ag^+ ions are precipitated by the addition of dilute HCl.

Chloride	K_s *at 25°C*
$PbCl_2$	1.6×10^{-5}
Hg_3Cl_2	1.3×10^{-15}
AgCl	1.8×10^{-16}

Group II: The sparingly soluble sulphides derived from Hg^{2+}, Pb^{2+}, Cu^{2+}, Cd^{2+} and Bi^{3+} ions are precipitated by the passage of H_2S through an acidic solution whose concentration of hydrogen ions is small.

Sulphide	K_s *at 25°C*
HgS	3×10^{-32}
PbS	8×10^{-28}
CuS	8×10^{-36}
CdS	7×10^{-27}
Bi_2S_3	1×10^{-96}

Group III: The sparingly soluble hydroxides derived from Fe^{2+}, Cr^{3+} and Al^{3+} ions are precipitated by the addition of ammonium hydroxide solution in the presence of excess ammonium chloride

Hydroxide	K_s *at 25°C*
$Fe(OH)_3$	6×10^{-28}
$Cr(OH)_3$	7×10^{-31}
$Al(OH)_3$	1.4×10^{-24}

Group IV: A further group of sparingly soluble sulphides, those derived from CO^{2+}, Ni^{2+}, Mn^{2+} and Zn^{2+} ions are precipitated by the passage of H_2S through a basic solution.

Sulphide	K_s *at 25°C*
CoS	8×10^{-23}
NiS	2×10^{-21}

MnS	1×10^{-11}
ZuS	8×10^{-25}

Group V: The sparingly soluble carbonates derived from Ca^{2+}, Sr^{2+} and Ba^{2+} ions are precipitated by the addition of ammonium carbonate solution in the presence of NH_4OH and NH_4Cl.

Carbonate	*Ks at 26°C*
$CaCO_3$	4.7×10^{-9}
$SrCO_3$	7×10^{-10}
$BaCO_3$	1.6×10^{-9}

The most common cations remaining in solution after the precipitation fo group V are Na^+, K^+ (all the simple salts of the alkali metals are water-soluble) and Mg^{2+} ions. The order of group precipitation is such that the precipitating reagents for later groups also precipitate in most cases cations in preceding groups, which must therefore the completely removed in the above prescribed order if they are not to interfere. For example,

(i) K_S of silver sulphide is 7×10^{-50} at 25°C. It means that silver sulphide will be precipitated in group II unless it has been completely removed as AgCl in group I.

(ii) Any of the cations normally precipitated as their sulphides in group II, will reappear in the group IV precipitate unless completely removed in group II.

(ii) K_S for $ZnCO_3$ at 25°C is 2.1×10^{-11}. Hence $ZnCO_3$ will be precipitated in group V unless it has been completely removed as ZnS in group IV.

2. The Separation of the Sulphides into Groups II and IV

The values of K_S for group II sulphide are smaller than those for group IV sulphides. Hence, it is possible to precipitate the sulphides of those cations which fall in group II selectively in the presence of those cations which fall in group IV if the $[S^{2-}]$ from the reagent H_2S is regulated such that it exceeds that required to reach the solubility products of the group II sulphides but is less than that required to reach the solubility produces of the group IV sulphides. This is done as follows:

H_2S is a weak diprotic acid for which:

(a) $$H_2S \rightleftharpoons H^+ + HS^-$$

$$\frac{[H^+][HS^-]}{[H_2S]} = K_{S_1} = 9.1 \times 10^{-3} \text{ at } 18°C$$

(b) $$HS^- \rightleftharpoons H^+ + S^{2-}$$

$$\frac{[H^+][S^{2-}]}{[HS^-]} = K_{S_2} = 1.2 \times 10^{-15} \text{ at } 18°C$$

Hence, on multiplying:

$$\frac{[H^+]^3[S^{2-}]}{[H_2S]} = 10^{-22}$$

$$\therefore \quad [S^{2-}] = \frac{[10^{-22}][H_2S]}{[H^+]^2}$$

If H_2S at 1 atmosphere is bubbled through water, for the saturated solution:

$$[H_2S] \sim 0.1 \text{ mole/litre}$$

$$\therefore \quad [S^2] = \frac{10^{-22} \times 10^{-1}}{[H^+]^2} = \frac{10^{-22}}{[H^+]^2}$$

i.e., the $[S^{2-}]$ is dependent on the $[H^+]$ and may be varied by varying the acidity of the solution.

If $$pH = 0(1 \text{ MHCl})$$

$$\therefore \quad [H^+] = 1 \text{ mole/litre}$$

or $$[S^{2-}] = 10^{-22} \text{ mole/litre}$$

If $$pH = 12(0.01 \text{ M NaOH}),$$

$$\therefore \quad [H^+] = 10^{-12} \text{ mole/litre}$$

or $$[S^{2-}] = 10 \text{ mole/litre}$$

Hence if the acidity is too low, the corresponding $[S^{2-}]$ is large enough to exceed the requirements of the solubility products of both the group II and the group IV sulphides which are precipitated together in group II.

Conversely, if the acidity is too high, the $[S^{2-}]$ ion is suppressed to a degree when it is lower than that required to reach the solubility products of even the group II sulphides. These, therefore, remain in solution and are precipitated in group IV where the [S^{2-}] in the basic solution is too high.

In practice, if the [H^+] of the solution, after the precipitation of the group I chlorides is adjusted to 0.3 M prior to passing H_2S, the group II sulphides are precipitated selectively.

(a) Selective precipitation of the group III hydroxide: Three hydroxides only are precipitated in group III, those of Fe^{2+}, Cr^{3+} and Al^{3+} whose K_s values lie between 10^{-30} to 10^{-33}. However, the K_s values for the hydroxides of those cations which are normally precipitated in group IV as their sulphide are also relatively small, though of a higher order than the above there:

Hydroxide	K_s *at 25°C*
$Co(OH)_2$	2×10^{-16}
$Ni(OH)_2$	2×10^{-15}
$Mn(OH)_2$	1.6×10^{-13}
$Zn(OH)_2$	7×10^{-18}

Now, for a solution fo NH_3 in water the following equilibria exist:

$$NH_3 + H_2O \rightleftharpoons NH_4OH \rightleftharpoons NH_4^+ + OH^-$$

for which $\dfrac{[NH_4^+][OH^-]}{[NH_4OH]} = K_b = 1.8 \times 10^{-5}$ at 25°C

$$\therefore \quad [NH_4^+][OH^-] = 1.8 \times 10^{-5}\,[NH_4OH]$$

But $\quad [NH_4^+] = [OH^-]$

$$\therefore \quad [OH^-]^2 = 1.8 \times 10^{-5}\,[NH_4OH]$$

For 1 M NH_4OH :

$$[OH^-]^2 = 1.8 \times 10^{-5}$$

or $\quad [OH^-] = 4.24 \times 10^{-3}$ mole/litre

For 0.1 M NH_4OH:

$$[OH^-]^2 = 1.8 \times 10^{-5} \times 10^{-1} = 1.8 \times 10^{-6}$$

or $\quad [OH^-] = 1.34 \times 10^{-2}$ mole/litre

Hence, if a solution of NH_4OH alone is used to precipitate the hydroxides in group III, the [OH^-] is high enough to exceed the requirements of the solubility products for the hydroxides of the cations which fall in group IV. These are therefore precipitated together with the normal group III hydroxides.

It is therefore necessary to suppress the $[OH^-]$ concentration to such an extent where it exceeds the requirements of the solubility products for the group III hydroxides but is lower than is necessary to reach the solubility products for the hydroxides of the group IV cations.

This is done by the addition of an excess of NH_4Cl, when a buffer solution is produced consisting of the weak base, NH_4OH, and a salt of the weak base and a strong acid, NH_4Cl. NH_4Cl, being a salt, is a strong electrolyte:

$$NH_4Cl \rightarrow NH_4^+ + Cl^-$$

The excess NH_4^+ ions depress the ionisation of NH_4OH in order that K_b for NH_4OH may be maintained. Hence the $[OH^-]$ is suppressed to a point where only the group III hydroxides are precipitated.

(b) Quantitative Analysis: The solubility products of the silver halides are small:

$$\text{At } 25°\text{C}, K_s \text{ of AgCl} = 1.8 \times 10^{-10}$$

$$K_s \text{ of AgBr} = 5.2 \times 10^{-13}$$

$$K_s \text{ of AgI} = 8.3 \times 10^{-17}$$

Hence, if a solution fo the soluble $AgNO_3$ is added to a solution containing a soluble halide, the halide ions are quantitatively removed from solution as the sparingly soluble silver salt. The concentration of halide ions remaining in solution when precipitation is just complete is very low and may be neglected.

For example, if $AgNO_3$ solution is added to NaCl solution, AgCl is precipitated. When precipitation is just complete, for the ions remaining in solution:

$$[Ag^+][Cl^-] = K_s = 1.8 \times 10^{-10}$$

$$\text{But} \quad [Ag^+] = [Cl^-]$$

$$\therefore \quad [Cl^-] = (1.8 \times 10^{-10})^{1/2} = (1.34 \times 10^{-5}) \text{ mole/litre}$$

Thus precipitation of sparingly soluble silver halides is made the basis of quantitative determinations of the concentrations of solution of soluble halides. A solution of the halides of unknown concentration is titrated with a standard solution of $AgNO_3$. The end point of the reaction occurs when precipitation fo the sparingly soluble silver halide is just complete.

3. Using Excess of Reagent

The use of excess of reagent is essential to ensure complete precipitation in gravimetric analysis. The insoluble substances like AgCl, $BaSO_4$, etc., have very slight solubility but by adding excess of precipitant their solubility product is further exceeded. Thus, the ionic product $[Ba^{2+}][SO_4^{2-}]$ becomes still higher than the solubility product of $BaSO_4$, by using large excess of H_2SO_4.

$$BaSO_4 \rightleftharpoons Ba^{2+} + SO_4^{2-}$$

$$H_2SO_4 \rightleftharpoons 2H^+ + SO_4^{3-}$$

4. Use of K_2CrO_4 as Indicator in $AgNO_3$ Chloride Titrations

In silver nitrate-chloride titration, the $AgNO_3$ is added from the burette into the chloride solution to which some K_2CrO_4 solution is added. At first the added $AgNO_3$ precipitates AgCl, but not Ag_2CrO_4 because AgCl has a lower solubility (0.00145 g per litre) than Ag_2CrO_4 (0.0285 g per litre). When sufficient $AgNO_3$ has been added to precipitate all the AgCl, subsequent addition causes the formation of brick-red Ag_2CrO_4 precipitate, which serves to note the end-point.

IONIC PRODUCT OF WATER

Pure water is essentially a covalent compound. Nevertheless, it ionises very slightly and the following equilibrium is established:

$$H_2O \rightleftharpoons H^+ + OH^- \qquad ...(1)$$

where H^+ is a hydrogen ion and OH^- is a hydroxyl ion. The removal fo the one orbital electron from a hydrogen atom yields the positive hydrogen ion or hydrogen cation, which is in fact a bare proton. The bare proton has no separate existence in solution and owes its stability to solvation by a water molecule to give the hydronium ion, H_3O^+. The ionisation equilibrium of water is thus represented more accurately as:

$$2H_2O \rightleftharpoons H_3O^+ + OH^-$$

However, as many qualitative and quantitative considerations remain essentially the same for the unsolvated and the solvated proton, it is usual for convenience to refer to the unsolvated proton, H^+.

On applying the law of mass action to equation (1), we get

$$K = \frac{[H^+][OH^-]}{H_2O} \qquad ...(2)$$

where, K is called the ionisation constant. As water is ionised to very slight extent (about 1 in 10^7 water molecules), the concentration of water molecules is very large as compared with H^+ and OH^- ions so that it can be regarded as practically constant.

Therefore, equation (2) may be written as

$$[H^+][OH^-] = K[H_2O] = k_w \qquad ...(3)$$

where constant k_w is constant called the *ionic product of water.*

Conductivity measurements show that in pure water,

$$[H^+] = 10^{-7} \text{ mole/litre at } 25°C$$

Furthermore in pure water,

$$[H^+] = [OH^-]$$

Therefore, equation (3) may become as

$$[10^{-7}][10^{-7}] = k_w$$

or

$$k_w = 10\text{–}14 \text{ at } 25°C \qquad ...(4)$$

Significance of Equation (3) and (4)

For all aqueous solutions, the product of $[H^+][OH^-]$ may be taken to be equal to 10^{-14}.

If extra H^+ ions are introduced, the ionisation of H_2O is suppressed to the point at which $[H^+][OH^-] = 10^{-14}$. Similarly, if extra OH^- ions are introduced, the ionisation of water is again suppressed to maintain the constancy of $[H^+][OH^-]$.

From equation (3), it follows that the ionic product of water may be defined as *the product of concentration of H^- and OH ions expressed in gram ions per litre.*

pH VALUE

From equation (3) and (4), we have

$$[H^+][OH^-] = 10^{-14} \qquad ...(5)$$

In case of pure water, $[H^+] = [OH^-]$

Therefore, equation (5) becomes as

$$[H^+][H^+] = 10^{-14}$$

or
$$[H^+] = 10^{-7}$$

In means if the concentration of H^+ ions in solution is 1×10^{-7} gm, ion per litre, it is natural and if more, it is acidic and if less it is alkaline. Thus the acidity or alkalinity of the solution can be expressed in terms of its H^+ ion concentration.

It is not very convenient to express acidity and basicity in terms of H^+ ions as its value are usually very small, especially in case of weakly ionised substances. In order to overcome this difficulty, *Sorensen,* a Danish biochemist in 1909 introduced a new notation to express the H^+ ion concentrations. According to his the pH of a solution is,

"*numerically equal to the negative power to which 10 must be raise din order to express the H^+ ion concentration*".

$$[H^+] = 10 \quad \text{or} \quad \log[H^+] = -\text{pH} \log 10$$

or
$$\log [H^+] = -\text{pH} \qquad [\because \log 10 = 1]$$

or
$$\text{pH} = -\log [H^+]$$

Hence pH is the *negative logarithm of hydrogen ion concentration.* For pure water or a neutral solution in which

$$[H^+] = 1 \times 10^{-7}$$

$$\text{pH} = -\log[H^+] = -\log[10^{-7}] = 7$$

The pH of neutral solution is 7. In acidic solution $[H^+] > 10^{-7}$ and therefore pH < 7. In alkaline solution $[OH^-] > [H^+]$ and therefore $[OH^-] > 10^{-7}$ and $[H^+] < 10^{-7}$ and, therefore, solution is acidic, and if it is more than 7, the solution is alkaline. A pH scale ranges from 0 to 14. This is shown in Fig. 2.17.

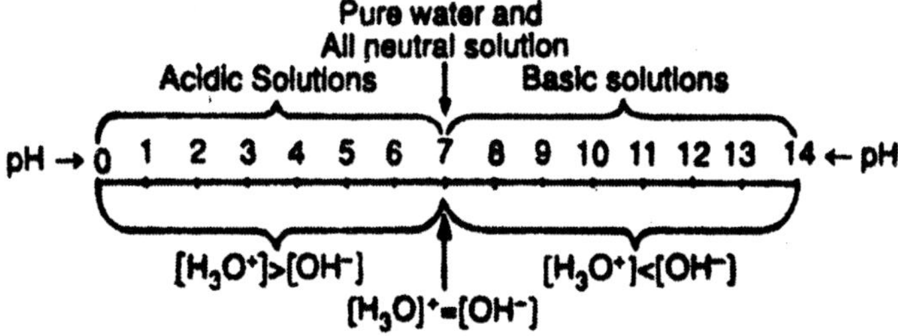

Fig. 2.17 pH scale.

Determination of pH Value

(1) *Colorimetric Method (For Colourless Solutions Only):* If the same amount of the same indicator gives the same colour in the same amount of two different solutions, then the pH of the two solutions must be the same. The colorimetric method (or indicator method) is the simplest method requiring least expensive equipments and is applied where both speed and moderate accuracy are expected.

The method consists in adding a definite quantity of *universal indicator* (a mixture of certain dyes which give different colours at different pH) to measured quantity of solution under test contained in a glass tube. The colour so developed is compared with a series of colours obtained by adding same amount of indicator to standard buffer solutions of known but varying pH values. If a complete match is obtained, the pH of the unknown solution will be the same as that of the standard solution. The matching of colours can be best done by using a colorimeter. The use of a colorimeter with a set of coloured glasses provides a quick and simple method for measuring pH values.

(2) *Emf Method:* This is the most accurate method for determining the pH value (or H^+ ion concentration). The solution whose pH is to be determined is taken in a vessel of suitable shape as shown in Fig. 2.18. In it is dipped a plantinised platinum electrode and a steady stream of moist hydrogen gas is passed through the solution. This half cell so formed is connected to a saturated

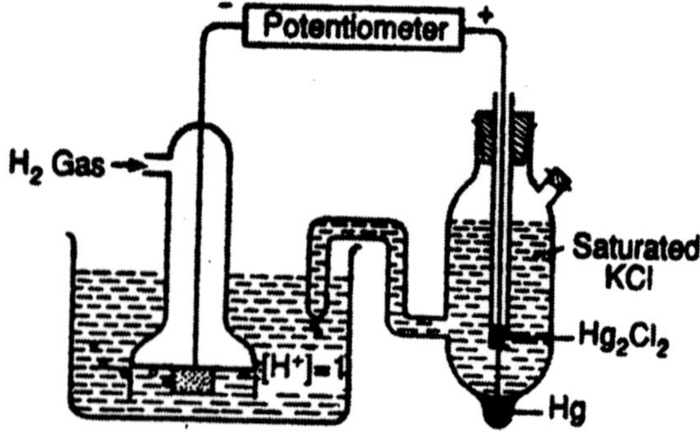

Fig. 2.18 : E.m.f. method for determining pH value.

calomel electrode through a salt bridge and the emf of the complete cell is determined by means of a potentiometer. The pH value is then calculated using the formula

$$pH = \frac{\text{Emf} - 0.2422}{0.0591}$$

INDICATORS

Indicators *are organic substances the presence of very small amount of which indicates the termination of a chemical reaction by a change of colour.* Indicators are of various types, *e.g.*, acid-base indicators, redox indicators, adsorption indicators, etc. *Acid-base indicators* are organic substances which have one colour in acid solution while an altogether different colour in alkaline solution. Various acid-base indicators show colour changes in a definite pH range, *e.g.*,

Table 2.3

Indicator	*pH range*	*Colour in acid*	*Colour in alkali*
Photophthalein	8.3—10.5	Colourless	Red
Litmus	5.5—7.4	Red	Blue
Methyl red	4.5—6.5	Red	Yellow
Methyl orange	3.2—4.5	Pink	Yellow

Theories of Indicators

1. Ostwald's Theory

According to this theory:

(1) Acid-base indicators are *weak organic acids or bases.*

(2) *They possess different colours in ionised and un-ionised states i.e.,*

$$\underset{\text{(one colour)}}{HIn} \rightleftharpoons H^+ + \underset{\text{(different colour)}}{In^-}$$

(3) *The colour of the indicator depends on the relative proportions of the unionised indicator molecules and its ions.*

Thus, *phenolphthalein* is a weak acid whose unionised molecules are colourless, while ions are red in colour *i.e.*

$$\underset{\text{(colourless)}}{HPh} \rightleftharpoons \underset{\text{(red)}}{Ph^-} + \underset{\text{(colourless)}}{H^+}$$

In presence of an acid (*i.e.*, H^+ ions) the equilibrium is forced backward (due to common H^+ ion), whereby resulting in the formation of colourless undissociated molecules. Addition of a strong alkali (*i.e.*, OH^- ions) results in the removal of H^+ ions.

$$\underset{\text{(of indicators)}}{H^+} + \underset{\text{(of added alkali)}}{OH^-} \rightleftharpoons \underset{\text{(nearly unionised)}}{H_2O}$$

This results in more and more dissociation of indicator. Therefore, in presence of an alkali the concentration of pink coloured Ph^- ions is far greater than colourless HPh molecules, and hence the solution will be pink.

If a weak base (*e.g.*, NH_4OH) is added, the OH^- ions furnished by it are very small in number, and hence the equilibrium is not shifted sufficiently to produce a large number of coloured Ph^- ions, *i.e.*,

$$HPh \rightleftharpoons H^+ + Ph^-$$

$$\underset{\text{(feebly ionised)}}{NH_4OH} \rightleftharpoons OH^- + NH_4^+$$

$$OH^- \downarrow H_2O \qquad NH_4^+ \rightleftharpoons \underset{\text{(Readily ionised)}}{NH_4Ph}$$

Moreover, NH_4Ph molecules formed are highly ionised to yield back NH_4^+ ions, which further reduces the number of Ph^- ions. Therefore, pink colour does not appear until a large excess of weak base is added to get visible colour change (or end-point). Hence, this explain why *phenolphthalein is not a good indicator when weak base is used.*

Methyl orange is a weak base which ionises to yield red ions, *i.e.*,

$$\underset{\text{(Yellow)}}{MeOH} \rightleftharpoons \underset{\text{(red)}}{Me^+} + \underset{\text{(colourless)}}{OH^-}$$

If a base (*i.e.*, OH^- ions) is added to the indicator, the OH^- ions will suppress the ionization of the indicator. Hence, the indicator will remain yellow in an alkali. However, if a small excess of acid (say, HCl) is added, the latter will force the equilibrium to the right by removing OH^- ions to form H_2O.

$$\underset{\text{(of indicator)}}{OH^-} + \underset{\text{(of added acid)}}{H^+} \rightleftharpoons \underset{\text{(practically unionised)}}{H_2O}$$

This will result in the formation of red coloured Me^+ ions in the solution.

It may be pointed that methyl orange is not a suitable indicator for titrating a *strong base against weak acid* like CH_3COOH. After the end point, the weak acid does not produce sufficient H^+ ions to shift the equilibrium appreciably to the right. Moreover, the slat formed (CH_3COOMe) although it is highly ionised, yet it does not produce sufficient Me^+ ions due to hydrolysis.

$$CH_3COOMe + H_2O \rightleftharpoons CH_3COOH + MeOH$$

Thus, the colour does not appear until sufficient excess of CH_3COOH has been added. Consequently the end point will not be correct.

2. Modern Quinoid Theory

According to it:

1. *An acid-base indicator is a dynamic equilibrium mixture of two alternative tautometric forms*; ordinarily one form in *benzenoid* while the other is *quinoid.*
2. *The two forms have differents colours.*
3. *Out of these one form exists in acidic solution, while the other in alkaline solution.*
4. *Change in pH causes the transition of benzenoid form to quinoid form and vice-versa* and consequently a change in colour.

Indicators and acid-base titrations: A process of acid-base titration is accompanied by a change in pH when successive amounts of base are added to a solution of an acid (or *vice-versa*). Indicators have the property of changing colour in dilute solutions when hydrogen ion concentration of the solution attains a definite value. The colour charge in the indicator occurs within a certain pH range. Let us consider an indicator such as HIn, which is a weak acid and its ionisation is represented as

$$\underset{\text{Colour A}}{HIn} \rightleftharpoons H^+ + \underset{\text{Colour B}}{In^-}$$

Application of law of mass action to this reversible reaction gives

$$\frac{[H^+][In^-]}{[HIn]} = K_1$$

The colour exhibited by the indicator is determined by the ratio of the concentrations of the two species HIn and In^-.

Thus, $$\frac{[In^-]}{[HIn]} = \frac{K_1}{[H^+]}$$

$$\log K_1 - \log[H^+] = \log\frac{[In^-]}{[HIn]}$$

$$pK_1 + pH = \log\frac{[In^-]}{[HIn]}$$

(i) When pH = pK_1, the ratio $\frac{[In^-]}{[HIn]}$ becomes equal to 1 *i.e.* 0.5 indicator is present in the acid form and 0.5 in the alkaline form.

(ii) When pH = $pK_1 - 1$, then [HIn] will be ten times and at pH = $pK_1 - 2$, 100 times $[In^-]$. At these pH values the colour of the undissociated indicator will predominate.

In general the useful range of pH is $pK_1 \pm 0.5$ to 1 unit.

Titration Curves

A plot between pH of the solution during titration and the amount of acid/base added is called titration curve. Plots of pH vs. volume of acid or base added in an acid-base titration are useful in that they show the equivalence point graphically and help in the choice of a proper indicator.

Titration curves for various acid-base pairs of varying strength: The nature of the titration curve depends on the ionization constants of the acid and base employed in a titration, *i.e.*, on the strength of the acids and bases. We shall discuss now the nature of titration curves of various acid-base parts of varying strength.

(i) *Titration of a strong acid against a strong base:* Fig. 2.19 shows the pH-titration curve of a solution of a strong acid (say HCl) against a solution fo a strong base (say NaOH). In the beginning when alkali is introduced into the acid solution, there is not much change in the pH value of the solution. This is in conformity with our earlier experience that strong acid solutions are good buffers at low pH. So the curve is almost flat until the end point is reached by successive additions of alkali when the sudden rise in pH of the solution takes place. The curve at this stage is almost vertical and parallel to pH axis.

The end point is also called the inflection point. On adding alkali after the inflection point, the sudden rise continues and soon the curve again becomes flat. This method can therefore, be used for determining the end point in case of the titrations involving solutions whose pH can be found out directly. In such titrations the pH is determined after successive additions of alkali and a graph is drawn between pH and the amount of alkali (volume of alkali) added to the acid solution. The inflection point (the point of the rapid rise in the curve) is the equivalence mid-point for the titration and represents the stage of titration where equivalent quantities of acid and base are present in the titration mixture. For the titration of a strong acid and a strong base, the equivalence point occurs at a pH 7.

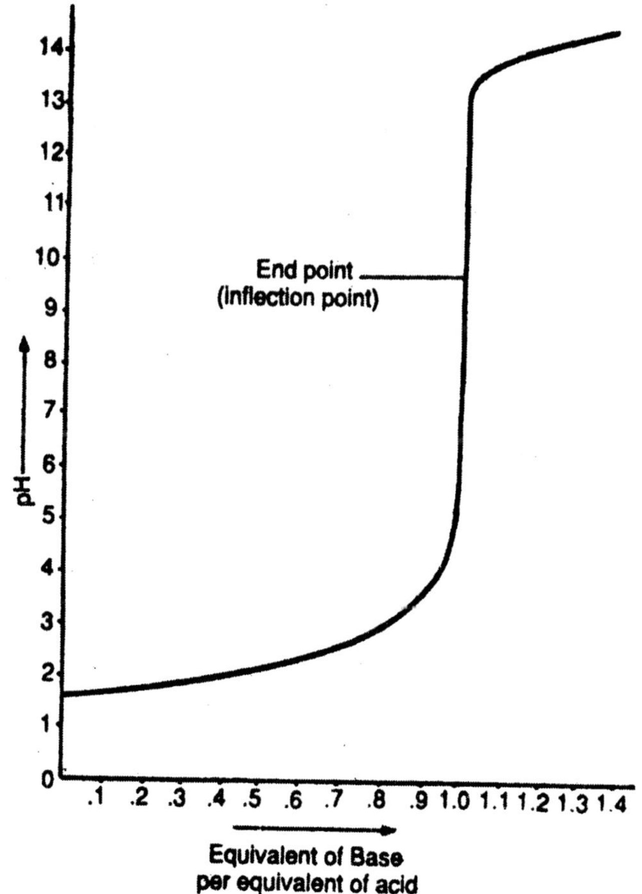

Fig. 2.19 : Titration fo a strong acid against a strong base.

In the titration of a strong base against strong acid the titration curve begins at high pH and remains flat till the end point is reached when there is a sudden fall in pH. Further addition of acid does not bring about an appreciable change in the pH of the solution and the curve is again flat. Here again the inflection point of the curve gives the end point.

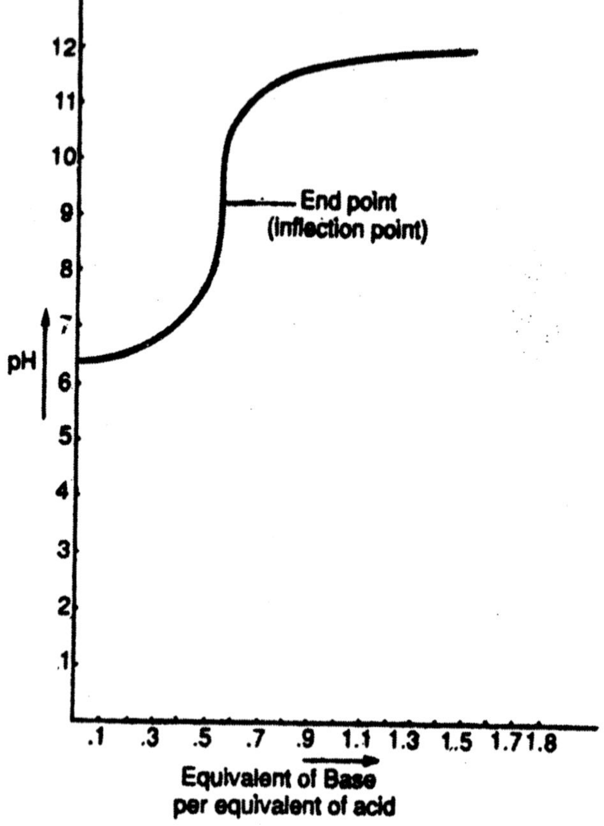

Fig. 2.20 : Titration of weak acid against a strong base.

(ii) *Titration of a weak acid against a strong base:* Fig. 2.20 shows the pH titration curve of solution of a weak acid (say acetic acid) against the solution of a strong base (say NaOH). Before adding alkali to the acid solution, its pH can be calculated from its ionization constant and its concentrations On adding alkali in the beginning, the free hydrogenion concentration decreases due to the neutralisation by hydroxyl ions furnished in the solution by strong base and there is a big change in pH.

End point in this case also is given by the inflection point of the curve but the jump in pH in this case is smaller than that in the case of pH-titration curve when a strong acid and strong base are involved in the titration. The jump in pH depends on the pK_a of the acid. Higher the pK_a of the acid, lower is the jump in pH at the end point it will be seen that vertical portion now begins beyond pH 7 and end point lies between pH 8 and 10. This is due to the hydrolysis of the salt which gives OH^- ions in aqueous solution.

In case of the titration of a solution fo strong base against a solution fo a weak acid, the curve starts with a higher pH and ends at a lower pH and the inflection gives the end point.

The nature of this curve is the same otherwise.

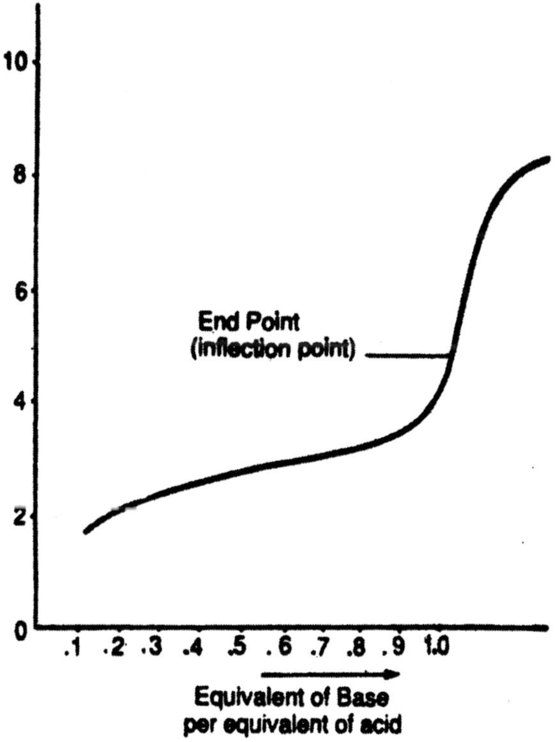

Fig. 2.21 : Titration of a strong acid against a weak base.

(iii) *Titration of a strong acid against a weak base:* There is not much difference in the titration curve of a strong acid against a weak base from the titration curve of a strong acid against a strong base till the end point is reached.

Here the vertical portion of the curve will now begin below pH 7 and end point lies between 4 and 6 (Fig. 2.21). This is again due to the hydrolysis of the salt which gives H^+ ions in aqueous solution.

(iv) *Titration of a weak acid against a weak base:* The nature of the curve depends on the ionisation constants of the weak acid (pK_a) and the weak base (pK_b) involved in the titration. If both the acid and base have very large pK_a and pK_b, the rise in pH at the end point is very small and may not be even observed.

Choice of an Indicator

The indicator selected for acid-base titrations should have the following characteristics:

(1) *It should change its colour at the pH, corresponding to the end point of the reaction, indicating that the reaction is complete.*

(2) *The colour change should be sharp.*

(3) *The change should be between the contrasting colours which are easily distinguished.*

Thus, in order to choose an indicator, it is necessary to know the pH of a solution at the end point. For example, when a strong acid is titrated against a strong base or *vice-versa* the resulting solution is very nearly neutral at the end point. Therefore, the ideal indicator will have a sharp colour change at about pH = 7. Thus, any indicator which undergoes colour change within the range 4—10 can be used. In the titration fo a weak acid with a strong base, the pH at the end point will be above 7 and any indicator changing colour between pH of 8 and 10 will be satisfactory.

Table 2.4 : Choice of indicators

Titration	*Marked pH range*	*Indicator*
Strong acid and strong base	4—10	Any indicator
Weak acid and strong base	7.5—10.5	Phenolphthalein (8.3—10)
Strong acid and weak base	3.5—6.5	Methyl red (4.4—6.3) or Methyl orange (3.1—4.5)
Weak acid and weak base	No marked change	End point cannot be detected accurately by any indicator.

In the titration of a strong acid and weak base, the pH at the end point will be below 7 and any indicator, changing colour between pH 3 and 6 will be satisfactory. In the titration of a weak acid with weak base, no sharp change in pH is obtained, so no indicator will give a satisfactory colour change. This means the choice of indicator is limited as follows:

Universal Indicators

These are mostly employed for determining the approximate pH of solutions. It is a mixture of methyl red, methyl orange, bromothymol blue and phenolphthalein. It covers a pH range of 3–11 and gives colour changes at different pH values as shown below:

Colour	*Red*	*Orange*	*Yellow*	*Yellowish green*	*Blue*	*Violet*
pH	3	4-5	6	7	9	10

BUFFER SOLUTIONS OR BUFFERS

We know that the pH of neutral water is 7. If we add 1 cc of 1N HCl to 1 litre of water, the $[H^+]$ concentration in the solution becomes 10^{-3} gm ions and the pH of the solution becomes 3. Similarly, the addition of 1 cc of 1N NaOH in a litre of pure water changes the pH of the solution to 11. Thus, we see that the pH of water changes by large amounts on the addition of small quantity of a strong acid or a strong base. But for certain biological and analytical purposes, we require solution whose pH should not change on keeping or on addition of small amounts of a strong acid or a strong base. *Such solutions which oppose the change in their pH on the addition of small amount of an acid or a base are called buffer solutions*. Some characteristics of buffer solution are:

(i) Their pH should not change on keeping.

(ii) Their pH should remain same on dilution.

(iii) Their pH should remain same on the addition of small amounts of strong acid and base.

Buffer Capacity

Often it is necessary to know the effectiveness of a buffer on a quantitative basis. To do so, we employ the term *buffer capacity* (β), first introduced by van Slyke in 1922. Buffer capacity is defined as the amount of acid or base that must be added to the buffer to produce a unit change of pH. Hence

$$\beta = \frac{d[B]}{d\,pH} \qquad ...(1)$$

where d[B] is the increase (in mol $litre^{-1}$) of strong base B. If a strong is added, Eq. (1) becomes

$$\beta = \frac{-d[B]}{-d\,pH}$$

The buffer capacity always has a positive value, however, since addition of base increases the pH and addition of acid decreases the pH. Thus, d[B] and dpH always have the same signs.

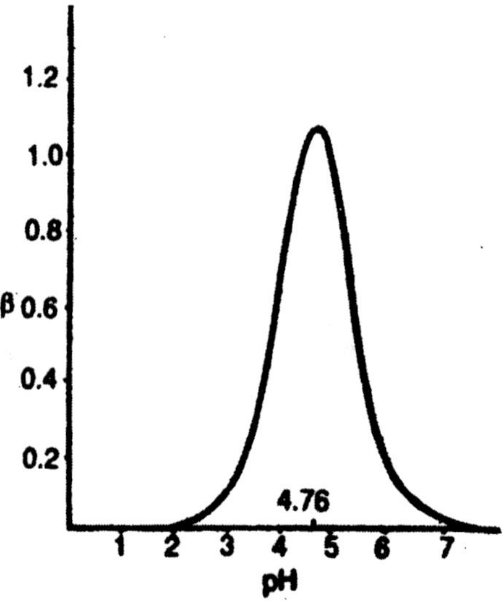

Fig. 2.22 : Buffer capacity of the 1 M CH_3COOH/1 M CH_3COONa buffer system. The maximum of the peak occurs at pH = 4.76, which is also equal to pK_a.

The value of β depends not only on the nature of the buffer, but also on the pH, which is determined by the relative concentrations of the acid and its conjugate base. Fig. 2.22 shows plots of buffer capacity versus pH for the CH_3COOH—CH_3COONa system. We see that the buffer functions best around its pKa value of 4.76.

Action of Buffers

A buffer solution invariably consists of a weak acid and its salt solution (acidic buffer) or a weak base and its salt solution (basic buffer). Some examples of buffer solutions are:

Constituents	*pH Range*
Glycine + Glycine hydrochloride	1.0 — 3.7
Phthalic acid + Potassium hydrogen phthalate	2.2 — 3.8
Acetic acid + Sodium acetate	3.7 — 5.6
Disodium citrate + Trisodium citrate	5.0 — 6.3
Monosodium phosphate + Disodium phosphate	5.8 — 8.0
Boric acid + Borax	6.8 — 9.2
Borax + Sodium hydroxide	9.2 — 11.2
Disodium phosphate + Trisodium phosphate	11.0 — 12.0

Action of Acidic Buffer

Let us illustrate its action by considering a mixture sodium acetate and acetic acid in water.

$$CH_3COOH \rightleftharpoons CH_3COO^- + H^+ \text{ (feebly ionised)}$$
$$CH_3COONa \rightleftharpoons CH_3COO- + Na^+ \text{ (highly ionised)}$$

As CH_3COONa is almost fully ionised, the acetate ions so produced suppress the ionisation of the acetic acid, so that the mixture contains unionised acetic acid molecules and a large number of acetate ions. When a few drops of an acid, say HCl, are added the H^+ ions from the acid added react with the CH_3COO^- ions to form unionised or feebly ionised acetic acid:

$$CH_3COO^- + H^+ \rightarrow CH_3COOH$$

On the other hand, when a few drops of an alkali, say, NaOH are added, the OH^- ions react with CH_3COOH to form unionised H_2O.

$$CH_3COOH + OH^- \rightarrow CH_3COO^- + H_2O$$

Thus, the addition of small amounts of the acid or alkali to a buffer solution does not alter the H^+ ions concentration of the buffer. In other words, the pH of the buffer remains unchanged.

Action of a Basic-buffer

Let us illustrate the action of a basic buffer by considering a mixture of NH_4OH and NH_4Cl in water. They exhibit the following equilibria

$$NH_4OH \rightleftharpoons NH_4^+ + OH^- \text{ (weakly ionised)}$$

$$NH_4Cl \rightleftharpoons NH_4^+ + Cl^- \text{ (completely ionised)}$$

The presence of the excess of NH_4^+ ions from completely ionised NH_4Cl suppresses the ionisation of NH_4OH, so that the mixture solution contains unionised NH_4OH molecules and a large number of NH_4^+ ions. When a few drops of alkali is added to this buffer solution, the OH^- ions produced from the alkali react with NH_4^+ present in solution to form unionised NH_4OH.

$$NH_4^+ + OH^- \rightarrow NH_4OH$$

On adding a few drops of the acid say HCl, the H^+ ions provided by the acid react with NH_4OH to give unionised H_2O.

$$NH_4OH + H^+ \rightarrow NH_4^+ + H_2O$$

Thus, the addition of small amounts of the alkali or acid to the above solution does not alter its H^+ ion concentration. In other words, the pH of buffer remains unchanged.

Mathematical Expression for the pH of an Acidic Buffer

Consider a solution having a weak acid HA and its salt BA, with a strong base. The ionisation of the weak acid is represented as:

$$HA \rightleftharpoons H^+ + A^-$$

or

$$K_a = \frac{[H^+][A^-]}{[HA]} \quad \text{or} \quad [H^+] = K_a \frac{[HA]}{[A^-]} \quad ...(1)$$

where K_a is the dissociation constant of the weak acid., The salt BA is almost completely ionised. Thus, high concentration of the A^- ions provided by almost complete ionisation of the salt BA suppresses the ionisation of the weak acid, so that it may be assumed that the free A^- ions are entirely due to the salt. Thus

$$[A^-] = [\text{Salt}] \text{ and } [HA] = [\text{Acid}] \quad ...(2)$$

Substituting Eq. (2) in (1), we get

$$[H^+] = K_a \frac{[\text{Acid}]}{[\text{Salt}]}$$

Taking logarithm of both sides, we get

$$\log [H^+] = \text{loh } K_a + \log \frac{[\text{Acid}]}{[\text{Salt}]}$$

or $$-\log [H^+] = -\text{loh } K_a - \log \frac{[\text{Acid}]}{[\text{Salt}]}$$

or $$pH = pK_a - \log \frac{[\text{Acid}]}{[\text{Salt}]} \quad ...(3)$$

$$= PK_a + \log \frac{[\text{Salt}]}{[\text{Acid}]}$$

where $pK_a = -\log K_a$. Equation (3) is known as Henderson-Hasselbalch equation.

Mathematical Expression for the pH of a Basic Buffer

Consider a weak base BOH and its salt AB. The dissociation of the weak base may be written as

$$BOH \rightarrow B^+ + OH^-$$

or $$K_a = \frac{[B^+][OH^-]}{[BOH]} \text{ or } [OH^-] = \frac{K_b[BOH]}{[B^+]} \quad ...(1)$$

where K_b is the dissociation constant of the weak base.

The high concentration of B^+ ions produced by the complete ionisation of the salt, which thus suppresses the ionisation of the weak base so that it may be assumed that free B^+ ions in the solution are entirely due to the salt. Thus,

$$[B^+] = \text{Salt and } [BOH] = \text{Base}$$

Thus, equation (1) may be put as

$$[OH^-] = K_b \frac{[\text{Base}]}{[\text{Salt}]}$$

Taking the logarithm of the above equation, we obtain

$$\log [OH^-] = \log K_b + \log [\text{Base}] - \log [\text{Salt}]$$

or $$-\log [OH^-] = -\log K_b - \log [\text{Base}] + \log [\text{Salt}]$$

or $$pOH = pK_b + \log \frac{[\text{Salt}]}{[\text{Base}]} \quad ...(2)$$

But pH = 14 – pOH, Equation (2) becomes as

$$pH = 14 - \left(pK_a + \log \frac{[\text{Salt}]}{[\text{Base}]} \right) \quad ...(3)$$

Now $[H^+][OH^-] = k_w = 10^{-14}$ or $\log k_w = -\log 10 = -14$

or $$14 = -\log k_w \qquad ...(4)$$

From Eqs. (3) and (4), we get

$$pH = -\log k_w - \left(pK_b + \log \frac{[Salt]}{[Base]}\right)$$

$$= -\log k_w - \left(-\log K_b + \log \frac{[Salt]}{[Base]}\right)$$

$$= -\log \frac{k_w}{k_b} + \log \frac{[Base]}{[Salt]} \qquad ...(5)$$

HYDROLYSIS

Pure water is neutral and is feebly ionised into H^+ and OH^- ions as:

$$H_2O \rightarrow H^+ + OH^-$$

The concentration of H^+ and OH^- ions is same. Now when a salt is dissolved in water, it ionises into ions, *e.g.*, the salt AB will ionise as:

$$AB \rightarrow A^+ + B^-$$

The ions of the salt may react with the ions of water to give

$$A^+ + OH^- \rightarrow AOH$$

$$B^- + H^+ \rightarrow HB$$

Now, the solution will be acidic or alkaline depending upon the relative strengths of the acid HB and the base AOH. If the acid is strong and the base is weak, then the solution will be acidic in nature. If the base is strong and the acid is weak, the solution will be basic in nature. Thus, this phenomenon in which a salt interacts with water to produce either an acidic or alkaline solution is termed as hydrolysis. Actually, hydrolysis involves the interaction of the ions of a salt with the water to produce H^+ or OH^- ions in solution. Thus, it may be defined as:

"*The interaction between the ions of salt and the ions of water to given free H^+ or OH^- ions in solution.*"

Example of Hydrolysis

Salt may be divided into four categories:

(a) Salts of Strong Acid and Strong Bases

These salt, when dissolved in water, produce a strong acid and a strong base, so that the relative proportions of H^+ and OH^- ions in solution remain unchanged and the solution remains neutral. IN simple words, it can be said that the salts of strong acids and bases do not undergo hydrolysis. Let us consider the dissolution of sodium chloride in water.

$$NaCl \rightarrow Na^+ + Cl^-$$

$$H_2O \rightarrow OH^- + H^+$$

$$Na^+ + OH^- \rightarrow NaOH \text{ (highly ionised)}$$

$$H^+ + Cl^- \rightarrow HCl \text{ (highly ionised)}$$

But the acid, HCl and the base, NaOH being equally strong are almost completely ionised. Thus, the concentration of H^+ and OH^- is not altered and the solution remains neutral. Examples of such salts are NaCl, KCl, Na_2SO_4, K_2SO_4, etc.

(b) Salts of Weak Acids and Strong Bases

These salts (like Na_2CO_3; CH_3COONa, KCN, etc.), when dissolved in water undergo hydrolysis in water to give an alkaline solution. For example,

$$CH_3COONa \rightarrow CH_3COO^- + Na^+$$

$$H_2O \rightarrow H^+ + OH^-$$

$$CH_3COO^- + H^+ \rightarrow CH_3COOH$$

$$Na^+ + OH^- \rightarrow NaOH$$

NaOH being a strong base is almost completely ionised and the concentration of OH^- ions in solutions is not altered. However, CH_3COOH being a weak acid is feebly ionised. Thus the concentration of H^+ ions in solution falls. Thus, the solution has more of OH^- ions than the H^+ ions. Hence, it is alkaline in nature.

(c) Salts of Strong Acids and Weak Bases

Examples are NH_4Cl, $CuSO_4$, etc. When such salts are dissolved in water, they produce acidic solutions. Consider the dissolving of ammonium chloride.

$$NH_4Cl \rightarrow NH_4^+ + Cl^-$$

$$H_2O \rightarrow H^+ + OH^-$$

$$NH_4^+ + OH^- \rightarrow NH_4OH$$

$$H^+ + Cl^- \rightarrow HCl$$

HCl being a strong acid is almost completely ionised and the concentration of the H^+ ions in solution is not affected. NH_4OH, however, is a weak base and feebly ionises to give only few OH^- ions in solution. Thus, the solution has a large concentration of the H^+ ions that the OH^- ions. Hence the solution is acidic in nature.

(d) Salts of Weak Acid and Weak Bases

Examples are CH_3COONH_4, $(NH_4)_2CO_3$, etc. Such salts when dissolved in water yield acidic or basic or even neutral solutions, depending upon the nature of the acid and the base produced. Consider the dissolving of ammonium acetate in water.

$$CH_3COONH_4 \rightarrow CH_3COO^- + NH_4^+$$

$$H_2O \rightarrow H^+ + OH^-$$

$$CH_3COO^- + H^+ \rightarrow CH_3COOH$$

$$NH_4^+ \; OH^- \rightarrow NH_4OH$$

Both CH_3COOH and NH_4OH are weak and are equally but feebly ionised. Hence the solution contains practically the same number of the H^+ and OH^- ions. Therefore, the solution is almost neutral. In case of $(NH_4)_2CO_3$, however, the solution is slightly basic because NH_4OH is relatively a stronger base than the H_2CO_3 acid.

MATHEMATICAL TREATMENT OF HYDROLYSIS

(a) Salt of Strong Acids and Weak Bases

(i) *Hydrolysis constant.* The hydrolysis of a salt of a strong acid and a weak base may be represented as:

$$A^+B^+ + H_2O \rightarrow HOH + B^- + H^+$$

or

$$A^+ + H_2O \rightarrow AOH + H^+$$

Applying the law of mass action to the above equilibrium,

$$K = \frac{[AOH][H^+]}{[A^+][H_2O]}$$

As water is taken in large excess, it means that its active mass remains practically constant. Therefore,

$$K[H_2O] = \frac{[AOH][H^+]}{A^+}$$

$$K_h = \frac{[AOH][H^+]}{A^+}, \text{ where } K_h = K[H_2O] \quad ...(1)$$

The constant K_h is called the hydrolysis constant of the salt.

(ii) *Relation between K_h, K_b and K_w.* The dissociation of the weak base produced in the hydrolysis of a strong acid and a weak base may be represented as:

$$AOH \rightarrow A^+ + OH^-$$

Applying the law of mass action,

$$K_b = \frac{[A^+][OH^-]}{[AOH]} \quad ...(2)$$

where K_b is the dissociation constant of the weak base.

Multiplying equations (1) and (2), we obtain

$$[H^+][OH^-] = K_h \times K_b$$

But $[H^+][OH^-] = K_w$, the ionic product of water

$$\because \quad K_w = K_h \times K_b$$

or

$$K_h = \frac{k_w}{K_b}. \quad ...(3)$$

From equation (3), it is evident that the *weaker the base, the greater will be the hydrolysis constant of the salt.*

(iii) *Degree of hydrolysis:* Degree of hydrolysis of salt may be defined as:

"The fraction of the total salt hydrolysed when the equilibrium has been established is called the degree of hydrolysis."

It is denoted by the letter '*h*'. In simple words, the extent to which the salt has been hydrolysed is known as the degree of hydrolysis. If *c* is the initial concentration of the salt in moles per litre and *h* is the degree of hydrolysis on the attainment of equilibrium, then the concentrations of various species will be represented as:

$$\underset{c(1-h)}{A^+} + \underset{\text{Excess}}{H_2O} \rightarrow \underset{ch}{AOH} + \underset{ch}{H^+} \quad ...(4)$$

Applying the law of mass action,

$$K_h = \frac{(ch)(ch)}{c(1-h)} \qquad ...(5)$$

$$= \frac{ch^2}{(1-h)} = \frac{ch^2}{(1-h)}$$

If the base is very weak, the value of h is very small as compared to unity. It means that 1 – h in the equation (5) may be replaced by 1, *i.e.*,

$$K_h = ch^2$$

or

$$h = \sqrt{\left(\frac{K_h}{c}\right)} \qquad ...(6)$$

From equation (3), we know

$$K_h = \frac{k_w}{K_c} \qquad ...(7)$$

Replacing the above value of K_h in equation (6), we obtain

$$h = \sqrt{\left(\frac{k_w}{cK_b}\right)} \qquad ...(8)$$

From the relation (8), it follows that the degree of hydrolysis of a salt of a weak base and a strong acid is:

(i) *directly proportional to the ionic product of water*

(ii) *inversely proportional to the value of* K_b, *and*

(iii) *inversely proportional to the value of c.*

From the above it follows that degree of hydrolysis increases if the base is weaker *i.e.*, if K_b is smaller, and also increases if the concentration decreases or the dilution increases.

As k_w increases much more than K_b, it means that the degree of hydrolysis at a given concentration increases with the increase in temperature.

(iv) Relation between pH, pK_b and c. From equation (4), it follows:

$$[H^+] = ch$$

$$= c\sqrt{\left(\frac{k_w}{cK_b}\right)} \qquad \text{[From equation (8)]}$$

$$= \sqrt{\left(\frac{ck_w}{K_b}\right)} \quad ...(9)$$

Taking logarithms and changing the signs throughout, Eq. (9) becomes as:

$$-\log [H^+] = -\frac{1}{2}\log k_w + \frac{1}{2}\log K_b - \frac{1}{2}\log c$$

$$pH = \frac{1}{2} pk_w - pK_b - \frac{1}{2}\log c \quad ...(10)$$

From equation (10), it follows that the pH of the solution must be less than $\frac{1}{2}pK_w$ and is less than 7.0 and so solutions of salts of the type under consideration will be acid solutions.

(b) Salts of Weak Acids and Strong Bases

(i) *Hydrolysis constant*: The hydrolysis of a salt of a weak acid and a strong base may be represented as:

$$(A^+ + B^-) + H_2O \rightarrow HB + A^+ + OH^-$$

$$B^- + H_2O \rightarrow HB + OH^-$$

Applying the law of mass action,

$$K = \frac{[OH^-][HB]}{[B^-][H_2O]} \quad ...(11)$$

As water is taken in large excess, its active mass remains practically constant. Therefore,

$$K[H_2O] = \frac{[OH^-][HB]}{[B^-]}$$

or

$$K_h = \frac{[OH^-][HB]}{[B]} \quad ...(12)$$

where K_h is known as the hydrolysis constant of the acid.

(ii) *Relation between K_h, K_a and k_w*: The dissociation of weak acid, HB, produced during hydrolysis may be represented as:

$$HB \rightleftharpoons H^+ + B^-$$

Applying the law of mass action,

$$K_a = \frac{[H^+][B^-]}{[HB^-]} \quad ...(13)$$

where K_a is known as the dissociation constant of the weak acid.,

Multiplying Eq. (12) by (13), we get

$$[H^+][OH^-] = K_a \times K_h$$

But $$[H^+][OH^-] = k_w$$

$\therefore$ $$k_w = K_a \times K_b$$

or $$K_b = \frac{k_w}{K_a} \quad ...(14)$$

From eq. (14), it follows that the hydrolysis constant, K_h, of the salt varies inversely as the dissociation constant, K_a, of the weak acid. Therefore, the weaker the acid, the greater is the hydrolysis constant of the salt.

(iii) *Degree of hydrolysis:* If c is the initial concentration fo the salt in moles per litre and h is the degree of hydrolysis on the attainment of equilibrium, the concentration of various species may be represented as:

$$\underset{c(1-h)}{B^-} + \underset{\text{excess}}{H_2O} \rightarrow \underset{ch}{HB} + \underset{ch}{OH^-} \quad ...(14A)$$

Applying the law of mass action,

$$K_h = \frac{(ch)(ch)}{c(1-h)} = \frac{ch^3}{1-h} \quad ...(15)$$

If the acid is very weak, h is negligible as compared to unity, *i.e.*, $1 - h \approx 1$. Thus, equation (15) becomes as:

$$K_h = ch^2$$

or $$h = \sqrt{\left(\frac{K_h}{c}\right)}$$

From equation (14), we have

$$K_h = \frac{k_w}{K_a}$$

$\therefore$ $$h = \sqrt{\left(\frac{k_w}{cK_a}\right)} \quad ...(16)$$

From equation (16), it follows that weaker the acid, greater is h, *i.e.*, the degree of hydrolysis. As value of k_w is also increased by a rise in temp., it means that the degree of hydrolysis 'h' will considerably increase with rise of temperature It is also seen that degree of hydrolysis 'h' also increases if the concentration decreases *i.e.*, dilution increases.

(iv) *Relation between pH, pK_a and c:* From Eq. (14A), the concentration of $[OH^-]$ ion is given by

$$[OH^-] = ch$$

But $$[H^+][OH^-] = k_w$$

$$[H^+] = \frac{k_w}{ch} \qquad \text{...(17)}$$

From equation (16), we have

$$h = \sqrt{\left(\frac{k_w}{cK_a}\right)}$$

Substituting the value of h in (17), we have

$$[H^+] = \frac{k_w}{c}\sqrt{\left(\frac{cK_a}{k_w}\right)} = \sqrt{\left(\frac{k_w K_a}{c}\right)} \qquad \text{...(18)}$$

Taking logarithms and changing the signs throughout, equation (18) becomes as

$$-\log[H^+] = -\frac{1}{2}\log k_w - \frac{1}{2}\log K_a + \frac{1}{2}\log c$$

or $$pH = pk_w + \frac{1}{2}pK_e + \frac{1}{2}\log c \qquad \text{...(19)}$$

From equation (19), it follows that pH or alkalinity of a solution of the salt of a weak acid and strong base increases with decreasing and strength, *i.e.*, increasing pK_a and increasing concentration.

(c) Salt of Weak Acids and Weak Bases

(i) *Hydrolysis constant*: The hydrolysis of the salt of a weak acid and a weak base may be represented as:

$$AB + H_2O \rightleftharpoons AOH + HB$$

Ionically, the above equilibria can be represented as:

$$A^+ + B^- + H_2O \rightleftharpoons \underset{\substack{\text{weak}\\\text{base}}}{AOH} + \underset{\substack{\text{weak}\\\text{acid}}}{HB} \qquad ...(19A)$$

Applying the law of mass action

$$K = \frac{[AOH][HB]}{[A^+][B^-][H_2O]}$$

$$K[H_2O] = \frac{[AOH][HB]}{[A^+][B^-]}$$

$$K_h = \frac{[AOH][HB]}{[A^+][B^-]} \qquad ...(20)$$

where K_h is the hydrolysis constant of salt.

(ii) Relation between K_h, k_w, K_a and K_b: The dissociation of the weak acid and weak base may be represented as:

$$HB \rightarrow H^+ + B^- \qquad ...(21)$$

and

$$AOH \rightarrow A^+ + OH^- \qquad ...(22)$$

Applying the law of mass action to the above equilibria (21) and (22),

$$K_a = \frac{[H^+][B^-]}{[HB]} \qquad ...(23)$$

$$K_b = \frac{[A^+][OH^-]}{[AOH]} \qquad ...(24)$$

where K_a and K_b are the dissociation constants of the weak acid and base respectively. Multiplying Eqs. (20), (23) and (24),

$$[H^+][OH^-] = K_a \times K_b \times K_h$$

But

$$[H^+][OH^-] = k_w$$

$\therefore$

$$K_w = K_a \times K_b \times K_h$$

or

$$K_h = \frac{k_w}{K_a \times K_b} \qquad ...(25)$$

(iii) *Degree of hydrolysis* If c is the initial concentration of the salt in moles/litre and 'h' is the degree of hydrolysis, then the concentration of various species of equilibria (19A) may be represented as:

$$\underset{c(1-h)}{A^+} + \underset{c(1-h)}{B^-} + \underset{\text{excess}}{H_2O} \rightarrow \underset{ch}{AOH} + \underset{ch}{HB} \qquad ...(26)$$

Applying the law of mass action,

$$K_h = \frac{[ch][ch]}{c(1-h)c(1-u)} = \frac{h^2}{(1-h)^2} \qquad ...(27)$$

If h being very small, h may be neglected and 1 – h may be taken as unity. Thus, equation (27) becomes as:

$$K_h = h^2 \qquad ...(28)$$

From Eq. (28) it follows that degree of hydrolysis of a weak acid and a weak base is independent of the dilution or concentration.

(iv) *Relation between pH, pK_w, pK_a and pK_b:* From equation (23), it follows that

$$[H^+] = K_a \frac{[HB]}{[B^-]} \qquad ...(29)$$

From equation (26), we get

$$B^- = c(1-h) \text{ and } HB = ch \qquad ...(30)$$

Substituting equation (30) in (29), we have

$$[H^+] = K_a \frac{ch}{c(1-h)} = K_a \frac{h}{1-h}$$

$$= K_a(K_h)^{1/2} \qquad \text{[from equation (27)]}$$

$$= K_a\left(\frac{k_a}{K_a K_b}\right)^{1/2} \qquad \text{[from equation (25)]}$$

$$= \left(\frac{k_w K_a}{K_b}\right)^{1/2}$$

$$= K_4(K_h)^{1/2} \qquad \text{[from equation (27)]}$$

$$K_a = \left(\frac{K_w}{K_b}\right)^{1/2} \qquad \text{[from equation (25)]}$$

$$= \left(\frac{k_w K_a}{K_a K_b}\right) \qquad ...(31)$$

Taking logarithm of both sides obtain, we get

$$-\log [H^+] = -\frac{1}{2}\log k_w - \frac{1}{2}\log K_a + \frac{1}{2}\log K_b$$

$$pH = \frac{1}{2}pk_a + \frac{1}{2}pK_a - \frac{1}{2}pK_b \qquad ...(32)$$

Now three different cases may arise:

(i) When $K_a = K_b$, it follows from equation (32), $pH = \frac{1}{2}pk_w = 7$, *i.e.*, the solution will be neutral.

(ii) When $K_a > K_b$; $pH < 7$, *i.e.*, the solution will be acidic.

(iii) When $K_a < K_b$; $pH > 7$, *i.e.*, the solution will be alkaline.

Determination of Degree of Hydrolysis

(a) From Colligative Properties

The colligative property of a solution number of the solute particles. Consider, 1 gm mole of the salt of weak acid and strong base like CH_3COONa which gets hydrolysed in the solution as:

$$\underset{1-x}{CH_3COO^-} + Na^+ + \underset{1-x}{H_2O} \rightarrow \underset{x}{CH_2COOH} + \underset{x}{Na^+} + \underset{x}{OH^-}$$

where x is the degree of hydrolysis. Then,

total number of particles before hydrolysis = 1 + 1 = 2

and the total number of particles after hydrolysis

$$= 1 - x + 1 - x + x + x + x$$

$$= 2 + x$$

$$\because \quad \frac{\text{Observed freezing point depression}}{\text{Calculated freezing point depression}} = \frac{2 + x}{2}$$

From the above relation, x can be calculated.

(b) From Conductance Measurements

The method is based upon the simple assumption that the equivalent conductance of an aqueous solution of a salt of a weak acid and strong base (or a weak base and strong acid) is due to the sum of equivalent conductances of strong base (or a weak base and strong acid) is due to the sum of equivalent conductances of

(a) the unhydrolysed salt and

(b) the free OH^- (or the H^+) ions produced by hydrolysis.

The weak acid or weak base formed during hydrolysis do not contribute anything towards the equivalent conductance of the salt solution at infinite dilution.

Procedure: The procedure involves the following steps:

(i) An aqueous solution fo salt of known dilution is prepared. Now determine the equivalent conductance at a certain dilution, say, V, *i.e.* λ_V.

(ii) Now mix the aqueous solution of the salt with excess of the weak acid or weak base as the case may be or the same which is produced during hydrolysis. This addition of weak acid or weak base checks the hydrolysis of the salt. Now determine the equivalent conductance of the mixture at the same dilution as before. Let it be λ'_V. Actually, this is the equivalent conductance of the unhydrolysed salt.

(iii) The equivalent conductance of the strong base or acid at infinite dilution is also determined by Kohlrausch's law. Let it be λ_∞.

Calculation: From the above values of the various equivalent conductances, the degree of hydrolysis may be calculated as follows:

Consider a salt AB of a strong acid and a weak base. Its hydrolysis may be represented as:

$$A^+B^- + H_2O \rightleftharpoons AOH + H^+ + B^-$$

or
$$\underset{1-h}{A^+} + H_2O \rightleftharpoons \underset{h}{AOH} + \underset{h}{A^+}$$

If h is the degree of hydrolysis, then

Concentration f unhydrolysed salt = 1 – h

Concentration of free H^+ ions = h

Equivalent conductance due to unhydrolysed salt = $\lambda'_V - (1-h)$ and equivalent conductance due to H^+ ions = $\lambda_\infty \times h$

∴ Equivalent conductance of hydrolysed salt,

$$\lambda_V = \lambda_\infty \times h + \lambda'_V(1-h)$$

or
$$h = \frac{\lambda_V - \lambda'_V}{\lambda_\infty - \lambda'_V}$$

This is known as Bredig's method.

(c) pH Determination Method

In case of the hydrolysis of a salt of a strong acid and a weak base,

$$A^+ + H_2O \rightleftharpoons AOH + H^+$$

and the value of $[H^+]$ is given as $[H^+] = ch$ ...(1)

where 'C' is the concentration of the salt in gm moles/litre and h is the degree of hydrolysis.

Taking logarithm of both sides of equation (1), we obtain

$$\log [H^+] = \log ch$$

or $$-\log [H^+] = -\log ch$$

or $$pH = -\log ch \quad ...(2)$$

In actual practice the aqueous solution of the salt is taken and its pH is determined with the help of pH meter. Thus, one can calculate the degree of hydrolysis.

(d) From the Dissociation Constants of the Weak Acids or Weak Bases

As we know in the case of

(i) a salt of a weak acid and a strong base

$$h = \sqrt{\left(\frac{k_w}{K_a \times C}\right)}] \quad ...(1)$$

(ii) a salt of a weak base and a strong acid

$$h = \sqrt{\left(\frac{k_w}{K_b \times C}\right)} \quad ...(2)$$

(iii) a salt of a weak acid and a weak base

$$h = \sqrt{\left(\frac{k_w}{K_a K_b}\right)} \quad ...(3)$$

Knowing dissociation constant of acid or base (K_a or K_b) or both as the case may be, it is a simple matter to calculate the degree of hydrolysis of a given salt at a given concentration.

(e) Distribution Method

This method can also be used provided the free acid or base formed during hydrolysis is soluble in a immiscible solvent like benzene. Let us illustrate this principle by considering the hydrolysis of aniline hydrochloride.

When aniline hydrochloride is hydrolysed, the following equilibria are established in solution.

$$C_6H_5NH_3^+Cl^- + H_2O \rightarrow C_6H_5NH_3^+OH^- + HCl$$
$$\downarrow$$
$$C_6H_5NH_2 + H_2O$$

Applying the law of mass action, the hydrolysis constant K_b is given as:

$$K_h = \frac{[\text{Concentration of free base}]\,[\text{Concentration of free acid}]}{[\text{Concentration of unhydrolysed salt}]}$$

If c is the initial concentration of the salt in moles per litre, then the degree of hydrolysis is given by:

$$h = \left(\frac{K_h}{c}\right)^{1/2} \quad ...(1)$$

The concentration of aniline formed by hydrolysis of aniline hydrochloride can be determined as follows:

(i) A known amount of aniline is distributed between benzene and water. The concentration of aniline in the benzene layer is measured by passing dry HCl gas through it and weighing the amount of aniline 'hydrochloride precipitated. Difference from the total weight of aniline taken gives the concentration of aniline in water layer. When distribution law is applied, distribution coefficient 'K' is given by

$$K = \frac{\text{Aniline in benzene}}{\text{Aniline in water}}$$

(ii) Now shake a known weight of aniline hydrochloride with a known weight of benzene and water. The free aniline formed distributes itself between benzene and water. Amount of the aniline in the benzene layer is determined by precipitating as aniline hydrochloride by passing dry HCl gas.

Then, calculate the amount of aniline in the water layer by using the distribution coefficient determined in the first step.

The sum of the weights of aniline in the benzene and water layers gives the total weights of aniline formed by hydrolysis. The concentration of free HCl (in moles per litre) will be the same as of aniline in moles/litre.

The concentration of unhydrolysed salt (c)

= [Initial concentration of the salt taken]

– [the loss of concentration due to hydrolysis]

= [Initial concentration of the salt]

– [Concentration of aniline formed by hyrolysis]

Substituting the various values in (1), the hydrolysis constant or the degree of hydrolysis of aniline hydrochloride can be determined.

(f) From E.M.F. Data

For details see the 'Applications of E.M.F. Measurements.'

MIGRATION OF IONS

Take a U-tube of uniform cross-section. Fill the lower limb of A, bend and the greater portion of the limb B with hot 4% solution of gelatin containing a few cc's of saturated KCl (to make it conducting) and a few drops of KOH plus a drop of phenolphathalein (to make it pink coloured). Allow it to cool, when it sets to pink coloured jelly. Then fill about 3/4th of the remaining portion of limb A with hot solution of gelatin + KCl + a drop of phenolphthalein. Allow it to cool when it sets to form a colourless jelly.

Then fill the remaining upper portions of the limb A and B with KOH and HCl solutions respectively. Finally, dip in these solutions two platinised platinum electrodes in such a way that – ve electrode is in contact with KOH solutions, while + ve electrode is in contact with HCl solution shown in Fig. 2.23.

On passing electric current, the H^+ ions move towards cathode; while OH^- ions move towards anode. The movement of these ions is indicated by decolorisation of pink colour in limb B and appearance of pink colour in limb A respectively. From the lengths of changed columns, due to the movement of H^+ and OH^- ions, it is found that speed of H^+ ions is nearly double that of OH^- ions. *Hence, it proves that different types of ions move with different speeds.*

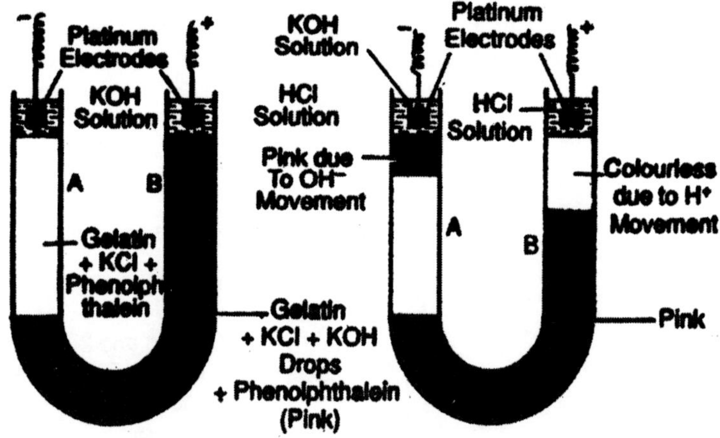

(a) Before passing electric current. (b) After passing electric current

Fig. 2.23 : Qualitative evidence for different speeds of ions.

SPEEDS OF IONS AND AMOUNTS LIBERATED AT THE ELECTRODES

When an electrolyte is submitted to slow electrolysis using non attackable electrodes it is found that *two types of ions are discharged in equivalent amounts,* but *loss of electrolyte around the two electrodes is not the same.* This clearly shows that *oppositely charged ions migrate with different speeds* and the *ions which migrate with greater speed produce a greater loss in amount of electrolyte around the electrode it leaves.* The same may theoretically be illustrated as follows:

(i) Consider an electrolytic cell divided into three imaginary compartments as presented by dotted lines. Let us suppose that there are 4 pairs each of cations and anions in anodic and cathodic compartments; while 6 pairs in the middle compartment before electrolysis (Fig. 2.24).

(ii) *Let only 2 anions move* from the cathodic to anodic compartment. The surplus unpaired ions in each compartment are two, which are discharged. As a result of this the *loss of number of molecules in the cathodic compartment is 2 molecules* (*i.e.*, 2 + ions and 2 – ions), while there is *no change in the anodic compartment.*

(iii) *Let cations and anions move with same speed:* Suppose in a particular time 2 anions and 2 cations move towards opposite electrodes.

The discharge of cations and anions at the two electrodes is again the *same* (*i.e.*, 4 ions each) and the *loss of electrolyte in the two compartments is also same, i.e.*, 2 molecules each.

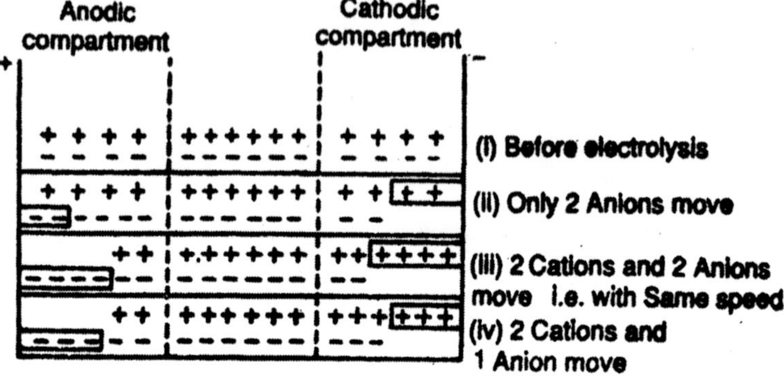

Fig. 2.24

(iv) *Let cations move at double the speed of anions:* Suppose in a particular time one anion and two cations move to the opposite electrodes. Again, the total number of ions discharged in each compartment is *same*, *i.e.*, 3 ions each. However, the *loss of electrolyte around the anode* (*i.e.*, 2 molecules) *is double the loss of electrolyte around the cathode* (*i.e.*, 2 molecules).

For the above it is clear that:

(a) *Loss of electrolyte around an electrode is proportional to the speed of the ion leaving it i.e.,*

$$\frac{\text{Loss around cathode}}{\text{Loss around anode}} = \frac{\text{Speed of cation}}{\text{Speed of anion}}$$

(b) *Whatever be the relative speeds of the two ions, the discharge of ions at the two electrodes is always equal.*

TRANSPORT NUMBER

The fraction of the total current carried by any one of the ionic species is known as transport number, transference number or Hittorf number of the species and may be denoted by sets of symbols like t_+ and t_-, t_a and t_c or n_c are n_a.

From the above definition, it follows that the transport number of an anion is

$$t_- = \frac{\text{Current carried by an anion}}{\text{Total current passes through the solution}} \quad ...(1)$$

and the transport number of cation is

$$t_+ = \frac{\text{Current carried by the cation}}{\text{Total current passed through the solution}} \quad ...(2)$$

But we know that the quantity of electricity carried by each ion is directly proportional to the speed of the concerned ions.

So, amount of electricity carried by anion $\propto$ speed of anion (u_a) amount of electricity carried by cation $\propto$ speed of cation (u_c).

$\because$ Total amount of electricity carried $\propto$ speed of anion + speed of cation

$$\propto u_a + u_c$$

Thus, the transport number of anion [From eq. (1), t is]

$$t_- = \frac{u_a}{u_a + u_c} \quad ...(3)$$

and the transport number of cation [From eq. (2)], t_+, is

$$t_+ = \frac{u_c}{u_a + u_c} \quad ...(4)$$

It follows from equation (3) and (4) that the transport number of an ion may be defined as the ratio of speed of the ion to the sum of the speeds of both cation and anion.

By adding Eqs. (3) and (4), we get

$$t_- + t_+ = \frac{u_a}{u_a + u_c} + \frac{u_c}{u_a + u_c} = \frac{u_a + u_c}{u_a + u_c} = 1 \quad ...(5)$$

$$t_- + t_+ = 1$$

From eq. (5), it follows that

(i) The sum of the transport numbers of all the species taking part in the transport of electricity is equal to unity, and

(ii) If the transport number of one of the ionic species is known, then that of the other can be easily calculated by using eqn. (5).

Experimental Determination of Transport Number

There are several methods for determining the transport numbers of ions, viz.

(i) Hittorf's Method

Principle : This method is based on the principle that "*During electrolysis, the fall of concentration around the electrode is proportional to the speed of ions moving away from that electrode.*"

Thus, the fall of concentration around the cathode ∝ speed of the anion

$$(u_a), \qquad ...(6)$$

and the fall of concentration around the anode ∝ speed of the cation

$$(u_c) \qquad ...(7)$$

Dividing eq. (6) by (7), we get

$$\frac{\text{Fall of concentration around the cathode}}{\text{Fall of concentration around the anode}} = \frac{u_a}{u_c}$$

Adding 1 to both sides, we get

$$1 + \frac{\text{Fall of concentration around the cathode}}{\text{Fall of concentration around the anode}} = 1 + \frac{u_a}{u_c}$$

or

$$\frac{\text{Fall of concentration around the anode + Fall of concentration around the cathode}}{\text{Fall of concentration around the anode}} = \frac{u_c + u_a}{u_c}$$

or

$$\frac{\text{Total fall of concentration}}{\text{Fall of concentration around the anode}} = \frac{u_c + u_a}{u_c}$$

Taking reciprocals of the abode equation,

$$\frac{\text{Fall of concentration around the anode}}{\text{Total fall of concentration}} = \frac{u_c}{u_c + u_a} \qquad ...(8)$$

From equation (4), we know $t_+ = \frac{u_c}{u_a + u_c}$

$$\frac{\text{Fall of concentration around the anode}}{\text{Total fall of concentration}} = t_+ \qquad ...(9)$$

Similarly,

$$\frac{\text{Fall of concentration around the cathode}}{\text{Total fall of concentration}} = t_- \qquad ...(10)$$

Discussion of the Principle

(i) Equations (9) and (10) are valid only if the electrodes are not attacked by the ions in solution.

(ii) If the anode is attacked by the anions, there will be an increase in the concentration around the anode instead of decrease. However, this increase in concentration at anode is equal to decrease in concentration that would have taken place if the anode were inert. Example is in the case of electrolysis of $AgNO_3$ using Ag anode.

Apparatus: Suppose one wants to determine the transport number of Ag^+ and NO_3^- ions in a solution of silver nitrate. For this an apparatus of the type shown in Figure 2.25 is set up.

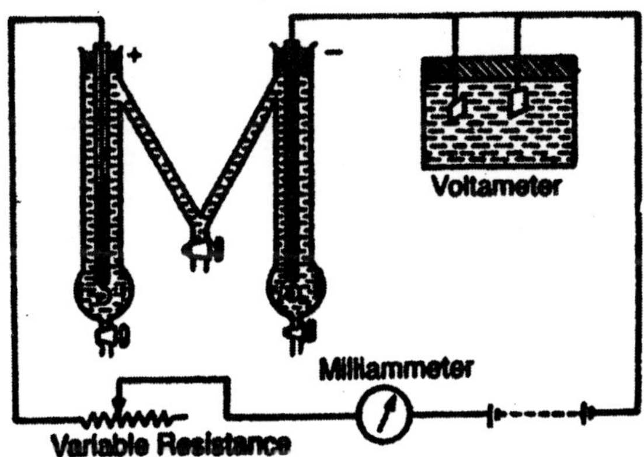

Fig. 2.25

(i) It consists of two vertical glass tubes connected by a V-tube in the middle. The end tubes containing the anode and cathode constitute the anodic and cathodic compartments respectively and V-tube constitutes the central compartment.

(ii) The tubes are provides with stop cocks at the bottom.

(iii) The electrodes are made of a suitable metal (attackable or non-attackable) and sealed in glass tubes which pass through rubber stoppers as shown in Fig. 2.25. Some mercury is also placed in the glass tubes to ensure proper contact.

(iv) The apparatus is filled up with a standard solution of silver nitrate.

(v) In addition to this apparatus, the circuit also contains a silver voltameter or an ammeter for determining the total electricity passed. The voltameter contains standard $AgNO_3$ solution and platinum electrodes.

Working

(i) A dilute solution fo known concentration (N/10) of the salt containing the ions, of which the transport numbers are to be determined, is filled in the apparatus.

(ii) The apparatus is then connected to a battery, variable resistance and voltameter in series as shown in Fig. (2.15). A low current of the order of 1 – 20 milliamperes is passed for 2 to 3 hours.

If high current density is used, diffusion sets in and the solution sin the various compartments are mixed and thus leading to wrong results.

(iii) After the electrolysis the stop cocks connecting the U-tube with the side tubes are closed and the solutions are withdrawn from the various compartments separately. The concentration is determined by titration against KCNS solution. These should be no change in the concentration of the solution from the V-tube, *i.e.*, central compartment. Knowing the original concentration of the solution taken, the change in concentration is determined separately.

(iv) The weight of Ag(or Cu) deposited in the voltameter is also determined separately. This represents the total quantity of electricity passed during electrolysis.

Calculations: Calculation involves the following steps:

Case I : *When the electrodes are unattackable.* This condition can be obtained by using platinum electrodes. Suppose that,

(i) The weight of $AgNO_3$ in the a g of anodic solution before electrolysis = X gm.

(ii) The weight of $AgNO_3$ in the b g of anodic solution after electrolysis = Y gm.

(iii) The weight of $AgNO_3$ in the a g of anodic solution after electrolysis $= \frac{Y}{b} \times a = Z$

$\therefore$ Full in concentration around anode = (X – Z) gm.

Let the weight of Ag deposited in the voltameter = 2 gm

$$\frac{\text{Total weight of } AgNO_3 \text{ electrolysed}}{\text{Weight of Ag deposited in voltameter}}$$

$$= \frac{\text{Eq. wt. of } AgNO_3}{\text{Eq. wt. of Ag}}$$

$$\therefore \quad \frac{\text{Total weight of } AgNO_3 \text{ electrode}}{w} = \frac{170}{108}$$

or $$\text{Total weight of } AgNO_3 \text{ electrolysed} = \frac{w \times 170}{108} = W \text{ (say)}$$

Now the transport No. of Ag^+ ion is given by

$$t_{Ag^+} = \frac{\text{Loss in wt. of } AgNO_3 \text{ around anode}}{\text{Total weight of } AgNO_3 \text{ electrolysed}}$$

$$= \frac{X - Z}{W} \qquad ...(11)$$

and the transport No. of NO_3^- ions is given by

$$t_{NO_3^-} = 1 - t_{Ag^+} = 1 - \frac{X - Z}{W} \qquad ...(12)$$

Case II : *When the electrodes are attackable.* This condition can be obtained by using Ag anode in $AgNO_3$ solution.

In this case $Z \geq X$, *i.e.*, the concentration of $AgNO_3$ in the anodic compartment after electrolysis is more than that before electrolysis. This is due to the fact that NO_3^- ions after losing charge at the anode attack the silver anode to form $AgNO_3$. Thus, the concentration of $AgNO_3$ in the anodic compartment increases.

The fall in concentration of $AgNO_3$ in the anodic compartment due to the migration of the Ag^+ ions may be calculated as follows:

The amount of $AgNO_3$ formed by the discharge of NO_3^- ions at the anode = No. of the NO_3^- ions discharged.

= total $AgNO_3$ electrolysed W (say)

$\therefore$ The actual increase in conc. of anodic compartment = W.

But observed increase in conc. of anodic compartment = Z – X.

$\therefore$ Fall in concentration due to the migration of Ag^+ ions

$$= \text{Actual increase} - \text{observed increase}$$

$$= W - (Z - X)$$

Hence
$$t_{Ag^+} = \frac{W - (Z - X)}{W} \quad ...(13)$$

and
$$t_{NO_3^-} = 1 - t_{Ag^+} = 1 - \frac{W - (Z - X)}{W} = \frac{Z - X}{W} \quad ...(14)$$

(2) Moving Boundary Method

Principle: It is based on measuring the rate of migration of one or both of the ionic species of the electrolyte, away from the similarly charged electrodes. This method was first of all suggested by Lodge and Whetham. It was modified by Masson (1899), Steel (1904) and Denison (1906). Finally it was the work of Macinnes (1923) whose work made the procedure more easy and simple.

Apparatus: The apparatus used in shown in the Fig. 2.26.

(i) Suppose one is interested in measuring the transport number of any cation, say Na^+ in NaCl. To do it, one has to select another suitable electrolyte called the *indicator-electrolyte* which has a common ion with the electrolyte under investigation. It is the usual practice to select such an electrolyte as indicator electrolyte whose cation must be slow as compared to the cation of the electrolyte of which one is interested in determining the transport number.

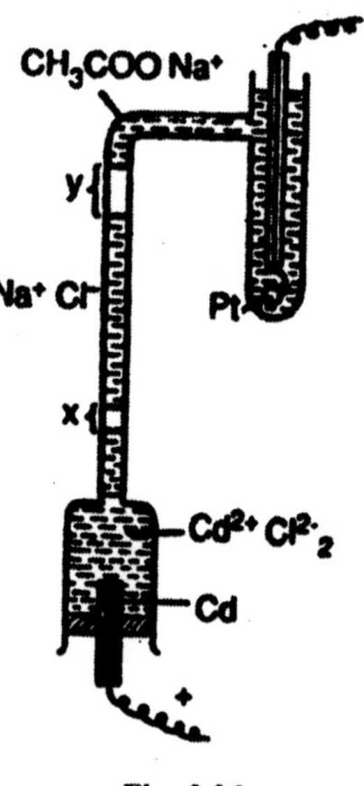

Fig. 2.26

For example, LiCl is used as an indicator electrolyte while determining the transport number of Na^+ because Li^+ ions move at a slower rate than Na^+. If one takes indicator ion moving faster than the experimental ion, the boundary line gets blurred.

In most of the methods, LiCl is not used as an indicator but instead of this cadmium is used as an anode which on electrolysis yields cadmium chloride which in turn itself acts as an indicator electrolyte.

(ii) The electrolytic cell consists of the shape as shown, the middle part of the left hand portion is vertical tube of uniform area of cross section.

(iii) The experimental solution NaCl is floated on top of a denser solution of $CdCl_2$.
At the bottom of the tube is fixed an anode of cadmium metal. Over the solution of NaCl a less dense solution of sodium acetate is floated.

(iv) The cathode used is of platinum. This is placed in a side tube so that the evolution of hydrogen gas at the cathode does not disturb the main liquid column.

Working

(i) A constant current is passed for 5 to 6 hours. The Na^+ ions move upwards towards the cathode closely followed by the Cd^{2+} ions. The sharp boundary moves gradually upwards. The movement of the boundary between NaCl and $CdCl_2$ can be easily followed on account of a difference in the refractive indices of the two solutions.

(ii) The rate of migration of Cl^- ions can also be observed as the acetate anions are slow moving than the chloride ions.

(iii) At the end of experiment, the distance through which the boundary has moved is noted. The time for which the current has passed is noted.

Calculations: Suppose a current of I amperes is passed for t seconds, the quality of electricity carried by Na^+ ions is given by

$$= t_+ \text{ It coulombs}$$

where t_+ is the transport number of the Na^+ cation

Suppose the concentration of NaCl solution = c gm eq. eq/c.c.

Suppose the area of cross section of the moving boundary = a cm^2.

Therefore, the amount of Na^+ ion that has migrated upwards

$$= \frac{t_+ It}{F} \text{ where F is Faraday.} \qquad ...(15)$$

Distance by which NaCl—$CdCl_2$ boundary moving during t seconds = x cm.

Volume of the solution cleared by the migration of Na^+ ions = ax

Number of gm equivalent of Na^+ ions transported = axc ...(16)

From Eqs. (15) and (16), we have $axc = \frac{t_+ It}{F}$

or
$$t_+ = \frac{F\,axc}{It} \qquad ...(17)$$

Similarly,
$$t_- = \frac{F \cdot ayc}{It} \qquad ...(18)$$

where y is the distance by which NaCl—CH_3COONa boundary moves during t send and t_- is the transport number of chloride anion.

Also
$$\frac{t_+}{t_-} = \frac{x}{y}$$

But
$$t_- = 1 - t_+$$

$\therefore$
$$\frac{t_+}{1-t_+} = \frac{x}{y} \quad \text{or} \quad t_+ = \frac{x}{x+y} \qquad ...(19)$$

Similarly
$$t_- = \frac{x}{x+y} \qquad ...(20)$$

Thus, t_+ and t_- can be evaluated.

(3) From Ionic Mobility

The ionic mobility or conductance of an ion is directly proportional to the speed of the ion. Thus,

$$\lambda_+ \propto u_+$$

or
$$\lambda_+ = Ku_+ \qquad ...(21)$$

Similarly,
$$\lambda_- \propto u_-$$

or
$$\lambda_- = Ku_- \qquad ...(22)$$

In Eqs. (21) and (22), u_+ and u_- are the speeds while λ_+ and λ_- are the ionic conductivities of the cation and anion respectively.

Now the transport number of cation is given by

$$t_+ = \frac{u_+}{u_+ + u_-}$$

$$= \frac{\frac{1}{K}\lambda_+}{\frac{1}{K}(\lambda_+ + \lambda_-)} \quad \text{[From Eqs. (21) and (22)]}$$

$$= \frac{\lambda}{\lambda_+ + \lambda_-} \quad ...(23)$$

But $\lambda_+ + \lambda_- = \lambda_\infty$ (Kohlrausch's Law)

$$\therefore \quad t_+ = \frac{\lambda_+}{\lambda_\infty} \quad ...(24)$$

Similarly

$$t_- = \frac{\lambda_-}{\lambda_\infty} \quad ...(25)$$

From Eqs. (24) and (25), it follows that the transport number of cation and anion can be determined if one knows the values of λ_+, λ_- and λ_∞.

In this case, oxidation takes place at the zinc electrode and reduction at the hydrogen electrode, as shown below:

Zinc electrode $Zn\ (s) \rightleftharpoons Zn^{2+} + 2e^-$

Hydrogen electrode $2H^+ + 2e^- \rightleftharpoons H_2(g)$

$\therefore$ *The net reaction is* $Zn(s) + 2H^+ \rightleftharpoons Zn^{2+} + H_2(g)$ (1 atm)

Nernst's Equation for Single Electrode Potential

From the Nernst theory it follows that the potential difference of the double layer formed at one electrode is termed as single electrode potential.

Let us consider a metal of valency 'n' dipping in a solution

$$M \rightleftharpoons M^{+n} + ne$$

Let P_1 be the osmotic pressure of its ions, P_3, be the solution pressure of the metal and E be the electrode potential, *i.e.* the difference of potential between the metal and solution. Now pass the current through the electrode reversible until 1 gm. ion of the metal gets dissolved. It means that the electricity required for the dissolution of 1 gm. ion of the metal will be nF* coulombs. Hence,

the electrical work done = NF*E volt-coulombs

i.e. $w_1 = nF^*E$ volt-coulombs

where F* is one faraday whose value is equal to 96500 coulombs and w_1 is the electrical work done .Go on diluting the solution till the osmotic pressure gets reduced from P_1 to $P_1 - dP_1$. At the same time the difference of potential between the metal and the solution is now changed from E to E – dE. At the potential difference E – dE, the electrical work done (w_2) to cause the dissolution of 1 gm. ion of the metal is given by

$$w_2 = (E - dE)\ nF^*\ \text{volt-coulombs}$$

The difference in electrical energy of work = $w_1 - w_2$

or
$$w_1 - w_2 = nEF^* - (E - dE)nF^*$$
$$= nF^*\ dE\ \text{volt coulombs.}$$

The different in electrical energy is equal to the osmotic work done in transferring 1 gm. ion of the metal from P1 to P_1—dP_1 which is equal to the product of volume and change in pressure *i.e.*,

$$nF^*dE \rightleftharpoons VdP_1 \qquad ...(1)$$

where V is the volume of the solvent containing 1 gm. ion of the metal. For ideal solutions, osmotic pressures are directly proportional to the activities of ions, then $V = RT/P_1$. Thus, equation (2) reduces to

$$nF^*dE = \frac{RT\,dP_1}{P_1} \qquad ...(2)$$

$$nF^*\int dE = RT\int \frac{dP_1}{P_1}$$

$$nF^*E = 2.303\ RT \log P_1 + K \qquad ...(3)$$

where 'K' is the integration constant whose value will be evaluated as follows: when the solution pressure (P_2) is equal to osmotic pressure (P_1) of the ions, it means that no potential difference exists between the metal and the solution, *i.e.*, E = 0 and $P_1 = P_2$. Hence equation (3) becomes as

$$0 = 2.303\ RT \log P_2 + K$$

or
$$K = -2.303\ RT \log P_2 \qquad ...(4)$$

Substituting the value of K from eq. (4) in (3), we get

$$nF^*E = 2.303\ (RT \log P_1 - RT \log P_2)$$

$$E = 2.303 \frac{RT}{nF^*} \log \frac{P_1}{P_2}$$

Equation (5) is termed as the Nernst equation for electrode potential. Let us now consider tow similar electrodes dipping in two different solutions. In this case, the potential difference between the electrodes will be given by

$$E_1 - E_2 = \frac{2.303\,RT}{nF^*}\left(\log\frac{(P_1)_A}{P_2} - \log\frac{(P_1)_B}{P_2}\right)$$

$$= \frac{2.303\,RT}{nF^*}\log\frac{(P_1)_A}{(P_2)_B} \qquad ...(6)$$

where $(P_1)_A$ and $(P_1)_B$ denote the osmotic pressure of the metal ions in two solutions A and B respectively. As we know that the osmotic pressure is proportional to concentration, it means, that equation (6) may be written as

$$E_1 - E_2 = \frac{2.303\,RT}{nF}\log\frac{(c_1)_A}{(c_2)_B}$$

STANDARD ELECTRODE POTENTIAL

When all the substances taking part in a reaction in a reversible cell are maintaining at unit activity, *i.e.*, in their standard states, the E.M.F. of any electrode in these conditions is known as standard electrode potential of the given cell (E°). If a reaction occurs by the passage of n Faradays in a cell in which the reactant and product have unit activities, the standard free energy change ΔG° is equal to $-$ nE F*. Hence

$$-\Delta G^\circ = nE^\circ F^* \qquad ...(1)$$

where F* is the Faraday of electricity. Let us now consider a reversible process (chemical reaction between A and B to form the products L and M)

$$aA + bB + ... \rightleftharpoons lL + mM + ...$$

When we apply the law of mass action to the above equilibrium, we get

$$K = \frac{a_L^l\, a_M^m}{a_A^a\, a_B^b \cdots} \qquad ...(2)$$

As the increase in free energy ΔG° at constant temperature is governed by the Vant Hoff's reaction isotherm, it, therefore, follows from this isotherm that

$$-\Delta G = RT\log_e K - RT\log_e\frac{a_L^l\, a_M^m}{a_A^a\, a_B^b \cdots} \qquad ...(3)$$

If in the process the reactants and products are taken in their respective standard states, *i.e.*, at unity, then eq. (3) becomes as

$$-\Delta G^{\circ} = RT \log_e K \qquad ...(4)$$

Eliminating ΔG° between equations (1) and (4), we get

$$nE^{\circ} F^{\circ} = RT \log_e K \qquad ...(5)$$

If the concentrations of reactants and products are taken at any arbitrary activities, then E.M.F. of the cell (E) is given by

$$\Delta G = -nEF^* \qquad ...(6)$$

Substituting equations (6) and (5) in (3), we obtain

$$nEF^* = nE^{\circ}F^* - RT \log_e \frac{a_L^l a_M^m}{a_A^a a_B^b \cdots}$$

$$E = E^{\circ} - \frac{RT}{nF^*} \log_e \frac{a_L^l a_M^m}{a_A^a a_B^b \cdots} \qquad ...(7)$$

Expression (7) is the general equation for the E.M.F. of any reversible cell in which the reactants and products are taken at any arbitrary activities. If a_A, a_B, a_C, ... and a_L, a_M, a_N, ... are unity, it means that equation (7) becomes as

$$E^{\circ} = E$$

So normal or standard electrode potential may be defined as:

"*The potential of the electrode when the activities of the reactants and products are unity.*"

The earliest tabulation of standard electrode potential was mainly performed by Wilsmore and later revised by Abegg, Luther and Auerbach (1911).

However, Lewis and Randall recalculated the standard electrode potential from activity data. Let us now consider a metal of valency z_+, reversible with respect to M^{z+} ions, the electrode reaction is

$$M \rightarrow M^{z+} + z_+e^-$$

The equation for the electrode potential is taken as:

$$E_+ = E^{\circ} - \frac{RT}{z_+F^*} \log_e \frac{a_M^+}{a_M} \qquad ...(9)$$

where a_M^+ is the activity of the cations in the solution and a_M is the activity of the solid metal. As the solution state of metal is generally taken as unity, it means that equation (9) takes the form

$$E_+ = E^\circ - \frac{RT}{z_+F^*} \log_e a_M^+ \qquad ...(10)$$

Again consider an electrode A which is reversible to A^- ions. The electrode reaction is

$$M^{z-} \rightarrow A + z^-e^-$$

Proceeding in the same manner as we have done earlier, we can prove that

$$E_- = E^\circ + \frac{RT}{z_-F^*} \log_e a_A^- \qquad ...(11)$$

From expressions (10) and (11), it follows that the general expression for any electrode reversible with respect to single ion of valency z± may be given as:

$$E_\pm = E^\circ \pm \frac{RT}{z \pm F^*} \ln a_l \qquad ...(12)$$

where a_l is the activity of a particular species. In equation (12), the upper signs are taken for a positive ion while the lower ones for a negative ion. Putting R = 8.313 Joules, F* = 96500 coulombs and by converting Neperian to Briggsian logarithms, the factor 2.3026 is utilised, it means that equation (12) becomes as

$$E_\pm = E^\circ \pm 1.9835 \times 10^{-4} \frac{T}{z\pm} \log a_l$$

$$= E^\circ \pm \frac{0.05915}{z\pm} \log a_l \text{ [when T = 298. 16°K]} \qquad ...(13)$$

Effect of concentration and valence on electrode potential. We have already proved that

$$E_l = E^\circ + \frac{0.0591}{n} \log a_l$$

From the above expression it is evident that a ten fold change in activity of an ion causes the electrode potential change by 0.0591/z± volts at 25°C. In a generalised manner, a change of activity by a factor of 10* changes the potential by $x \times 0.0591/z\pm$ volts at 25°. For monovalent ions, a ten fold change in concentration of solution will cause a change of 0.591 volts in monovalent and for bivalent ions, a change of 0.591/2 volts at 25°.

SIGN OF ELECTRODE POTENTIAL

For years two different signal conventions for electrode potentials have been in common use.

Without real justification, they are usually referred to as

(i) *European convention.*

(ii) *American convention.*

American convention was designated by Lewis and Randall and has found popularity among the physical chemists. To illustrate the two conventions, let us consider the following reaction taking place at the zinc electrode:

$$Zn \rightarrow Zn^{2+} + 2e^-$$

According to European convention, one would assign a value of – 0.763 V for the standard potential of this electrode with respect to hydrogen electrode. On the other hand, the value according to the American convention, would be + 0.763 V. Thus, it would appear that the difference between the two conventions is the sign only. Let us now consider the reverse reaction f zinc electrode as:

$$An^{2+} + 2e^- \quad \rightarrow Zn$$

According to the European convention, the standard electrode potential will be the same, *i.e.*, – 0.763 V but according to the American convention it is not + 0.763 V but is – 0.763. Thus, according to American convention it is evident that

(i) If the reaction is an oxidation, the electrode potential is positive as in $Zn \rightarrow Zn^{2+} + 2e^-$ and

(ii) If it is a reduction, the electrode potential is negative in $Zn^{2+} + 2e^- \rightarrow Zn$.

Thus, the electrode potential is a *bivariant* quantity according to the American convention, *i.e.*, both oxidation and reduction, whereas in the European convention it is an invariant quantity.

In an attempt to reconcile the two conventions, IUPAC have recommended changes in terminology that

(i) Electrode potential will be reserved for the European convention and,

(ii) E.M.F. of a half cell, in the American convention.

According to *latest convention* adopted by IUPAC (International Union of Pure and Applied Chemistry), *the electrode potential is given a positive sign if the electrode reaction involves* reduction (*i.e.* taking up of electrodes from the electrode) *when connected to the standard hydrogen electrode and a* negative sign *if the electrode reaction involves* oxidation (*i.e. liberation of electrons*) *when connected to the standard hydrogen electrode taken arbitrarily as zero.*

Thus, when copper electrode (copper rod dipping in a solution of a copper salt) is connected with a standard hydrogen electrode, *reduction* takes place at the copper electrode.

The electrode reactions are represented below:

Hydrogen electrode $H_2\,(g) \rightleftharpoons 2H^+ + 2e^-$ (Oxidation)

Copper electrode $Cu^{2+} + 2e^- \rightleftharpoons Cu(s)$ (Reduction)

Here, according to the above convention, the potential of copper electrode is taken as *positive*. Thus, $E(Cu^{2+}, Cu)$ is positive.

However, if zinc electrode is connected with the standard hydrogen electrode, *oxidation* takes place at the zinc electrode.

The electrode reaction are represented below:

Zinc electrode $Zn(s) \rightleftharpoons Zn^{2+} + 2e^-$ (Oxidation)

Hydrogen electrode $2H^+ + 2e^- \rightleftharpoons H_2(g)$ (Reduction)

Hence, the potential of the zinc electrode is taken as *negative*. Thus, $E(Zn^{2+}, Zn)$ is negative.

3

PHASE RULE

INTRODUCTION TO PHASE RULE

The phase rule of Gibbs may be expressed in the simple form :

$$P + F = C + 2 \quad ...(1)$$

where P is the number of *phases*, F is the number of *degrees of freedom* and C is the number of *components* in a system at equilibrium. Equation (1) does not involve the mass of phase as it has no influence on the state of equilibrium. In words, the phase rule (Eq. 1) may be stated as:

"In a heterogeneous system in equilibrium, the number of degrees of freedom plus the number of phases are equal to the number of components plus 2."

Before deriving the phase rule, it is necessary to explain the meaning of phase, component and degree of freedom. *Phase :* As stated earlier, a heterogeneous system consists of different parts of each of which is homogeneous in itself and is separated from others by bounding surfaces or interfaces. The different parts of heterogeneous system can be separated by mechanical methods. *These homogeneous, physically distinct and mechanical separable parts of the heterogeneous system in equilibrium are called phases. A phase may be gas, liquid or solid.*

Examples : (i) In a system consisting of ice, water and vapour, there are three phases; ice (solid phase), water (liquid phase) and vapour (gas phase). Each phase is homogeneous in itself and can be separated from the other phases by mechanical methods.

(ii) *All gases are completely miscible :* Therefore, a gaseous mixture irrespective of the number of gases present, consists of one phase only.

(iii) Two or more liquids which are miscible with one another constitute a single phase as there is no bounding surfaces separating the different liquids. Example is the alcohol and water which exist in one phase only as two are completely miscible. As there will be vapour in contact with the liquid, there will be two phases, *i.e.*, liquid and gas in miscible liquids.

(iv) A solid mixture consists of many as phases as are solids present. A heterogeneous equilibrium,

$$CaCO_3 \rightleftharpoons CaO + CO_2$$

involves three phases, solid ($CaCO_3$), solid (CaO) and gas (CO_2).

Thus, there are three phases, two solids and one gaseous.

Component : A component is an element or a compound present in a system. The concentration of this can undergo variation independently. But in the phase rule the component does not stand for the total number of constituents of the system. This term, is however, defined as :

"The number of components of a system at equilibrium is defined as the minimum number of molecular species in terms of which the composition of all the phases may be expressed quantitatively:

The meaning of this term will be made clear by considering some actual cases.

(i) Consider a water system consisting of the three phases :

$$\text{Ice} \rightleftharpoons \text{water} \rightleftharpoons \text{Vapour.}$$

Thus, water system is one component system as each phase is essentially a different physical form of the same chemical compound represented by H_2O.

(ii) Consider the dissociation of $CaCO_3$ by heat

$$CaCO_3 \rightleftharpoons CaO + CO_2$$

This equilibrium consists of three phases, two solids and one gas. At first sight, it seems that it is a three component system. This cannot be so as three constituents are not independent of each other as demanded by the definition. For example :

$$CaCO_3 = CaO + CO_2$$

$$\text{or} \quad CaO = CaCO_3 - CO_2$$

or $CO_2 = CaCO_3 - CaO$.

Thus, the composition of all the three phases can be expressed by taking two of the constituents only. Hence it forms a two component system. For example :

(a) By taking CaO and CO_2 as the components

Phase	Component
$CaCO_3$	$= CaO + CO_2$
CaO	$= CaO + 0\ CO_2$
COa	$= 0\ CaO + CO_2$

(b) By taking $CaCO_3$ and CO_2 as components

Phase	Component
$CaCO_3$	$= CaCO_3 + 0\ CO_2$
CaO	$= CaCO_3 - CO_2$
CO_2	$= 0\ CaCO_3 + CO_2$

(c) By taking $CaCO_3$ and CaO as components

Phase	Component
$CaCO_3$	$= CaCO_3 + 0\ CaO$
CaO	$= 0\ CaCO_3 + CaO$
CO_2	$= CaCO_3 - CaO$

(iii) Consider the dissociation of ammonium chloride in a closed vessel which is in equilibrium with its products, ammonia and hydrochloric acid

$$\underset{\text{solid}}{NH_4Cl} \rightleftharpoons \underset{\text{gas}}{NH_4Cl} \rightleftharpoons NH_3 + HCl$$

The system is considered as a one-component system because the composition of both the solid and gaseous phases can be stated in terms of NH_4Cl only. This is shown as below:

Phase	*Component*
Solid	NH_4Cl
Gaseous	$x.NH_3 + x.HCl$ or $x\ NH_4Cl$

If a small amount of NH_3 or HCl is added to the equilibrium mixture then the number of components would become two. Suppose excess of hydrochloric acid is added, then the gaseous mixture will contain x moles of NH_3 + x moles of HCl + y moles of HCl. Tlherefore, the composition of two phases of ammonium chloride may be expressed as

Phase	*Component*
Solid NH_4Cl	NH_4Cl
Gas	.t.NHa +X.HCl –{ y.HCl
	= x NH_4Cl + y HCl

Thus, the minimum number of components by means of which the composition of both phases can be expressed arc the two components NH_4Cl and HCl. Thus, it is a two *component system*. Similarly it behaves as a two component system if ammonia is added to the equilibrium mixture.

Degree of freedom or variance of the system : The degree of freedom or variance of a system is defined as the least number of variable factors such as temperature, pressure and concentration which should be arbitrarily fixed in order to define the system completely.

(i) A gaseous mixture, say of carbon dioxide and nitrogen, is completely defined when three variables temperature, pressure and concentration are specified. Then, the system is said to have three degrees of freedom or to be trivariant.

(ii) If one consider a system consisting of unsaturated water vapour, two variables pressure and temperature are to be specified to define the system because pressure may have any value at a particular temperature. This system is said to have two degrees of freedom or bivariant.

PHASE REACTIONS

There are certain physical or chemical reactions which are accompanied by the appearance or disappearance of a phase. Such reactions are known as phase reactions.

A phase reaction is generally indicated in Fig. 3.1.

From Fig. 3.1, it follows that the examples of phase reactions are as follows:

(i) Change of solid to liquid.

(ii) Change of liquid to solid.

(iii) Change of liquid to gas.

(iv) Change of solid to gas.

(v) Change of gas to liquid.

(vi) Change of gas to solid.

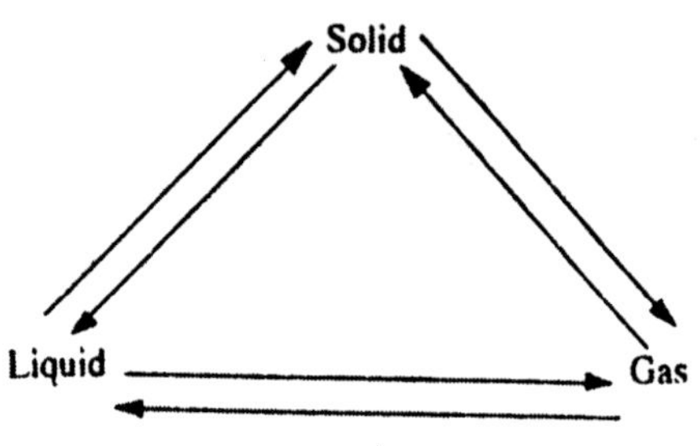

Fig. 3.1

From the above examples, it is evident that all processes of vaporisation, sublimation, freezing, melting, solidification, crystallisation, and precipitation are phase reactions.

All chemical processes may not be phase reactions, *e.g.*, the combination of gaseous ammonia and gaseous hydrogen chloride to form solid ammonium chloride is a phase reaction because the, reactants and products are in different phases.

But the reaction between gaseous hydrogen and gaseous chlorine to form gaseous hydrogen chloride cannot be regarded as a phase reaction because all the, substances are in one phase.

CONDITIONS FOR EQUILIBRIUM BETWEEN PHASES

In order for each phase to be in equilibrium in a multiphase system, the following equilibrium conditions are to be fulfilled :

(a) *Thermal Equilibrium :* It means that the temperature of all the phases must be identical, otherwise there would occur the flow of heat from one phase to another. One can prove it thermodynamically as follows :

Suppose we consider two phases α and β at temperature T^{α} and T^{β}. Suppose S^{α} and S^{β} are the entropies of the two phases. Suppose the heat, dq, is transferred from phase α to β at equilibrium. Due to this heat transfer, the change of entropy of the system at equilibrium will be as follows :

$$dS = dS^{\alpha} + dS^{\beta} = 0 \qquad ...(2)$$

We know, $$dS^{\alpha} = -\frac{dq}{T\alpha} \text{ and } dS^{\beta} = \frac{dq}{T\beta} \qquad ...(3)$$

In equation (3), –ve and +ve signs are used to signify the heat loss and heat gain by the phases α and β respective ly.

On substituting equation (3) in (2), we get

$$-\frac{dq}{T\alpha} + \frac{dq}{T\beta} = 0$$

or $$T\alpha = T\beta \qquad ...(4)$$

From the equation (4), *it follows that the temperature of all the phases in equilibrium is same.*

(b) *Mechanical Equilibrium :* By this equilibrium it is meant that all the phases in equilibrium must be under the same pressure. It is not so, one phase would increase its volume at the expense of another. This may be deduced as follows :

Suppose there occurs an expansion in volume, dV, of α-phase at the expense of β-phase. Then at constant temperature, the change in Helmholtz free energy will become as follows :

$$dA = dA^{\alpha} + dA^{\beta} = 0 \text{ (At equilibrium)} \qquad ...(5)$$

we know $dA^{\alpha} = P^{\alpha}\, dv$ and $dA^{\beta} = -P^{\beta}\, dv$ ..(6)

In equilibrium (6), positive sign indicates increase in volume of α-phase whereas negative sign indicates decrease in volume of β-phase.

On substituting equation (6) in (5), we get

$$P^{\alpha} = P^{\beta} \qquad ...(7)$$

From equation (2), it follows *that the phases in equilibrium are under the same pressure.*

(c) *Chemical Equilibrium :* By this equilibrium, it is meant that the chemical potential of any components is same in all the phases at equilibrium.

Let us consider a closed system of P phases indicated by α, β, γ, ... having u total of C components indicated by 1, 2, 3, ... P in equilibrium and each at the same pressure and temperature. We know that the Gibbs tree energy (G) of each phase is a function of temperature, pressure and composition, *i.e.*,

$$G^{\alpha} = f(T, P, n_i)^{\alpha} \quad ...(8)$$

$$G^{\alpha} = f(T, P, n_i)^{\beta} \quad ...(9)$$

$$\cdots \quad \cdots \quad \cdots$$

$$G^{P} = f(T, P, n_i)^{P} \quad ...(10)$$

where i = l, 2, 3, 4, ..., i, C The free energy change of the entire system will be the sum of change of free energies of each phase.

$$dG = dG^{\alpha} + + dG^{\beta} + dG^{v} + ...$$

For a multicomponent system, we know

$$dG = -S\ dT + V\ dp + \Sigma\mu_i\ dn$$

At constant temperature and pressure, dT = dP = 0 and therefore the above equation becomes as

$$(dG)r,\ p = \Sigma\mu_i\ dn_i$$

If there occurs a transfer of mass from one phase to another, we can write

$$dG = \sum_{i=1}^{0} \mu_i^{\alpha}\ dn_i^{\alpha} + \Sigma\mu_i^{\beta}\ dn_i^{\beta} + \Sigma\mu_i^{y}\ dn_i^{y} + ... \quad ...(11)$$

If we are considering a closed system at equilibrium (dG = 0), we can write

$$\Sigma\mu_i^{\alpha}\ dn_i^{\alpha} + \Sigma\mu_i^{\beta}\ dn_i^{\beta} + \Sigma\mu_i^{y}\ dn_i^{y} + ... = 0$$

At equilibrium total mass of each component is constant because we are considering a closed system. Therefore, we can write

$$\left.\begin{array}{l} dn_1^{\alpha} + dn_1^{\beta} + dn_1^{y} + ... + dn_1^{P} = 0 \\ dn_2^{\alpha} + dn_2^{\beta} + dn_2^{y} + ... + dn_2^{P} = 0 \\ \cdots \quad \cdots \quad \cdots \quad \cdots \quad \cdots \\ dn_i^{\alpha} + dn_i^{\beta} + dn_i^{y} + ... + dn_i^{P} = 0 \\ \cdots \quad \cdots \quad \cdots \quad \cdots \quad \cdots \\ dn_c^{\alpha} + dn_c^{\beta} + dn_c^{y} + ... + dn_c^{P} = 0 \end{array}\right\} \quad ...(12)$$

For all the equations in expression (12) to be zero for all possible variations of n, subject to the restrictions posed by equation (11), it is required that

$$\left.\begin{matrix} \mu_1^\alpha = \mu_1^\beta = \mu_1^\gamma = \dots = \mu_1^P \\ \mu_2^\alpha = \mu_2^\beta = \mu_2^\gamma = \dots = \mu_2^P \\ \dots \quad \dots \quad \dots \quad \dots \\ \mu_c^\alpha = \mu_c^\beta = \mu_c^\gamma = \dots = \mu_c^P \end{matrix}\right\} \quad \dots(13)$$

From expression (13), it is evident that for any component i in the system, the value of chemical potential (α_i) must be same in every phase provided the system is in equilibrium at constant T and P.

DERIVATION OF PHASE RULE

J. Willard Gibbs (1876) deduced a *mathematical* relation between the number of degrees of freedom (F), the number of components (C) and the number of phases (P) in a heterogeneous system at equilibrium. The relation is as follows :

$$F = C - P + 2$$

It was first derived by Gibbs from thermodynamic considerations and is known as *Gibbs phase rule.* Let us consider a system of 'C' components distributed between 'P' phases and let each phase contain all the 'C' components. To define the composition of each phase it is required to specify (C – l) concentration terms and hence the concentration of the remaining components will be defined by the difference. As there are 'P' phases the total number of variables for concentration alone is P (C – l).

As the process is being carried out at a constant temperature and pressure, it is required to add two more variables. Hence the total number of variables is becoming equal to P(C – 1) + 2.

We will now determine how many equations involving these variables are there when all the phases are in equilibrium. This can be shown in the following manner :

Consider a heterogeneous system of two phases (a and b) containing three components (1, 2, 3) in equilibrium. The chemical potential (μ) of these components in two phases can be represented as :

$$\mu_1 (a), \mu_2 (a), \mu_3 (a), \mu_1 (b), \mu_2 (b), \mu_3 (b)$$

As we are considering the closed system in equilibrium at a definite temperature, and pressure, the *Gibbs-Duhem* equation can be applied to it.

Thus, we get $\Sigma\mu \, dn = 0$.

Suppose a small amount δn_1 is transferred, under equilibrium conditions from phase a to b and therefore:

$$-\mu_1 (a) \, dn_1 + \mu_1 (b) \, dn_1 = 0$$

or $$\mu_1 (a) = \mu_1 (6)$$

This equation reveals that the equilibrium is between two phases when the chemical potential of a particular component is the same in each phase. This can be applied to all the components. It, therefore, follows that

$$\mu_2 (a) = \mu_2 (b); \mu_3 (a) = \mu_3 (b); \text{ etc.}$$

Now, imagine the system is having three phases. Therefore, the chemical potential of a particular component must be same in all the phases, *i.e.*,

$$\mu_1(a) = \mu_1(b) = \mu_1(c)$$

Hence the number of independent equations defining the equilibrium between three phases are two only. In general, the above conditions also hold good for 'P' phases and 'C' components to give a set of following relations :

$$\mu_1(a) = \mu_1(\&) = \mu_1(c) = \ldots = \mu_1(P)$$

$$\mu_1(a) = \mu_2 (b) = \mu_2(c) = \ldots = \mu_2 (P)$$

$$\ldots \quad \ldots \quad \ldots \quad \ldots \quad \ldots$$

$$\ldots \quad \ldots \quad \ldots \quad \ldots \quad \ldots$$

$$\mu_c(a) = \mu_c(b) = \mu_c(c) = \ldots = \mu_c(P)$$

These equations constitute C(P – l) independent equations. Thus the degree of freedom F must be equal to

$$F = [P(C - 1) + 2] - [C(P - 1)]$$

$$= C - P + 2$$

This is the phase rule of Gibbs.

Discussion : In the derivation of equation (14) or (15), it was assumed that each component is present in every phase. If a component is missing from a particular phase, however, the number of concentration variables is decreased by one. But at the same time the number of possible equations is also decreased by one. Hence the value of (C–P), and therefore F. remains the same whether each constituent is present in every phase or not. This means that the phase rule is not restricted by the assumption made, and is generally valid under all conditions of distribution provided that equilibrium exists in the system.

Advantages of Phase Role

(i) It gives a simple method of classifying equilibrium states of systems.

(ii) It confirms that the different systems having the same number of degrees of freedom behave in like manner.

(iii) It predicts the behaviour of systems when subjected to changes in the variables such as pressure, temperature and volume.

(iv) The phase rule is applicable to macroscopic systems. Therefore, it is not necessary to take into account about their molecular structures.

(v) It is applicable to physical as well as to chemical phase reactions.

(vi) Phase rule takes no account of nature of the reactants or products in phase reactions.

(vii) Phase rule predicts that a number of substances would remain in equilibrium or not in equilibrium if some of the substances have been transformed into the new substances.

Limitations of the Phase Role

(i) As the phase rule is applicable to heterogeneous systems in equilibrium, it is therefore of no use for such systems which are slow in reaching the equilibrium state.

(ii) As the phase rule is applicable to a single equilibrium state, it never tells about the number of other equilibrium possible in the system.

(iii) In phase rule various variables are temperature, pressure and composition. This phase rule does not consider the electric and magnetic influences. If such variables are considered, the factor 2 of the phase rule has to be adjusted accordingly.

(iv) All the phases in the system must be present under the same pressure, temperature and gravitational force.

(v) No liquid or solid phases should be finely divided otherwise their vapour pressures will differ from their normal values.

ONE-COMPONENT SYSTEMS

When one component is the smallest number by means of which the composition of each phase is expressed, it is known as a one component system.

Maximum Number of Phases : When F is minimum, P becomes maximum. The minimum number of degrees of freedom possible in a system is zero. Consider an equilibrium state of a one component system having zero degree of freedom. According to the phase rule,

we have $P = C - F + 2$ ...(1)

or $P = 1 - 0 + 2 = 3$ ...(2)

Thus, the *maximum number of phases in any equilibrium state of one component system having zero degree of freedom will be three.* It cannot be more then three.

Maximum Number of F : The maximum value of F is expected in the case when P is minimum. As the minimum number of phases in any system is one, it follows from phase rule equation that the maximum degree of freedom in one component system are two.

$$F = C - P + 2 = 1 - 1 + 2 = 2$$

THE WATER SYSTEM

It is one component system as H_2O is the only chemical compound involved. Water exists in three possible phases, namely, solid, liquid and vapour. The phase diagram of this system is given in Fig. 3.2 which is obtained by plotting pressure—temperature curves.

The diagram consists of:

(i) *Curve :* Three curves are OA, OB and OC.

(ii) *Triple point :* The above three curves meet at the point 0 which is known as triple point.

(iii) *Areas :* Three curves divide the diagram into three areas AOC, AOB and BOC.

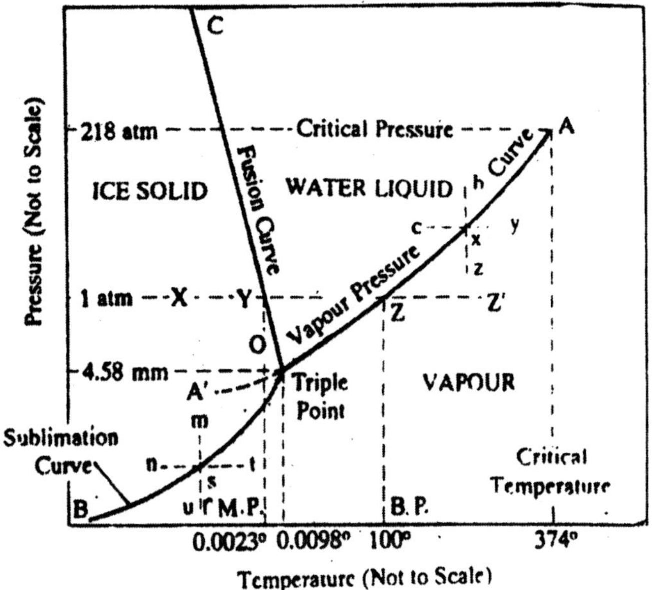

Fig. 3.2 : Water System.

Description of the Phase Diagram

Curves

AO : This curve is known as the *vapour pressure curve* because it gives the vapour pressure of water at different temperatures. The curve starts from O which is the freezing point of water and ends at A, the critical temperature of water (374°C). Beyond the point A, *i.e.,* critical temperature, the two phases liquid water and vapour merge into each other.

From this curve it is seen that for any given temperature, there exists a fixed value of pressure. Similarly for each vapour pressure, temperature has also a fixed value. Thus, the degree of freedom on any point of this curve is one or it is univariant. This is also predicted by the phase rule.

$$F = C - P + 2 = 1 - 2 + 2 = 1$$

The number of phases is taken as two because liquid water and vapour exist along the curve OA.

OB : It is the *vapour pressure carve* of ordinary ice and is known as the sublimation curve. Along this curve, solid ice is in equilibrium with its vapour. Thus, this curve shows the variation of vapour pressure of ice at various temperatures.

The curve starts at O and ends at B, *i.e.,* absolute zero (–273°C). At this temperature, no vapour can exist and, therefore only ice is left. But on the other points of the curve OB, ice is in equilibrium with vapour. Thus, there are two phases. Applying the phase rule, we have

$$F = C - P + 2 = 1 - 2 + 2 = 1$$

Thus, the system has one degree of freedom or is univariant. It means that each vapour pressure term can be maintained only at a fixed temperature. This is found to be so.

OC : It is the *melting point or fusion curve* of ice as it indicates the effect of pressure on melting point of ice. The inclination of the curve OC towards the pressure axis, *i.e.,* y-axis indicates that the melting point of ice is lowered by increase of pressure. At any point on the curve OC two phases ice and liquid water are in equilibrium. Applying the phase rule, we get

$$F = C - P + 2 - 1 - 2 + 2 \text{ or } F = 1$$

Thus, the system is univariant. It means that for any given pressure, melting point must have one definite value. This has been found in actual practice.

Triple point O : The three curves OA, OB and OC meet at a point O which is known as *triple point.* At this point all the three phases namely, ice. water and vapour co-exist. Thus, the value of P is three. Applying the phase rule to this point,

$$F = C - P + 2 = 1 - 3 + 2 \text{ or } F = 0$$

Thus, the degree of freedom at triple point is zero. It means that three phases can co-exist in equilibrium only at a definite temperature and pressure which correspond to the point 0. The values of pressure and temperature at the point O are 4.5 mm and 0.0075°C. If pressure and temperature are varied from the value for this point, three different cases may arise :

(i) *If pressure is raised without changing the temperature, only liquid phase will be there. Other phases ice and vapour will be converted into the liquid.*

(ii) *If pressure is lowered without changing the temperature, only vapour phase will be there. Two phases ice and liquid will also be converted into the vapour.*

(iii) *If pressure and temperature are varied together, one or the other curve (i.e., OA or OB or OC) will be followed depending upon values of pressure and temperature.*

Areas, AOB, BOC and COA : The areas between the curves OA OB and OC arc AOB, BOC and COA. In each area a single phase *i.e.*, solid ice, water and vapour is of stable existence Thus,

(i) in the area AOB, only vapour exists.

(ii) in the area AOC, only water exists.

(iii) and in the area BOC, only ice exists.

All these areas are bivariant. *i.e.*, in order to locate any point in the area, the temperature and pressure must be defined. This also follows from the phase rule,

$$F = C - P + 2 = 1 - 1 + 2 = 2$$

Metastable Equilibrium : Sometimes it is possible to cool water below its freezing point without the separation of solid ice. Thus, the curve AO can be extended to A' by cooling below its freezing point without separating ice. Along the curve OA' the liquid is in metastable equilibrium with the vapour. Therefore, this curve (OA') is said to be in metastable equilibrium. As soon as a small ice particle is kept in contact with the supercooled liquid, it at once changes into the solid ice and the curve merges in OB.

From the phase diagram, it is seen that the curve OB lies below the curve OA'. Thus the metastable system processes a higher vapour pressure than the stable system if same temperature is considered.

Effect of Change of Temperature and Pressure on Equilibrium

The real significance of phase diagram can be understood only if we follow the changes that are occurring on changing the pressure or temperature of the system. Suppose we are interested to study the effect of heating ice under a pressure of 1 atmosphere and at a temperature represented by the point X in the Fig. 3.2. At the point X, there is only one phase, *i.e.*, ice and therefore this point is said to be bivariant ($F = 1 - 1 + 2 = 2$). Due to the bivariant nature of X. The temperature can have any value at the same pressure. Therefore, heating the ice slowly at constant pressure will cause the system to move along XY. Up to point Y there is only ice phase. Beyond the Y, fusion sets in, *i.e.*, the liquid phase also appears. It means that at point beyond Y there arc two phases, *i.e.*, ice and water hence the system has one degree of freedom (F = 1 – 2 + 2 = 1). Due to one degree of freedom the temperature cannot alter without the alteration of pressure. The effect of continued heating will result the conversion of ice into water. As soon as the point Z is reached, there occurs complete conversion of solid into liquid. At point Z, there is only liquid phase and the system is bivariant. Beyond the point Z, vapourisation sets in. It means that the number of phases will become two and the system has only one degree of freedom. The continued heating will at change the temperature because the pressure is being kept constant. The only effect of further heating will be to convert more and more of the liquid into vapour. The temperature will rise along the line ZZ'.

Effect of Change of Temperature and Pressure on Equilibrium

It is very interesting to see the effect of heating and applying pressure on the system at the triple point 0 where all the three phases are coexisting. If we do the heating at the triple point 0, there occurs the melting of more and more of the solid into liquid but there occurs no increase in temperature or pressure until the whole of ice has completely converted into liquid water. When this is completed, the system will have now two phases, *i.e.*, liquid and vapour. This means that the system will change from nonvariant to univariant. Therefore, further heating will result a rise in temperature, causing the equilibrium to shift along the curve OA.

Now if we apply the pressure to the system at the triple point, no change in pressure or temperature will occur as long as all the three phases are present. The only effect of applying pressure will be to convert vapour into liquid or solid phase. Ultimately vapour phase will disappear completely and convert into liquid and solid phases, and the further application of pressure will cause increase of pressure with change of temperature along the curve OC.

DEUTERIUM OXIDE SYSTEM

Deuterium oxide occurs in ordinary water in the ratio of 1 : 4000. This is prepared from ordinary water by fractional electrolysis. The chemical properties of H_2O and D_2O are more or less similar but there is marked difference in the physical properties of D_2O and H_2O. The phase diagram of heavy water (deuterium oxide) is shown in Fig. 3.3. This diagram is Just similar to the water system but has some differences which are as follows :

(i) Triple point in heavy water system is 3.82°C but not 0.0075°C as in H_2O system.

(ii) Vapour pressure of the liquid heavy water becomes equal to 760mm of Hg at 111.42°C whereas in case of H_2O system it becomes at 100°C.

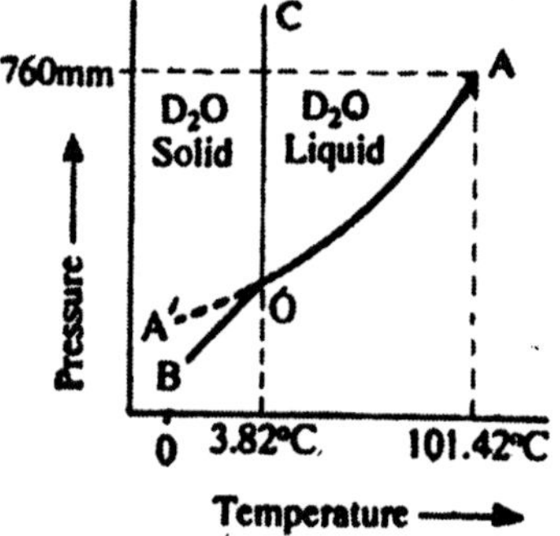

Fig. 3.3 : Deuterium oxide system.

(iii) On freezing, heavy water expands. The melting point of solid heavy water is lowered by 0.0069°C for each atmospheric pressure applied. Hence, the curve OC is sloping towards the left in D_2O system but this is less than the corresponding curve for H_2O.

The equilibrium diagram of heavy water contains three curves OA, OB and OC. These three curves divide the phase diagram into three regions AOC, BOC and AOB. Point 0 is the triple point. All the curves, areas and triple points can be explained on exactly the same lines as in water system. The salient features of heavy water system are outlined in Table 3.1.

Table 3.1 : Salient Feature of D_2O System

Curves/Reglong/ Triple point	*Name*	*Phase*	*Degree of Freedom*
Curve OA	Vaporisation curve of D_2O liquid	$D_2O(l) \rightleftharpoons D_2O(v)$	1
Curve OB	Vaporisation curve of D_2O solid	$D_2O(s) \rightleftharpoons D_2O(v)$	1
Curve OC	Fusion curve of D_2O solid	$D_2O(s) \rightleftharpoons D_2O(l)$	1

(Contd....)

Curves/Reglong/ Triple point	*Name*	*Phase*	*Degree of Freedom*
Region AOB	–	D_2O (vapour)	2
Region AOC	–	D_2O (liquid)	2
Region BOC		D_2O (solid)	2
Point 0	Triple point	$D_2O(s) \rightleftharpoons D_2O(l) \rightleftharpoons D_2O(v)$	0
Curve OA'	Metastable vaporisation curve of liquid D_2O	$D_2O(l) \rightleftharpoons D_2O$ (v)	2

ICE SYSTEM

Tammann ann Bridgmann observed that water under high pressure exists in seven different crystalline forms of ice which are shown in the phase diagram Fig. 3.4. The seven forms of ice are ice I, ice II, ice III, ice IV, ice V, ice VI and ice VII. Ice I is ordinary ice, whereas the other forms are different from ice I in density, crystal structure, and other physical properties and are stable at higher pressure with the exception of IV which is not shown on the diagram because its existence though indicated was not confirmed.

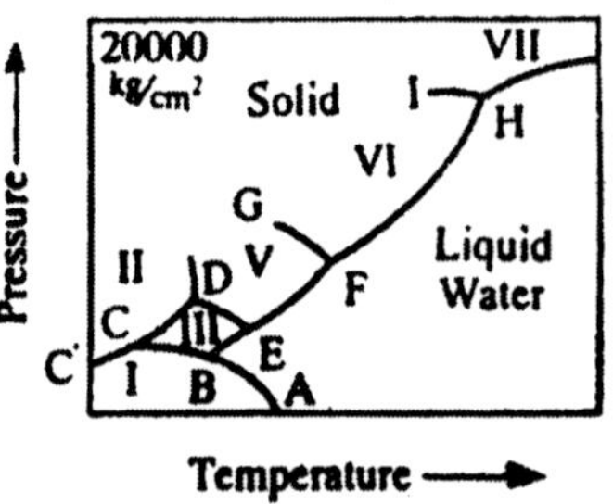

Fig. 3.4 : Ice System

There are ten stable curves in the system, viz., AB, BC, CD, BE, DE, EF. FG. FH, HI and HJ.

A is the normal triple point as it exists in water system. At A, ice I, liquid and vapour are in equilibrium. The degree of freedom of this point is one which follows from the phase rule :

$$F = C - P + 2 = 1 - 3 + 2 = 0$$

This point A is stable at 0.0075°C at a corresponding pressure of 4.58 mm. of Hg.

When the pressure is increased, the melting point of ice I falls until the triple point B is reached which corresponds to –22°C at 2110 kg cm^2. At the point, ice I, ice III and liquid are in equilibrium. The degree of freedom of this point is zero. If pressure is further applied to the state corresponding to point B, a value of –34°C at pressure of 2170 kg. cm^2 corresponding to point C is reached. At this triple point C, three phases in equilibrium are ice I, ice II and ice III.

Further increase in pressure results in the transformation of ice III into ice V, ice VI and ice VII. Thus, there are Seven triple points in this system which are all non-variant. Only three forms of ice viz. ice I, ice VI and ice VII are stable above 0°C. In fact, ice VII is solid at the normal boiling point of water (100°C at a pressure of 22400 kg cm^2). It is remarkable that melting of ice is quite hot. The melting point curve beyond the ice VII curve has been extended to 190°C and pressure of 40,000 kg. cm^2.

The various salient features of ice system are given in Table 3.2.

Table 3.2 : Salient Features of the Ice System

Triple point	*Temperature°C*	*Pressure*	*Phases*
A	+00075°	4.58 mm of Hg	Ice I $\rightleftharpoons$ liquid $\rightleftharpoons$ Vapour
B	–22°	2,110 kg cm^2	Ice I $\rightleftharpoons$ Ice III $\rightleftharpoons$ liquid
C	–34.7°	2,170 kg cm^2	Ice I $\rightleftharpoons$ Ice II $\rightleftharpoons$ Ice III
D	–24.3°	3,500 kg cm^2	Ice II $\rightleftharpoons$ Ice III $\rightleftharpoons$ Ice V
E	–17°	3,520 kg cm^2	Ice II $\rightleftharpoons$ Ice V $\rightleftharpoons$ Liquid
P	+01°	6,375 kg cm^2	Ice V $\rightleftharpoons$ Ice VI $\rightleftharpoons$ Liquid
G	+81.6"	22,400 kg. cm^2	Ice VI $\rightleftharpoons$ Ice VII $\rightleftharpoons$ Liquid

THE SULPHUR SYSTEM

It is a one component system consisting of four phases :

Rhombic sulphur S_R...............Solid phase

Monoclinic sulphur S_M. ....Solid phase

Liquid sulphur S_L,................. Liquid phase

Vapour sulphur S_V..................Vapour phase.

The phase diagram of sulphur system is represented in Fig. (3.5).

The diagram consists of :

(i) *Curves* AB, BC, CD, BE, CE and EG and for metastable curves.

(ii) *Areas* : The above curves divide the diagram into 4 areas ABEG, ABCD, GECD and BEC.

(iii) *Stable triple points* : Three stable triple points are B. C and E.

(iv) *Metastable point* : F is metastable point.

Description

Curve AB : This curve is the *sublimation* or *vapour pressure* curve of rhombic sulphur (S_R). This gives the variation of vapour pressure of S_R at different temperatures. Thus, along this curve, two phases rhombic sulphur and vapour are in equilibrium with each other. It follows from the phase rule.

$$F = C - P + 2 = 1 - 2 + 2 = 1.$$

Thus, the system is univariant. It means that there can be one vapour pressure value at one temperature.

Point B : At the point B, rhombic sulphur changes into monoclinic sulphur. This is known as transition temperature. Thus, there are three phases (two solids and one vapour) in equilibrium at this point (B) and it is non-variant.

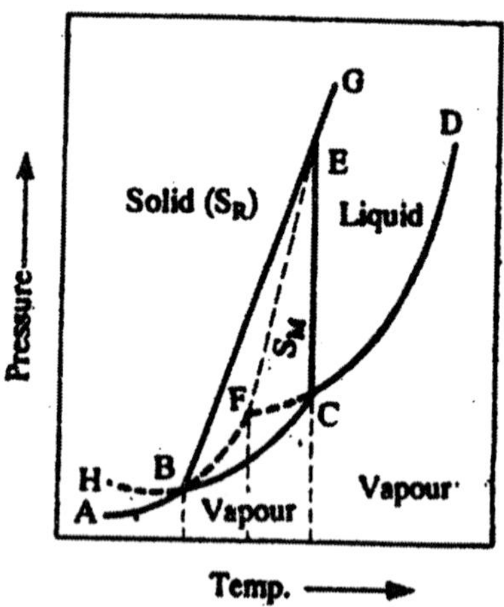

Fig. 3.5 : Sulphur system.

$$F = C - P + 2 = 1 - 3 + 2 = 0$$

Thus, B is a triple point.

Curve BC : It is the *vapour pressure* or *sublimation* curve for monoclinic sulphur. It gives the variation of vapour pressure of S_M (monoclinic sulphur) at different temperatures. At any point on this curve BC, the two phases SM and Sy are in equilibrium. The system is univariant. *i.e.,* it has one degree of freedom.

$$F = C - P + 2 = 1 - 2 + 2 = 1$$

Point C : It represents the *melting point* of monoclinic sulphur (120°). This is another triple point. At this point three phases (S_M. S_L, and S_V) are in equilibrium. Thus, this is a non-variant point.

Curve CD : It is the vapour pressure curve of liquid sulphur. This curve starts at point C which is melting point of S_M and terminate at point D, the critical temperature. Beyond this temperature there is one phase S_V.

Along this curve SL and, Sy are in equilibrium. Thus the system is univariant since at any point on the curve for the fixed value of pressure, temperature has a fixed value. This is also evident from the phase rule,

$$F = C - P + 2 = 1 - 2 + 2 \text{ or } F - 1$$

Curve BE : This is known as the *transition curve* of S_R to S_M. It gives the effect of pressure on the transition point. Both phases (S_R and S_M) are solids. The system is monovariant. This curve terminates at E beyond which SM disappears. As the transition point is raised With increase of pressure, the curve BE, therefore, slopes away from the pressure axis.

Curve CE : It is the melting or fusion curve for monoclinic sulphur. It shows the effect of pressure on melting point of S_M which is raised by increase of pressure, the curve CE, therefore slopes slightly away from the pressure axis.

At any point of the curve, the two phases S_M and S_L are in equilibrium with each other. This also has one degree of freedom.

Point E : The two curves BE and CE meet at the point E. Thus, it is another triple point which is nonvariant. At this point, three phases (S_R, S_M and S_L) are in equilibrium.

Curve EG : It is the fashion, curve for rhombic sulphur. Along this curve S_R and S_L are in equilibrium. Therefore, the number of phases is one and hence the system is monovariant.

Metastable Equilibrium

Curve BF : If the temperature of rhombic sulphur at 95.5°C is allowed to rise quickly, the transition of S_R or S_M does not take place at B But this curve is extended to F, the melting point of S_R. The curve BF so obtained is the metastable curve of rhombic sulphur. If system is disturbed slightly, the curve BE merges into BC.

Curve CF : If liquid sulphur is allowed to cool very carefully the solid phase will not separate out at C. Thus, the curve CD can be extended to F by supercooling the liquid carefully. Thus, the curve CF represents metastable equilibrium between S_L and S_V.

Curve EF . It is the metastable fusion curve of rhombic sulphur. Along this curve, rhombic sulphur is in metastable equilibrium with liquid sulphur. This curve gives the effect of pressure on the melting point of rhombic sulphur in metastable state.

Point F : It is the metastable triple point where S_R. S_L and S_V co-exist in the equilibrium ; like B, C and E, it is non-variant.

Areas. There are four areas :

(i) Area ABEG : It contains only rhombic sulphur.

(ii) Area ABCD : It contains only vapour sulphur.

(iii) Area GECD : It contains on ly liquid sulphur.

(iv) Area BEC : It contains only monoclinic sulphur.

Thus, all the above areas contain one phase only. It means that there are two degrees of freedom which is also evident from the phase rule equation.

$$F = C - P + 2 = 1 - 1 + 2 \text{ or } F = 2$$

It means that both pressure and temperature are required to define any point in all the above areas.

EXPERIMENTAL DETERMINATION OF TRANSITION POINT

The various methods are as follows :

1. *The Dilatometric Method :* This method is based on the simple principle that there occurs a change in volume due to the transition of one form into another because each form has a different density.

The determination is carried out in an apparatus known as *dilatometer* as shown in Fig. 3.6. It consists of a wide tube having a graduated capillary (in mm) side-tube A. The substance whose transition temperature is to be determined is filled in the lower half of the tube whereas the rest of the tube is filled with an inert solvent such as xylene. The stopper carrying the thermometer is inserted into the tube. The dilatometer is now kept in a thermostat. The level of the xylene in the capillary is noted. The temperature of the thermostat is increased degree by degree. As long as no change is taking place in the solid, the rise in the xylene level in the capillary is regular and uniform. As soon as the transition point is reached, there occurs a sudden change in volume of the solid which causes an abrupt rise or fall in the level. The same thing happens on cooling when the transition point is again reached. But the heating curve and cooling curve do not coincide as there is always a lag. Usually, a mean of the temperatures is taken as the transition temperature. This is shown in Fig. 3.7.

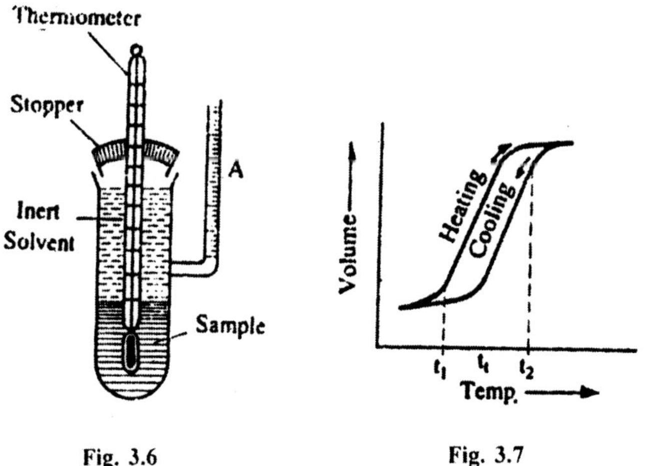

Fig. 3.6

Fig. 3.7

2. *The Solubility Method :* Each form possesses a different solubility at a given temperature. But both the forms have the same solubility at the transition temperature. In this method, the solubility curves for both the forms are plotted against temperature (Fig. 3.8). In

this figure, transition temperature corresponds to that point at which the two curves intersect each other.

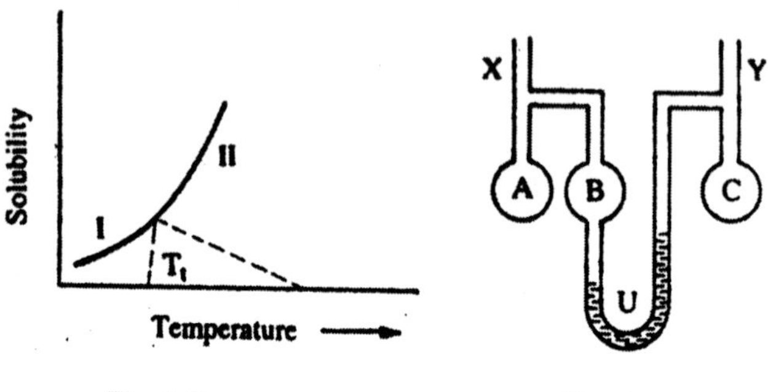

Fig. 3.8 Fig. 3.9

3. *The Vapour Pressure Method :* The principle of this method is that the vapour pressure of the two forms becomes equal at the transition temperature. Bremer and Frowein employed an apparatus shown in Fig. 3.9 known as tensiometer to measure the vapour pressure. It consists of a U-tube containing mercury. The two limbs of U-tube are connected to bulbs B and C. The bulb B is fused to another bulb A which has side-let X. Similarly the bulb C has side-let Y.

In bulb A, water is kept and then the bulb is sealed at X. The pure hydrated substance is kept in the bulb C. An exhaust pump is connected to Y and then the pump is started to reduce the pressure in the apparatus by removing all the air from it. Now the bulb C is also sealed at Y. Note the mercury level in the U-tube. After this, the bulb C is heated to start dehydration of the hydrate by keeping the apparatus in a thermostat. When the equilibrium is attained, the difference in the mercury level is accurately read out with the help of a travelling microscope. Then, the value of p_1 can be obtained by applying the following equation:

$$p_1 - p_2 = \text{Difference in mercury level}$$

where p_1 = Decomposition pressure of the substance kept in the bulb C, and

p_2 = Saturation pressure due to the water present in bulb A.

Various values of pi are obtained for different temperatures and are plotted against temperature to get a graph shown in Fig. 3.10. The sharp break in the curve corresponds to the transition temperature.

4. *Thermometric Method :* The principle of this method is that each form possesses a different energy content. If a substance which can undergo transition is heated, its temperature rises gradually until the transition temperature is reached at which the temperature remains constant until its transition into the other form is complete.

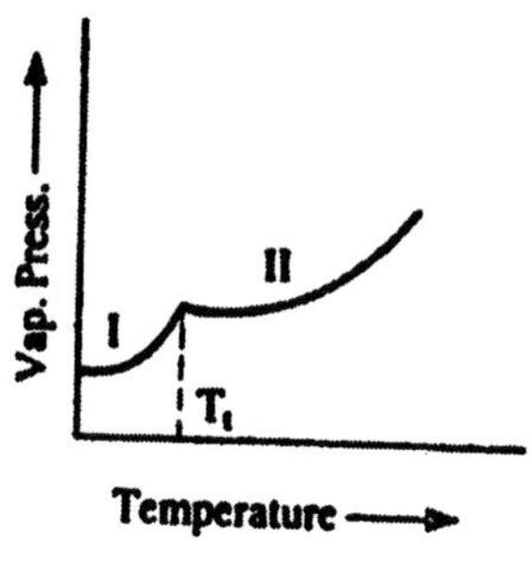

Fig. 3.10

TWO COMPONENT SYSTEMS

When two components denote the smallest number by means of which composition of each phase can be expressed, it is known as a two component system.

Maximum number of Phases : When F is minimum, P becomes maximum. The minimum number of degrees of freedom possible in any system is zero. Consider an equilibrium state of a two component system having zero degree of freedom. According to the phase rule, we have

$$F = C - P + 2 \quad ...(1)$$

$$0 = 2 - P + 2 \text{ or } P = 4$$

Thus, the maximum number of phases in any equilibrium state of a two component system having zero degree of freedom will be four.

Maximum number of F : The maximum value of F is expected in the case when P is minimum.

As the minimum number of phases in any system is one, it follows from equation (1) that maximum degree of freedom in a two component system are three.

$$F = C - P + 2 = 2 - 1 + 2 \text{ or } F = 3$$

Phase Diagram, As the value of F is three, it means that a phase is to be defined by three variables, *i.e.,*. pressure, temperature and composition. Pressure, temperature and composition could be represented by three co-ordinate axes at right angles, yielding models for two component systems. But the space models cannot be conveniently represented on paper.

In order to have simple phase diagram for two component systems, it is a usual practice to choose any two of three variables for graphic representation, assuming third to be constant. In this manner, we can have

(i) Pressure-temperature diagram (P–T)

(ii) Temperature-composition diagram (T–C)

(iii) Composition-pressure diagram (C–P)

In most of the studies, it is the usual practice to keep the pressure constant and thus, an T–C diagram is the one which is most often met.

Reduced phase rule : In all the above P–T. T–C and C–P diagrams, one of the variables is fixed. Thus the degree of freedom of the system is reduced by one. Therefore the phase rule equation (1) for the component system is written as

$$F - 1 = C - P + 1$$

$$F = C - P + 1 \qquad ...(2)$$

where F' gives the remaining degrees of freedom of the system. Equation (2) is known as *reduced phase rule.*

TYPES OF TWO COMPONENT SYSTEMS

We will only discuss the solid-liquid systems of two components.

Solid-liquid Systems of two Components

Condensed systems : The effect of pressure on the equilibrium between liquids and solids is very small. Therefore, the effect of pressure on such systems may be disregarded. Therefore, the experiments in liquid-solid equilibria are usually conducted under constant atmospheric pressure. *"Such a system in which only solid and liquid phase are considered and the experiments are carried out under atmospheric pressure, is called a condensed system."*

The graphical representation of condensed systems of two-components is simplified as there are only two variables, *i.e.*, temperature and composition, keeping pressure to be constant and these can be represented on a plane paper in the case of a condensed system, variables gets fixed. In other words, one of the degrees of freedom of the system is reduced by one. Therefore, the equation (1) becomes as:

$$F - 1 = C - P + 1$$

or $$F = C - P + 1 \quad ...(3)$$

where F' gives the remaining degrees of freedom of the system. Equation (3) is known as *condensed phase rule* which is similar to the equation (2). Solid-liquid equilibria of number of types are known. But the following types are more important:

Type I. Two components are completely miscible m liquid phase and the solid phase consist of pure components. Examples are : (a) Lead-antimony, (b) Lead-silver, (c) Potassium iodide-water system.

Type II. The components enter into chemical combination giving rise to one or more compounds. Such types of two-component systems are of two kinds :

IInd (A). Two components form a compound with congruent melting point. Examples are :

(i) Zinc-magnesium system, (ii) $FeCl_3$–water system.

IInd (B). Two components form a compound with incongruent melting point. Examples are :

(i) Picric acid-benzene system, (ii) Gold-antimony system.

(ii) Sodium sulphate-water system.

Simple Eutectic System or Solid-liquid Equilibrium of Type I.

Consider a two-component system having two components A and B. These components are completely miscible in liquid state. *They do not form any compound.* The phase equilibrium diagram for this type is shown in Fig. 3.11.

The phase diagram (3) consists of

(i) Curves. AC and CB

(ii) Points. A, B and C

(iii) *Areas* : ADC, D'CB, OO'D'D and they are above the curve ACB.

Description

(i) *Point A* : It represents the melting point of the pure component A.

(ii) *Point B* : It represents the melting point of the pure component B.

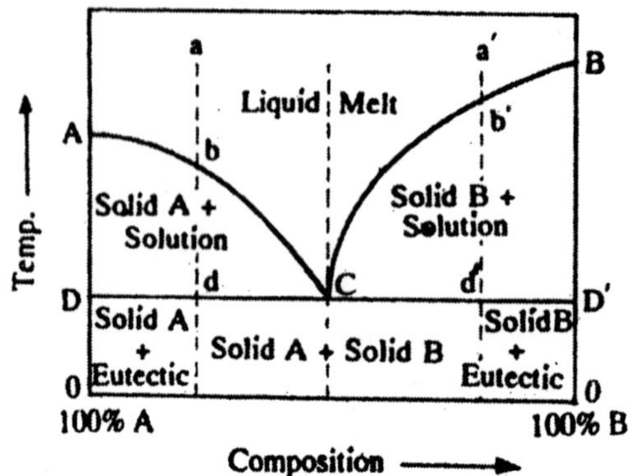

Fig. 3.11 : Solid-liquid system (Type I)

(iii) *Curve AC* : On gradual addition of small amounts of B to A, the freezing point of A is lowered along this curve AC. Therefore, this curve AC is known as the freezing point of the component A.

Along this curve AC, the solid A is in equilibrium with the solution of the component B in A (liquid) at different temperatures. Thus, the number of phases along this curve AC is two. As measurements are being made at constant atmospheric pressure, it follows from the condensed phase rule (3),

$$F = C - P + 1 = 2 - 2 + 1 = 1$$

Thus any point on the curve AC has one degree of freedom or is univariant. It means that only composition varies along the curve AC.

Curve BC : As increasing quantities of A are added to B, the freezing point of B falls along the curve BC. Along this curve, the solid B is in equilibrium with the solution of the component A in B at different temperatures. Again the number of phases along the curve BC like the curve AC is two. Again, applying the condensed phase rule equation, we get

$$F' = C - P + 1 = 2 - 2 + 1 = 1$$

Thus, any point on the curve BC is univariant, *i.e.*, only composition varies along BC.

Point C : At the point C, where the two curves AC and BC intersect, obviously both the sides A and B must be in equilibrium with the liquid phase. As three phases co-exist at this point, it follows from the modified phase rule (3).

$$F' = C - P + 1 = 2 - 3 + 1 = 0$$

Thus, the system at the point C is non-variant, *i.e.,* has no degree of freedom. This means that there is only one temperature at which the liquid phase can be in equilibrium with both solids A and B, and that the composition of the liquid phase must also be definite. From the figure, it is seen that the point C is the lowest temperature at which the liquid can exist. Therefore C has been called the eutectic point because it represents the lowest melting point of any mixture of solids A and B.

Area Above ABC : In this area the two components A and B exist as a homogeneous liquid solution because the temperature in it is above the melting point of any mixture. It means that there is only one phase in the area. Therefore, applying the condensed phase rule into this area, we obtain

$$F' = C - P + 1 = 2 - 1 + 1 s = 2$$

Thus, the system in the area above curve ACB is bivariant. It means that temperature and composition are required to define any point in the area above curve ACB.

Area OO'D'D : In this area only solid can exist because the liquid phase can not exist below the eutectic temperature.

Area ADC : Any point in the area ADC represents equilibrium between solid A and liquid. The composition of the phases in equilibrium can he obtained by drawing a horizontal line through the particular point. This line, where it cuts the curve AD, gives the nature of the solid and the AC to give the composition of the liquid.

Area CD'B : The above explanation is also true for this area.

Cooling of liquid mixtures : It is possible to predict the behaviour of any system on heating or cooling by using equilibrium diagram (3). This type of study is of special importance in the study of alloys in metallurgy.

Consider a liquid mixture of components represented by a point a. Allow it to cool at constant pressure. This fall in temperature continues without any change of composition till the point b on the curve AC is reached. At the point b, solid A separates out at a temperature t_1 which corresponds to the point b. Thus, the system at b involves two phases, *i.e.,* solid A and liquid solution.

Applying the condensed phase rule to this point b, it gives

$$F' = C - P + = 2 - 2 + 1 - 1$$

Thus, the system at point b is univariant. It means that there will be a fall in temperature only when the composition, of liquid phase changes. Therefore, if the solid A keeps on separating out along the curve bC and the solution becomes relatively richer in B. At the point y solid A is in equilibrium with its mixture of composition y at a temperature t_2 which corresponds to this point. Thus, in the area ACD solid A is in equilibrium with solution of different compositions depending upon different temperatures.

When the point d is reached which corresponds to the eutectic point C, the second solid B also starts separating out. Now the system will have three phases. Again, applying the phase rule to this system, we obtain

$$F' = C - P + 1 = 2 - 3 + 1 \text{ or } F' = 0$$

Thus, the system is univariant. It implies that on cooling further, the solid A and solid B separate out in the fixed ratio so that the composition of the liquid mixture remains constant as required at the point C. Here temperature does not change. This continues till the whole solution phase has been completely solidified. Thus, the system in the area OO'D'D consists of mixture of solid A and solid B. In that area, the system becomes univariant,

$$F' = C - P + 1 = 2 - 2 + 1$$

or $$F' = 1.$$

It means that on further cooling the temperature will only fall in the area BD'DA keeping the composition of two solids as constant. Similarly, one can explain if one considers a liquid of composition a' on the right side of eutectic point. Examples are : (i) Animony-lead system, (ii) Silicon-aluminium system. (Hi) Bismuth-cadmium system, (iv) $KCl–CaCl_2$ system, (v) Benzene-methyl chloride system, (vi) $KI-H_2O$ system, (vii) Lead-silver system.

We will now discuss some examples of type 1 of two-compond involving solid-liquid equilibria.

Lead-silver system : It is a two component system. The two metals lead and silver are completely miscible in the liquid state and do not form any compound. Therefore, the equilibrium diagram of this system is similar to that shown in the figure (3.11). The special features of lead-silver system are demonstrated in Fig. 3.12.

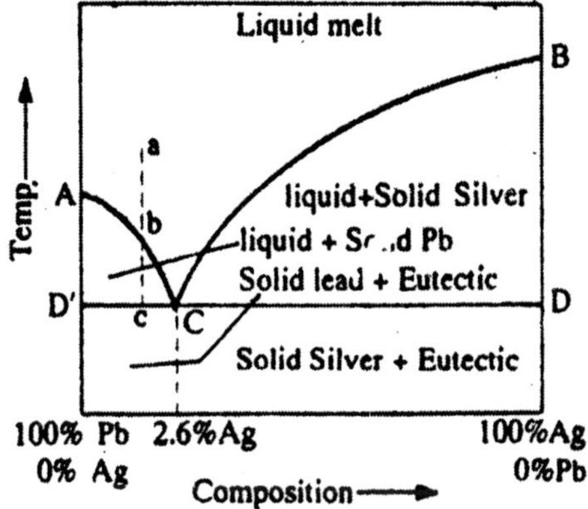

Fig. 3.12 : Lead-Silver System.

As the experiment is being carried out at constant pressure, the vapour phase is not considered and the condensed phase rule is applied.

$$F' - C - P + 1$$

Curve AC : A is the melting point of lead (327°). When increasing quantities of pure silver are added to lead metal, the freezing point of lead lowers along the curve AC. Therefore, AC is the freezing point curve of lead.

Along the curve AC, the solid lead is in equilibrium with the liquid melt. Thus, the number of phases along the curve is two. As all measurements are made at atmospheric pressure, on applying condensed phase rule, $F' = C - P + 1 = 2 - 2 + 1 = 1$

Therefore, the system is univariant. It means only composition varies along the curve AC.

Curve BC : Point B represents the melting point of pure silver (961°). When increasing quantities of pure lead are added, the freezing of silver lowers along the curve BC. Therefore, this curve may be called the *freezing point* curve of silver. Again, the solid silver is in equilibrium with the liquid melt along the curve BC. Thus, there are two phases. Applying the condensed phase rule equation, we obtain

$$F' = C - P + 1 = 2 - 2 + 1 = 1.$$

Again, the curve BC like AC is univariant and only composition varies along this.

Eutectic point C : The two curves AC and BC intersect at the eutectic point C. This point, being common to both the curves, represents the condition under which three phases lead, solid silver and liquid melt co-exist. The degree of freedom here is zero which also follows from reduced phase rule :

$$F' = C - P + 1 = 2 - 3 + 1 = 0$$

Thus, the point C is invariant, *i.e.*, has no degree of freedom. This point C (308°C) lies at a temperature which is lower than the melting points of silver as well as lead metal. The composition of an alloy of lead (97.6%) and silver (2.4%) corresponding to the point C is also fixed. Therefore, the point C is called the *eutectic point i.e., lowest temperature at which the liquid melt can exist. The temperature corresponding to this point is known as eutectic temperature and composition to this point is known as eutectic composition.*

Area above curves AC and BC : In this area, silver and lead are present as a homogeneous liquid solution. Thus, there is only one phase in this area. Applying the condensed rule to any point in this area, we obtain :

$$F' = C - P + 1 = 2 - 1 + 1 = 2$$

Thus, the system in this area is bivariant. It means that both temperature and composition are required to define any point in this area.

Area ACD' : In this area, solid lead is in equilibrium with the liquid melt.

Area BCD : In this area, solid silver is in equilibrium with the liquid melt,

Significance of lead-silver system : The phase diagram is utilised in the separation of silver from lead in Pattinson's Process for desilverisation of lead.

Suppose a represents the molten argentiferous lead containing a small amount of silver in it. Allow it to cool. Only temperature of the liquid melt falls with no change in concentration till the point b on the curve AC is reached. On further cooling lead begins to crystallise out and the solution becomes richer in silver. Further cooling will continue along the curve bC. Lead continues to separate out and is constantly removed by means of ladels. The liquid melt continues to be richer and richer in silver till the point C is reached where an alloy containing 2.6% of silver is obtained. This is then subjected to cupellation process.

The above process of increasing the concentration of silver in argentiferous lead is known as *Pattinson's process.*

Potassium Iodide-water System

The potassium iodide-water system is a typical example of a two-component system which forms a eutectic mixture. Hence its phase diagram is similar to the Fig. 3.13. The various features of this system are illustrated in the T–C diagram, pressure being considered as one *atm* pressure.

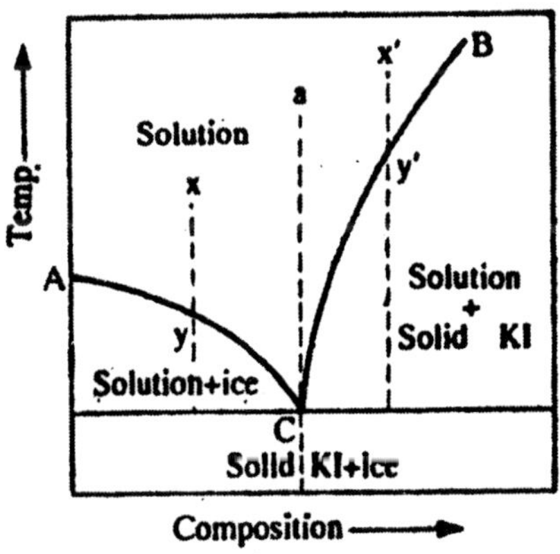

Fig. 3.13 : KI-H_2O System.

In the above diagram, the melting point of potassium iodide is not shown because it is very high.

Curve AC : Point A represents the freezing point of water at 1 atm pressure. This temperature is 0°C.

When increasing quantities of potassium iodide are added, the freezing point of water is lowered along the curve AC. Thus, AC may be called the freezing point curve of water.

Along the curve AC, the solution of KI in H_2O is in contact with ice. Thus, there are two phases. Applying the condensed phase rule to any point on this area.

$$F' = C - P + 1 = 2 - 2 + 1$$

or $$F' = 1.$$

Thus, the system is univariant, *i.e.*, has one degree of freedom. It means that there will be a definite composition of the solution corresponding to each temperature. When further quantities of potassium iodide are added, the lowering of freezing point continues along the curve AC till the point C is reached. Here the solution becomes saturated with respect to solid potassium iodide and freezes at a constant temperature (–23°C) and fixed composition (52% KI).

Curve BC : This is known as the solubility curve of potassium iodide in water at different temperatures. The solubility increases with rise of temperature and becomes maximum at the point B which is the boiling point of the saturated solution. Allow a saturated solution of potassium iodide at B to cool. The solution will freeze along the curve BC. During this, solid KI crystallises out.

This crystallisation of KI continues till the temperature and composition corresponding to the point C is reached. At this point, the solution freezes as a whole with fixed composition.

Along the curve BC, solid KI is in equilibrium with solution.

Again, applying the condensed phase rule to this system, we obtain

$$F' = C - P + 1 = 2 - 2 + 1 \text{ or } F' = 1$$

Thus the system is univariant, *i.e.*, has no degree of freedom. The steep rise of curve BC indicates that the solubility of KI increases slowly with the rise of temperature.

Point C : It is the eutectic or cryohydric point. At this point the two curves AC and BC intersect. This point corresponds to a definite temperature (–23°C) and definite composition (527% KI + 487% ice).

At this point C, ice, solution and solid KI are in equilibrium. This point has no degree of freedom which is evident from phase rule equation.

$$F' = C - P + 1 = 2 - 3 + 1 \text{ or } F' = 0$$

Thus, C is a non-variant point.

Area above AOB : If the point lies above the curve BC. It means that the dilute solution of KI is there. But the area above the curve BC represents the concentrated solution of potassium iodide.

Cooling : Three cases may arise :

(i) Consider a dilute solution represented by the point x. On cooling, the temperature will lower along the curve xy without any change in composition.

This continues till the point y is reached. At this point ice separates out. Here the system will become univariant and further cooling will take place along the curve yC. This continued till the point C is reached. At this point ice and K.I freeze as a whole with fixed composition, *i.e.,* eutectic mixture.

(ii) Consider a concentrated solution represented by the point x'. Allow it to cool. This cooling continues without any change in composition along the curve x'y'. At the point y', solid KI separates out.

Further cooling will proceed along the curve y'C until the point C is reached. At this point, ice and K[freeze as a whole to give the eutectic mixture [KI = 52% and ice = 48%].

(iii) Consider a solution which is represented by the point a. This point lies vertically above the eutectic point C. When the solution is allowed to cool, the temperature will lower along the curve aC until the eutectic point C is reached. At this point, ice and potassium iodide separate out simultaneously.

This mixture of potassium iodide and ice deposited at the eutectic point is known as *cryohydrate.*

Type IInd A.

Solid-liquid equilibrium of type IInd A :

A compound is said to hive a congruent melting point when it melts sharply at a constant temperature into a liquid of the same composition as that of solid from which it is derived.

Suppose A and B are two components and AB is the compound formed by their chemical combination. The compound AB has a sharp congruent melting point. The phase diagram for such a phase system is shown in Figure (3.14).

Curve AC : Point A represents the melting point of pure component A. By gradual addition of B to A, the freezing point of A is lowered along the curve AC. Thus, the curve AC represents the freezing point curve of the component A. Along this curve, the solid A is in equilibrium with the solution of the component B in A. Therefore, the number of phases along AC is two. Applying the condensed phase rule to the system, we obtain

$$F' = C - P + 1 = 2 - 2 + 1 = 1$$

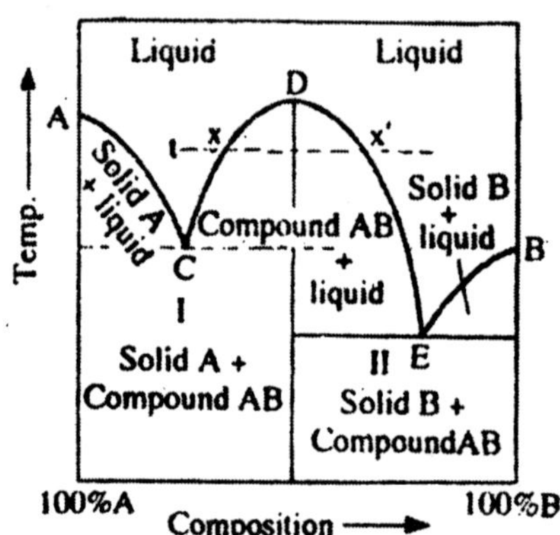

Fig. 3.14 : Solid-liquid equilibrium (Type II)

Thus, the system is univariant, *i.e.*, only the composition varies along AB.

Curve BE : Point B represents the melting point of pure component B. When A is added to B, the freezing point B is lowered along the curve BE. The curve is known as the *freezing point curve* of the component B.

Along this curve, the solid B is in equilibrium with the solution of the component A in B. Thus, the number of phases along BE is two. Applying the condensed phase rule to this system,

$$F' = C - P + 1 = 2 - 2 + 1$$

or $$F' = 1.$$

Thus, the system is univariant, *i.e.*, only the composition varies along BE.

Curve CDE : This curve is known as *freezing point carve* of the solid compound AB. Along this curve, the solid compound AB is in equilibrium with the liquid phase at different temperatures. The point D is known as *congruent melting* point of the compound because both the liquid and solid phases have the same composition at the point D. It means that two-component system becomes one component at this temperature as both the solid and liquid phases contain the same compound AB. Applying the phase rule to this point D, we get

$$F' = C - P + 1 = 1 - 2 + 1 = 0$$

Thus, the point D is non-variant : It means that this point D represents a definite melting point A.

Eutectic point C : The two curves AC and CDE intersect at the point C. At this point, solid A and compound AB are in equilibrium with liquid phase. Therefore, the number of phases is three. Now, applying the phase rule equation to this point C, we obtain

$$F' - C - P + 1 = 2 - 3 + 1 = 0$$

Thus, the system at C is invariant, *i.e.*, has no degree of freedom. It means that solid A, solid compound AB and liquid can only exist at a definite temperature and the composition of the liquid phase must also be definite.

Eutectic point E : The two-curves BE and CDE intersect at the point E where solid B, solid compound AB are in equilibrium with liquid phase. Again, the point E is invariant, *i.e.*, has no degree of freedom. The horizontal line passing through the eutectic point C and E is known as solid because only phase exists along it.

Areas I and II : The names of the various phases in areas I and II are shown in the Fig. 3.14.

General Discussion

(i) In Fig. 3.14 it is shown that the melting point (congruent) lies above the melting points of A and B. But it is not so. Certain cases are known in which the congruent melting point of the compound AB may be above, below or between those of two components A and B.

(ii) In Fig. 3.14 it is shown that the compound AB contains equimolecular amounts of A and B. This is not always so.

(iii) Draw a line XX_1. It means that the liquid phase has two compositions X and X_1 in equilibrium with the same solid AB at a certain temperature, say, t. In simple words it means that the compound AB has two solubilities at the same temperature* This paradox is explained by the diagram (3.14). AB is made up of two parts with DD' as the dividing line. Thus, the left half of the diagram will be a two component system of solid A and compound AB and the right half of the diagram shows the two component system involving solid B and compound AB. Also, the curve DC will represent the freezing point curve of solid compound AB when A is added to it. Similarly, the curve DE will represent the freezing point curve of compound AB when B is added to it.

(iv) When two components form more than one compound, the equilibrium diagram will have curve analogous to CDE for each compound.

Exa.nples are:

(i) Zinc-magnesium system,

(ii) Aluminium-magnesium system,

(ii) Mercury-thorium system,

(iv) Gold-tin system,

(v) Phenol-aniline system,

(vi) Ferric chloride-water system.

We will now discuss some examples.

Zinc-Magnesium System : It is an example of a two component system in which two metals form a compound having congruent melting point. The compound is an alloy having the formula $Mg(Zn)_2$ with a true melting point 590°C which lies between the melting points of two metals. The composition of $Mg(Zn)_2$ by weight is

$$Mg = 79\%$$

and $$Zn = 21\%$$

The melting point of zinc is 420°C and that of magnesium is 650°C. The melting point of $Mg(Zn)_2$ is 590°C. The compound is fairly stable and it melts without change of composition.

As the experiment is being carried out at constant pressure, vapour phase therefore, is not considered and the condensed phase rule will be applied :

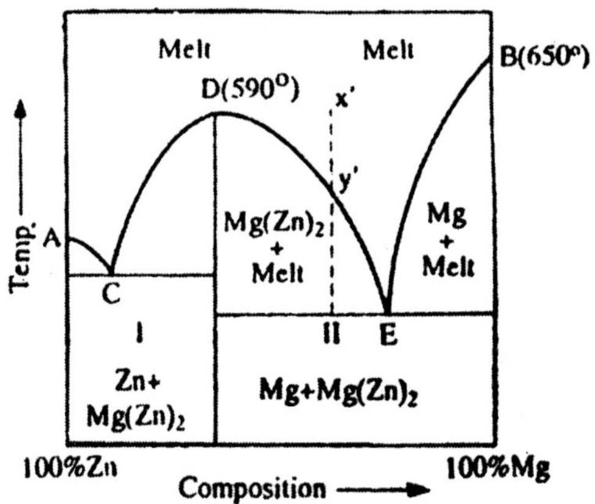

Fig. 3.15 : Zn—Mg system

$$F' = C - P + 1$$

The equilibrium diagram of such a system is represented in Fig. 3.15.

Point A : It represents the melting point of pure zinc (M.P. of zinc = 420°C],

Curve AC : By gradual addition of Mg to Zn, the melting point of zinc is lowered along the curve AC. Along this curve the solid zinc is in equilibrium with melt (solution). Thus, the number of phases is two. Applying the condensed phase rule equation*to this curve, we get

$$F' = C - P + 1 = 2 - 2 + 1 = 1$$

Thus, the system is univariant, *i.e.*, only the composition varies along the curve.

Point B : This represents the melting point of pure magnesium [M.P. = 65°C].

Curve BE : When zinc is added to Mg the melting point of Mg is lowered along the curve BE. Along the curve BE, the solid Mg is in equilibrium with melt at different temperatures and hence is regarded as the freezing or melting point curve of magnesium. Again, only composition varies along this curve which is evident from phase rule,

$$F' = C - P + 1 = 2 - 2 + 1 = 1 \qquad [\because P=2]$$

Curve DC : When further quantities of zinc are added, the melting point of the compound Mg $(Zn)_2$ lowers along this curve. Thus, the curve represents the lowering of melting point of compound Mg $(Zn)_2$ when zinc is added.

Curve DE : This curve represents the lowering of melting point of compound Mg $(Zn)_2$ when magnesium metal is added to it.

Point D : At this point, the liquid and solid have the same composition, *i.e.*, $Mg(Zn)_2$ and melt at this point without decomposition. Thus, the temperature corresponding to the point D is regarded as-the congruent melting point of $Mg(Zn)_2$.

Point C : The two curves AC and CD intersect at the point C. Thus, there will be three phases at this point. These three phases are solid zinc, compound $Mg(Zn)_2$ and melt. This point is invariant which is evident from the phase rule,

$$F' = C - P + 1 = 2 - 3 + 1 = 0$$

The point C is known as eutectic point.

Point E : This is the point where two curves BE and DE intersect at the point E. Thus, there are three phases which are in equilibrium at this point. These three phases are solid Mg, compound $[Mg(Zn)_2]$ and melt. Again, it is invariant and is known as eutectic point.

Cooling : Consider a melt of composition represented by a point x. Allow it to cool along the line xy. When it reaches y, the solid $Mg(Zn)_2$ separates out from the melt. At this point there are two phases and hence the system becomes univariant.

$$F' = C - P + 1 = 2 - 2 + 1 = 1$$

It means that the change in composition will also take place if the temperature is further lowered beyond the point y. This cooling will, therefore, take place along yC. At the point C, solid zinc will separate out. Thus, there are three phases at this point and hence it is nonvariant.

$$F' = C - P + 1 = 2 - 3 + 1 = 0$$

It means that there will not be any change in temperature as well as composition at this point if cooling is continued and as long as all the three phases are present. Similarly, consider a liquid melt of composition represented by a point x'. Allow it to cool. The temperature will, fall along x'y'. At point y'. solid $Mg(Zn)_2$ separates. Further cooling will therefore, take place along y 'E'. At the point E, magnesium also separates out.

Ferric Chloride-water System

It is an interesting example in which ferric chloride and water form a number of compounds. All compounds have congruent melting points.

There are in all 8 phases in the system out of which there are four stable hydrates.

Phases	*Composition*
(i) $Fe_2Cl_6.12H_2O$	$Fe_2Cl_6 + 12H_2O$
(ii) $Fe_2Cl_6.7H_2O$	$Fe_2Cl_6 + 7H_2O$
(iii) $Fe_2Cl_6.5H_2O$	$Fe_2Cl_6 + 5H_2O$
(iv) $Fe_2Cl_6.4H_2O$	$Fe_2Cl_6 + 4H_2O$
(v) Anhydrous Fe_2Cl_6	$Fe_2Cl_6 + H_2O$
(vi) Solution	$Fe_2Cl_6 + xH_2O$
(vii) Ice	$0\ Fe_2Cl_6 + xH_2O$
(viii) Vapour	$0\ Fe_2Cl_6 + xH_2O$

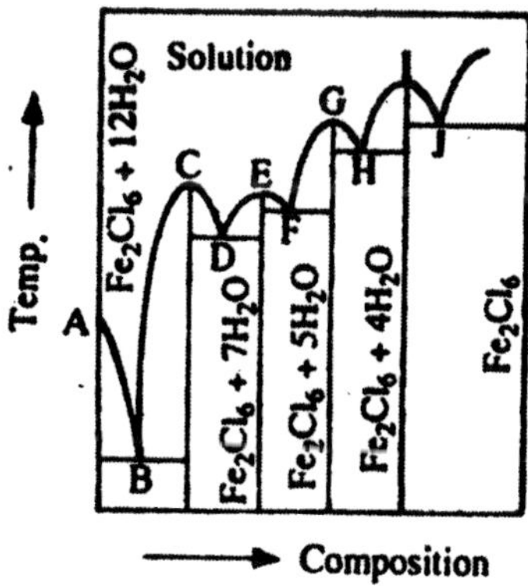

Fig. 3.16 : $FeCl_3$ – H–2–O system.

As the composions of all the above phases can be represented by Fe_2Cl_6 and H_2O it is therefore, regarded as a two component system. The complete diagram of this system is shown in the Fig. (3.16). As all the measurements are being carried out at constant pressure, vapour phase is, therefore, not considered and the condensed phase rule will be applied,

$$F' = C - P + 1.$$